STUDENT PO

BIMAN DAS

STATE UNIVERSITY OF NEW YORK, POTSDAM

FOURTH EDITION

PHYSICS

for

SCIENTISTS & ENGINEERS

with Modern Physics

GIANCOLI

PEARSON

Prentice
Hall

Upper Saddle River, NJ 07458

Sponsoring Editor: Christian Botting
Managing Editor, Science: Gina M. Cheselka
Supplement Cover Manager: Paul Gourhan
Supplement Cover Designer: Victoria Colotta
Operations Specialist: Amanda Smith
Senior Operations Supervisor: Alan Fischer
Cover photographs top left clockwise: Daly & Newton/Getty Images, Mahaux Photography/Getty
Images, Inc.-Image Bank, The Microwave Sky: NASA/WMAP Science Team, Giuseppe Molesini,
Istituto Nazionale di Ottica Florance

© 2009 Pearson Education, Inc.
Pearson Prentice Hall
Pearson Education, Inc.
Upper Saddle River, NJ 07458

Printed in the United States of America

10 9 8 7 6 5 4 3 2 1

ISBN-13:   978-0-13-227326-8
ISBN-10:   0-13-227326-8

Pearson Education Ltd., *London*
Pearson Education Australia Pty. Ltd., *Sydney*
Pearson Education Singapore, Pte. Ltd.
Pearson Education North Asia Ltd., *Hong Kong*
Pearson Education Canada, Inc., *Toronto*
Pearson Educación de Mexico, S.A. de C.V.
Pearson Education—Japan, *Tokyo*
Pearson Education Malaysia, Pte. Ltd.

CONTENTS

10 Rotational Motion   98

11 Angular Momentum; General Rotation   113

39   Quantum Mechanics of Atoms   491

40   Molecules and Solids   506

41   Nuclear Physics and Radioactivity   528

Preface

It was my pleasure to author the *Student Pocket Companion* to accompany *Physics for Scientists & Engineers with Modern Physics*, fourth edition, by Doug Giancoli. Although I had written a few Pocket Companions in the recent past, this was the first Pocket Companion that I wrote for a calculus-based physics text. This Pocket Companion contains the basic concepts, principles, equations, and applications from each section of the Giancoli text. It also provides tips for solving physics problems in each chapter. Students should find the mini-text useful for reviewing the basic concepts and equations, solving homework assignments, and pre-exam review.

This small paperback mini-text is easy to carry to the class for adding notes. Students will be able to use it as an alternative to highlighting material from the text. They will be able to pay more attention to the class lecture since all concepts have been summarized in this mini-text.

Each chapter of the *Student Pocket Companion* begins with a list of important concepts that the chapter contains. Each section of the mini-text follows the same order as the text, and it summarizes important concepts from each section in an easy-to-understand language. Learning how to solve real-world problems is an important part in any physics course. To master problem solving, I have given tips for problem solving at the end of each chapter based on my experience of working closely with students. I have mentioned the key concepts and equations students need to use for solving problems in different sections of each chapter. Wherever possible, I have pointed out common mistakes that students often make in solving problems.

Writing the *Student Pocket Companion* was a lot of work, but fun. I thank Mr. Christian E. Botting, Sponsoring Editor, Pearson Addison-Wesley, Physics & Astronomy, for his constant support and cooperation. Special thanks are due to my wife Indrani Das and my daughters Debapria Das and Deea Das for their support. I would not have found the time to write the book without their cooperation.

I hope students will find it useful.

Biman Das

To the Student: How to Use the Student Pocket Companion

This *Student Pocket Companion* (*SPC*) is a mini-text that is intended to help you understand the concepts, laws, principles, equations, and applications of physics, as outlined in the text by Doug Giancoli. It is a condensed version of Giancoli's text, but it is not meant to replace it. This small paperback mini-text is easy to carry and you can read it at home or even when you are on a bus, train, or airplane.

Some suggestions on how to use the *SPC*:

- Read the relevant chapters from the *SPC* before you attend class lectures.

- Carry the *SPC* to class to take brief notes in the margins where necessary.

- Carefully read the *Tips for Solving Problems* at the end of the chapter before solving problems. Be wary of the common mistakes as pointed out.

- Use the equations and concepts from the *SPC* to complete homework assignments.

- Review the chapters of the *SPC* before you take any quiz, test, or exam.

- Use the equations and the concepts from the *SPC* where necessary to analyze laboratory data.

- Review the concepts and important equations before you take other physics tests such as MCAT.

- Use the *SPC* as a reference for important concepts, laws, equations, and applications, even after you finish the course.

Biman Das

To my mother, Mrs. Renuka Rani Das

CHAPTER 1
Introduction, Measurement, Estimating

This chapter focuses on the nature of science, measurements, uncertainty, dimensions, different systems of units, standards, and dimensional analysis.

Important Concepts
> Model
> Theory
> Law
> Uncertainty
> Significant figures
> Scientific notations
> Length, mass, and time
> Unit
> MKS system
> British engineering system
> Cgs system
> Base and derived quantities
> Order-of-magnitude
> Dimensions and dimensional analysis

1–1 The Nature of Science

The role of physics is to search for orders among a wide range of natural phenomena. Observation and careful experimentation are one side of the scientific process, while its other side is the invention of new **theories** to explain the observations. Science is not a mechanical process of collecting facts. Like poetry and art, it is a creative activity that engages the emotion and the intellect. Science is not a copy of nature but a recreation of her by the act of discovery. Great theories of science are creative achievements and compared with the great works of art and literature.

The prediction of a theory is **tested** by experimentation. A new theory is often accepted when its predictions are in better agreement with experiment than the older theory, or when it explains a greater range of phenomena than does the older theory. For example, Copernicus's heliocentric model of the universe explains a greater variety of phenomena than Ptolemy's geocentric model of the universe. Einstein's theory of relativity, which differs very little from Newtonian theory for motion of ordinary objects, provides accurate explanations of phenomena for objects moving at speeds close to the speed of light. The theory unifies the concepts of mass and energy and shows that space and time are relative.

1–2 Models, Theories, and Laws

A **model** is a mental image of phenomena. It often gives us a deeper understanding and structural similarity to the phenomena from the analogy of a known system. Often a model is developed and then modified to fit the experimental phenomena. For example, there are models about the atomic structure of matter. The atomic models were developed and then modified.

A **theory** is broader, more detailed, and attempts to solve a set of problems. Models are helpful and they often lead to important theories. A theory is usually deeper and more complex than a simple model.

Laws are concise and general statements about how nature behaves. Often laws can be expressed by a mathematical relationship, such as $F = ma$ (Newton's second law).

Scientific laws are *descriptive* in that they tell how nature *does* behave. On the other hand, political laws are *prescriptic* and tell how we *ought* to behave.

1–3 Measurement and Uncertainty; Significant Figures

Uncertainty

Accurate and precise measurements are important aspect of physics. Accuracy refers to how close a result is to the accepted value. Precision refers to the repeatability of the measurement. Measured numbers usually have some degree of uncertainty or error. For example, when you use a centimeter ruler to measure a length, the result cannot be claimed to be accurate to about 0.1 cm, the smallest division on the ruler. A precise micrometer provides a more precise value of the measured length, but still the result will not be exact.

When giving a result of a measurement, it is important to state the **estimated uncertainty** in the measurement. For example, a measured length can be written as 5.2 ± 0.1 cm. Here ± 0.1 cm (plus or minus 0.1 cm) represents the estimated uncertainty in the measurement meaning that the actual length lies between 5.1 and 5.3 cm. The **percentage uncertainty,** which is the ratio of the uncertainty to the measured value times 100, is given by

$$\frac{0.1}{5.2} \times 100 \approx 2\%.$$

Significant Figures

The number of **significant figures** in a measurement is equal to the number of digits that are estimated or known with some reliability. When making measurements or writing results of a calculation (using a calculator), do not keep more digits in the answer than justified.

When multiplying and dividing numbers, leave as many significant figures in the answer as there are in the number with the fewest number of significant figures. Similarly, when adding or subtracting numbers, leave the same number of decimal places in the answer as there are in the number with the least number of significant figures.

Scientific Notation

In scientific notation, a number is written as:
(a decimal between 1 to 10) $\times 10^{\text{an integer exponent}}$.
For example, 36900 is written as 3.69×10^4 (if the number is known to an accuracy of three significant figures) or as 3.690×10^4 (if it is known to four).

1–4 Units, Standards, and the SI System

The measurement of a quantity is made relative to a standard or **unit**, which must be specified together with the numerical value of the quantity. Several systems of units are used including the Systeme International (SI) unit. The SI units (also called the mks units), of length, mass, and time (called base quantities) are *kilogram* (kg), *meter* (m), and *second* (s), respectively.

Length

One **meter** (m) is the length of path traveled by light in a vacuum in 1/299,792,458 of a second.

Time

One **second** (s) is the time it takes for radiation from a cesium-133 atom to complete 9,192,631,770 cycles of oscillation.

Mass

One **kilogram** (kg) is the mass of a particular platinum-iridium alloy cylinder at the International Bureau of Weights and Standards in Sevres, France. When dealing with atoms and molecules, mass is expressed in the unified atomic mass unit (u). 1 u = 1.6605×10^{-27} kg.

Unit Prefixes

In the metric system, prefixes are used to express very large and very small numbers. The following prefixes are commonly used: *kilo* (k), *mega* (M), *giga* (G), *milli* (m), *micro* (μ) and *nano* (n) for 10^3, 10^6, 10^9, 10^{-3}, 10^{-6} and 10^{-9}, respectively.

Systems of Units

The most important system of unit is the Systeme International (SI). Another metric system sometimes used is called the **cgs system**, in which centimeter, gram, and second are the standard units of length, mass, and time, respectively. The **British Engineering system**, which is often encountered in everyday use in the United States, takes as its standards the foot for length, the pound for force, and the second for time.

Base vs. Derived Quantities

Physical quantities are of two types: *base* and *derived*. The units of the corresponding quantities are called *base units* and *derived units*. Base quantities are seven in number and each of them are defined in terms of a standard. Derived quantities are defined in terms of base quantities.

1–5 Converting Units

Units can be converted from one set of units to another by multiplying by the appropriate conversion factors.

Conversion factors are ratios of the two units. Some important equivalents are listed here:

1 in = 0.0254 m	1 mi = 1.609 km
1 ft = 0.3048 m	1 m = 3.281 ft
1 rad = 57.3°	1 lb = 4.448 N
1 km/hr = 0.6214 mi/hr	1 hp = 746 W

1–6 Order-of-Magnitude: Rapid Estimating

An **order-of-magnitude** calculation is a rough estimate of a quantity that is accurate to within a factor of about 10. Such a calculation provides a quick idea of what order of magnitude should be expected from a complete and detailed calculation. Such an estimate is important when precise calculation is either impossible or when we are interested only in the approximate value of the quantity.

*1–7 Dimensions and Dimensional Analysis

The **dimension** of a physical quantity represents the type of units or base quantities it is made of. The dimensions of length, time, and mass are expressed as [L], [T], and [M], respectively. The velocity, v, has dimensions of length per time; $v = [L]/[T]$. The area of a triangle $A = \frac{1}{2} bh$ (b = base, h = height) and the area of a circle $A = \pi r^2$ (r = radius) have the same dimension $[L^2]$.

Any valid formula must be dimensionally consistent. That is, each term in the formula must have the same dimensions. Dimensions can be used in working out relationships; such a procedure is called the **dimensional analysis**. However, a dimension check is not a guarantee that the formula is correct because dimensionless factors (such as $\frac{1}{2}$, π) do not show up in the dimension check.

Tips for Solving the Problems

1. Recognize kg, m, s, and other SI symbols in a problem. Also, recognize different prefixes that may appear in a problem.

2. Use SI units whenever possible. Use conversion factors, if necessary. Do not use different sets of units (such as mm, m, and km) in the same problem.

3. Express the result using scientific notation (power of 10 notations). Keep only the desired number of significant figures in the final answer to a problem.

4. Each term in an equation has the same dimension. Use this fact to determine the dimension of a quantity that appears in an equation.

CHAPTER 2
Describing Motion: Kinematics in One Dimension

Kinematics is the study of motion without its cause. This chapter focuses on kinematics in one dimension.

Important Concepts

> Kinematics
> Position
> Displacement
> Average speed
> Average velocity
> Instantaneous speed
> Instantaneous velocity
> Derivative
> Average acceleration
> Instantaneous acceleration
> Free-fall motion
> Acceleration due to gravity
> Numerical integration

Mechanics is the study of motion of objects and the concepts of force and energy. It is divided into two parts:
1. **Kinematics** that describes how objects move, and,
2. **Dynamics** that studies motion with its cause.

2–1 Reference Frames and Displacement

The measurement of position or distance of an object is made with respect to a **frame of reference**. We often draw a set of **coordinate axes** to represent a frame of reference as shown.

For one-dimensional motion along, say, the x-axis, the position of an object is given by its x-coordinates.

The **displacement** of a moving object is the *change in position* of the object. If the initial and final positions are x_1 and x_2 respectively, the displacement is: $\Delta x = x_2 - x_1$.

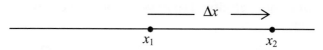

- Displacement is expressed in meters and it can be positive or negative. The distance traveled by an object is the length of travel and is always positive.

- Quantities, such as displacement, that have both magnitude and direction are called **vectors**.

2–2 Average Velocity

The **average speed** of a moving object is the ratio of the distance traveled over the time elapsed to travel this distance:

$$\text{average speed} = \frac{\text{distance traveled}}{\text{time elapsed}} \qquad (2\text{--}1)$$

The **average velocity** is the average speed with direction.

$$\text{average velocity} = \frac{\text{displacement}}{\text{time elapsed}}.$$

$$\bar{v} = \frac{\Delta x}{\Delta t} = \frac{x_2 - x_1}{t_2 - t_1} \qquad (2\text{--}2)$$

Here $\bar{v}$ is the average velocity and x_1 and x_2 are the positions of the object at time t_1 and t_2, respectively.

The SI unit for both the average speed and the average velocity is meters per second (m/s). Average speed does not provide information about the direction of motion, however.

2–3 Instantaneous Velocity

Velocity at a particular instant of time is called **instantaneous velocity**. The instantaneous velocity is given by

$$v = \lim_{\Delta t \to 0} \frac{\Delta x}{\Delta t} \qquad (2\text{–}3)$$

The preceding limit as $\Delta t \to 0$ is called the derivative of x with respect to t and is written as

$$v = \lim_{\Delta t \to 0} \frac{\Delta x}{\Delta t} \quad v = \frac{dx}{dt} \qquad (2\text{–}4)$$

Graphically, the slope of the straight line P_1P_2 obtained by joining the points P_1 and P_2 on the position (x) vs. time (t) curve of a moving object represents the average velocity of the object between the points P_1 and P_2, that is, between the time interval $(t_2 - t_1)$.

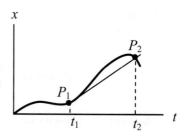

Also, *the slope of the tangent to the position (x) vs. time (t) curve at a point P is the instantaneous velocity of the object at that time.* Instantaneous velocity (that is, the slope) can be positive, negative, or zero. Its

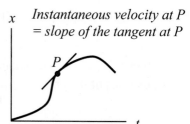

Instantaneous velocity at P = slope of the tangent at P

sign indicates the direction of motion.

When an object moves with a constant velocity, its average velocity and instantaneous velocity are the same.

2–4 Acceleration

The rate of change of velocity is called acceleration. The **average acceleration** is defined as

$$\text{average acceleration} = \frac{\text{change of velocity}}{\text{time elapsed}}$$

$$\bar{a} = \frac{\Delta v}{\Delta t} = \frac{v_2 - v_1}{t_2 - t_1} \qquad (2\text{–}5)$$

Here $\bar{a}$ is the average velocity and v_1 and v_2 are the velocity of the object at time t_1 and t_2, respectively.

The acceleration at a particular time is called **instantaneous acceleration**. Instantaneous acceleration is

$$a = \lim_{\Delta t \to 0} \frac{\Delta v}{\Delta t} \qquad (2\text{–}6)$$

In the notation of calculus,

$$a = \frac{dv}{dt} = \frac{d}{dt}\left(\frac{dx}{dt}\right) = \frac{d^2 x}{dt^2}$$

d^2x/dt^2 is called the second derivative of x with respect to t.

The SI unit of acceleration is meters per second per second, m/s^2.

The acceleration of a moving object can be obtained from its velocity (*v*)-versus-*time* (*t*) graph.

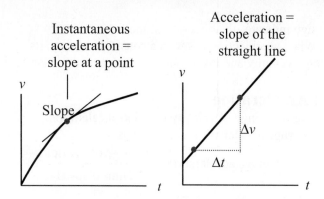

The average acceleration in a v-versus-t graph is the slope of the straight line connecting the two points corresponding to the two different times. The instantaneous acceleration is the slope of the tangent line at a particular instant of time. For an object moving with a constant acceleration, its acceleration is the slope of the straight line representing its v-versus-t graph.

2–5 Motion at Constant Acceleration

In many practical situations, the acceleration of a moving object (such as, the acceleration of a thrown ball) is constant or can be approximated to be constant. For an object starting from a position x_0 with an initial velocity v_0 and moving at a constant acceleration a, the following equations can be used to describe its motion:

Variables	Equation	
v, a, and t	$v = v_0 + at$	(2–7) (also 2–12a)
x, x_0, and t	$x = x_0 + \bar{v}\,t$	(2–8)
v_0, v, and v_{av}	$\bar{v} = \dfrac{1}{2}(v_0 + v)$	(2–9) (also 2–12d)

| x, v_0, a, and t | $x = x_0 + v_0t + \frac{1}{2}at^2$ (2–10) (also 2–12c) |
| x, v, and a | $v^2 = v_0^2 + 2a(x - x_0)$ (2–11) (also 2–12d) |

2–6 Solving Problems

To learn physics, you need to practice by solving problems. Although there is no recipe for problem solving, the following guidelines have been found to be effective in problem solving:

1. Read and reread the problem carefully.
2. Draw a diagram of the situation.
3. List known data and what are unknown(s).
4. Think about the physics that apply.
5. Identify appropriate equations or definitions that relate to the problem and solve the equation(s).
6. Carry out the calculation and report result to the correct number of significant figures.
7. Check if the result is reasonable.
8. Keep track of units that may serve as a check on your solution.

2–7 Freely Falling Objects

The most common example of a uniformly accelerated motion is the motion of an object falling freely near the surface of the earth. Galileo showed that in the **absence of air resistance, all freely falling objects move with the same constant acceleration** of about 9.80 m/s² or about 32 m/s². The free-fall acceleration is denoted by the symbol g and is called the **acceleration due to gravity**. The value of g varies slightly from place to place on the surface of the earth.

The equations describing the motion of an object with a constant acceleration (Equations 2–7 through 2–11) can be applied to the free-fall motion, since g is approximately constant. Since free-fall motion is in the y direction, replace x by y and x_0 by y_0 in the equations. *It is arbitrary whether you choose y to be positive upward or downward, but be consistent throughout a problem.*

*2–8 Variable Acceleration; Integral Calculus

Kinematic equations for motion (Equation 2–7 and 2–10) at constant acceleration can be derived using calculus from the definitions $a = dv/dt$ and $v = dx/dt$, respectively. The procedure involves integrating the expressions $dv = a\,dt$ and $dx = v\,dt$ between the proper limits. Calculus can also be used to study motion when acceleration is not a constant.

*2–9 Graphical Analysis and Numerical Integration

If the velocity $v(t)$ of an object is given by a graph such as shown, the total displacement between any two times t_1 and t_2 is equal to the sum of the displacements over the

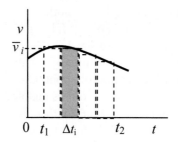

subintervals as given by the expresion:

$$x_2 - x_1 = \sum_{t_1}^{t_2} \overline{v}_i \, \Delta t_i \qquad (2\text{–}12\text{a})$$

In the limit Δt_i tends to zero, the total displacement is equal to the area between the velocity curve and the t axis between the times t_1 and t_2:

$$x_2 - x_1 = \lim_{\Delta t \to 0} \sum_{t_1}^{t_2} \overline{v}_i \, \Delta t_i, \; = \int_{t_1}^{t_2} v(t) dt. \qquad (2\text{–}12\text{b})$$

where $\overline{v}_i$ i represents the average velocity over the interval Δt_i.

Similarly, the total change in velocity for the same time interval is

$$v_2 - v_1 = \lim_{\Delta t \to 0} \sum_{t_1}^{t_2} \overline{a}_i \, \Delta t_i, \; = \int_{t_1}^{t_2} a(t) dt. \qquad (2\text{–}12\text{c})$$

If the velocity or acceleration is known at discrete intervals, the summation forms of the preceding equations can be used to determine the displacement or velocity, respectively. This procedure is called **numerical integration**.

Tips for Solving the Problems:

1. Read the problem carefully. Draw a sketch of the problem. Determine the subcategory to which the problem belongs, such as motion with constant velocity or motion with constant acceleration.

2. Recognize the symbols or notations, such as Δx, Δv, Δt, dx/dt, dv/dt, and d^2x/dt^2.

3. Make sure to note the difference between the two positions of an object to calculate its displacement, Δx. Also, subtract the two times to find the time interval, Δt. If there is no acceleration, instantaneous velocity and average velocity are the same.

4. For problems involving motion with constant acceleration (such as those in Sections 2–5 and 2–6), use the equations introduced in Section 2–5. Each of these equations relates four variables and usually all of them will be given except one. Find the right equation to determine the unknown. It may help to make a table like the following:

v	v_0	a	t
Given	Given	Given	?

5. Problems involving free-fall motion (such as in Section 2–7) can be solved using the equations from Section 2–5. In this case, $a = g = 9.80$ m/s^2. If upward direction is considered positive, g is negative.

6. Remember that the slope of a *position-versus-time* graph is velocit*y*, and the slope of a *velocity-versus-time* graph is acceleration.

7. To get instantaneous velocity, v, from an expression of the position, x, take the derivative of x with respect to t, and to get instantaneous acceleration, a, from an expression of v, take the derivative of v (that is, second derivative of x) with respect to t. On the other hand, to obtain v from an expression of a, integrate the expression of a with respect to t. Also, to obtain x from an expression of v, integrate the expression of v with respect to t.

8. To determine displacement and velocity by numerical integration, use the equations $x_2 - x_1 = \sum_{t_1}^{t_2} \overline{v}_i \, \Delta t_i$ and

    $v_2 - v_1 = \sum_{t_1}^{t_2} \overline{a}_i \, \Delta t_i$, respectively. As you use more and smaller subintervals, you will get a more accurate result.

CHAPTER 3
Kinematics in Two or Three Dimensions; Vectors

This chapter describes the rules of addition and subtraction of vectors. The chapter also explores kinematics in two dimensions, including projectile motion.

Important Concepts

Scalars and vectors
Addition and subtraction of vectors
Resultant
Tail-to-tip method
Parallelogram method
Components of a vector
Unit vectors
Projectile motion
Range for a projectile motion
Maximum height for a projectile motion
Time of flight for a projectile motion
Path of a projectile
Relative velocity

3–1 Vectors and Scalars

Physical quantities are generally of two types:

- A physical quantity that has both a magnitude and a direction is called a **vector**. Examples include displacement, velocity, acceleration, force, and momentum.

- A quantity that can be expressed by a number and its unit is called a **scalar**. Examples include mass, time, volume of a container, and temperature of air.

An arrow is drawn to indicate a vector on a diagram. The length of the arrow is proportional to the magnitude of the vector, and the direction of the arrow specifies the direction of the vector. **Boldface** with a tiny arrow over the symbol is used to indicate a vector in a written symbol (such as $\vec{\mathbf{v}}$ for velocity), and *italic* (such as *v* for velocity) is used for its magnitude.

3–2 Addition of Vectors—Graphical Methods

Simple arithmetic, which is used to add scalars, can be used to add vectors if the vectors are in the same direction. But simple arithmetic does not work if the two vectors do not act along the same line. For example, let $\vec{\mathbf{D}}_1$ represents the displacement vector of a person to the east and $\vec{\mathbf{D}}_2$ represent his subsequent displacement vector to the north.

In this case, the magnitude of the **resultant displacement** $\vec{\mathbf{D}}_R$ can be obtained from the Pythagorean theorem, since D_1, D_2, and D_R form a right angle with D_R as the hypotenuse. Thus, $D_R = \sqrt{D_1{}^2 + D_2{}^2}$.

Tail-to-tip method: To add two vectors $\vec{V}_1$ and $\vec{V}_2$ by the tail-to-tip method, place the tail of $\vec{V}_2$ at the head of $\vec{V}_1$. The resultant vector $\vec{V}_R$ is the vector joining the tail of $\vec{V}_1$ to the head of $\vec{V}_2$, or $\vec{V}_R = \vec{V}_1 + \vec{V}_2$. Measure the length of $\vec{V}_R$ and change the length to the magnitude by multiplying with the chosen scale factor. Measure the angle with the positive x-axis for the direction.

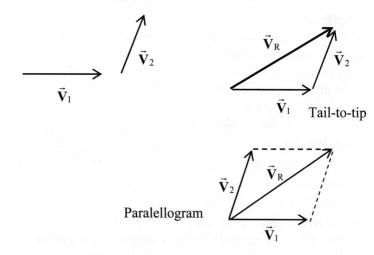

Tail-to-tip

Paralellogram

Parallelogram method: Draw $\vec{V}_1$ and $\vec{V}_2$ tail to tail and form a parallelogram as shown. The diagonal represents the resultant vector $\vec{V}_R$ whose magnitude and direction can be measured from the diagrams as mentioned in the tail-to-tip method.

Note that, $\vec{V}_1 + \vec{V}_2 = \vec{V}_2 + \vec{V}_1$.   [commutative law] (3–1a)

The tail-to-tip method can be extended to add more than two vectors. Place all the vectors head-to-tail one by one. The line joining the tail of the first vector to the head of the last vector represents the resultant vector.

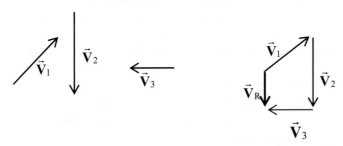

$\vec{V}_1 + (\vec{V}_2 + \vec{V}_3) = (\vec{V}_1 + \vec{V}_2) + \vec{V}_3$. [associative law] (3–1b)

3–3 Subtraction of Vectors and Multiplication of a Vector by a Scalar

An arrow having the same length as the original vector, but pointing in the opposite direction, represents the *negative* of a vector.

$$\overleftarrow{-\vec{V}} \qquad \overrightarrow{\vec{V}}$$

To subtract two vectors $\vec{V}_1$ and $\vec{V}_2$, add $\vec{V}_2$ with $-\vec{V}_1$.

Thus, $\vec{V}_2 - \vec{V}_1 = \vec{V}_2 + (-\vec{V}_1)$.

Multiplying $\vec{V}$ by a scalar c, we get a vector $c\vec{V}$ whose magnitude is c times that of $\vec{V}$. If c is positive, the direction of $c\vec{V}$ is the same as that of $\vec{V}$; if c is negative, the direction of $c\vec{V}$ is opposite to that of $\vec{V}$.

3–4 Adding Vectors by Components

The graphical method of addition of vectors is not very accurate and is not useful for vectors in three dimensions. It is necessary to resolve vectors into their components to add vectors precisely. To find the components of a vector, you need to set up a coordinate system. The components of a vector $\vec{V}$ are the projections (length of the perpendicular) of the vector on the x and y axes. The components are written as V_x and V_y.

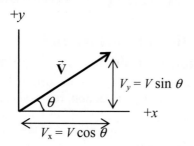

If a vector $\vec{V}$ makes an angle θ relative to the positive x axis, the x and y components of the vector are given by (respectively)

$$V_x = V \cos \theta \qquad (3\text{–}2a)$$

$$V_y = V \sin \theta \qquad (3\text{–}2b)$$

V_x and V_y can be positive, negative, or zero.

If the components of a vector are given, we can determine the magnitude of the vector using the Pythagorean theorem, and its direction using the inverse tangent of the ratio of the components:

$$V = \sqrt{V_x^2 + V_y^2} \qquad\qquad (3\text{–}3a)$$

$$\tan\theta = \frac{V_y}{V_x} \qquad\qquad (3\text{–}3b)$$

To add two vectors using components, the first step is to resolve each vector into its components. If $\vec{V}_1$ and $\vec{V}_2$ are the two vectors to be added, and the resultant vector is $\vec{V}$, then $\vec{V} = \vec{V}_1 + \vec{V}_2$. In this case, it can be shown that

$V_x = V_{1x} + V_{2x}$ and

$V_y = V_{1y} + V_{2y} \qquad\qquad (3\text{–}4)$

where V_x and V_y are the components of the resultant vector $\vec{V}$. The magnitude and direction of the resultant vector $\vec{V}$ can be obtained using Equations 3–4.

3–5 Unit Vectors

A **unit vector** have a magnitude exactly equal to one. It points along the coordinate axes. In a rectangular coordinate system, the unit vectors pointing along the positive x, y, and z axes are called $\hat{\mathbf{i}}$, $\hat{\mathbf{j}}$, and $\hat{\mathbf{k}}$, respectively.

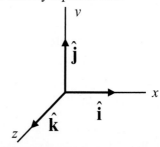

An arbitrary vector $\vec{V}$ can be conveniently expressed in terms of its components using unit vectors as:

$$\vec{V} = V_x\hat{\mathbf{i}} + V_y\hat{\mathbf{j}} + V_z\hat{\mathbf{k}}$$

In unit vector notation, vector addition (for the two-dimensional case) can be written as

$$\vec{V} = \vec{V}_1 + \vec{V}_2 = (V_{1x} + V_{2x})\,\hat{\mathbf{i}} + (V_{1y} + V_{2y})\,\hat{\mathbf{j}} \quad (3\text{--}5)$$

3–6 Vector Kinematics

A vector $\vec{r}_1$ drawn from the origin to a point P_1 (describing the position of a moving object at time t_1) is called a position vector. In unit vector notation,

$$\vec{r}_1 = x_1\hat{\mathbf{i}} + y_1\hat{\mathbf{j}} + z_1\hat{\mathbf{k}} \qquad (3\text{--}6a)$$

where x_1, y_1, and z_1 are the x, y, and z coordinates of the position, respectively. The position vector for the point P_2 is

$$\vec{r}_2 = x_2\hat{\mathbf{i}} + y_2\hat{\mathbf{j}} + z_2\hat{\mathbf{k}}$$

The **displacement vector** is given by the change in position:

$$\Delta\vec{r} = \vec{r}_2 - \vec{r}_1$$

$$= (x_2 - x_1)\,\hat{\mathbf{i}} + (y_2 - y_1)\,\hat{\mathbf{j}} + (z_2 - z_1)\,\hat{\mathbf{k}} \qquad (3\text{--}6b)$$

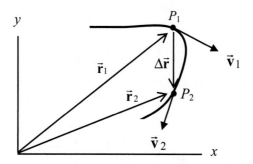

The **average velocity vector** is the ratio of the displacement vector, $\Delta\vec{\mathbf{r}}$, and the elapsed time, Δt.

$$\text{Average velocity} = \frac{\Delta\vec{\mathbf{r}}}{\Delta t} \qquad (3\text{--}7)$$

The **instantaneous velocity vector**, $\vec{\mathbf{v}}$, can be obtained in the limit $\Delta t \to 0$.

$$\vec{\mathbf{v}} = \lim_{\Delta t \to 0} \frac{\Delta\vec{\mathbf{r}}}{\Delta t} = \frac{d\vec{\mathbf{r}}}{dt} \qquad (3\text{--}8)$$

$$= v_x\hat{\mathbf{i}} + v_y\,\hat{\mathbf{j}} + v_z\hat{\mathbf{k}} \qquad (3\text{--}9)$$

The **average acceleration vector** is the ratio of the velocity vector $\Delta\mathbf{v}$ divided by the elapsed time.

$$\text{Average acceleration} = \frac{\Delta\vec{\mathbf{v}}}{\Delta t} \qquad (3\text{--}10)$$

The instantaneous acceleration vector is

$$\vec{\mathbf{a}} = \lim_{\Delta t \to 0} \frac{\Delta \vec{\mathbf{v}}}{\Delta t} = \frac{d\vec{\mathbf{v}}}{dt} \qquad (3\text{–}11)$$

$$= \frac{dv_x}{dt}\,\hat{\mathbf{i}} + \frac{dv_y}{dt}\,\hat{\mathbf{j}} + \frac{dv_z}{dt}\,\hat{\mathbf{k}}$$

$$= a_x\hat{\mathbf{i}} + a_y\,\hat{\mathbf{j}} + a_z\vec{\mathbf{k}} \qquad (3\text{–}12)$$

$$= \frac{d^2x}{dt^2}\,\hat{\mathbf{i}} + \frac{d^2y}{dt^2}\,\hat{\mathbf{j}} + \frac{d^2z}{dt^2}\,\hat{\mathbf{k}} \qquad (3\text{–}12a)$$

The velocity vector, $\vec{\mathbf{v}}$, is always in the direction of the motion of the object. The acceleration vector, $\vec{\mathbf{a}}$, can point in a direction other than the direction of motion.

Constant Acceleration

To study motion in two dimensions at constant acceleration, we apply the kinematic equations separately for the horizontal (x) and vertical (y) components of motion. The equations are listed below where v_x and v_y are velocity components and a_x and a_y are the components of the acceleration, respectively.

Table 3–1

x component (horizontal)	y component (vertical)
$v_x = v_{x0} + a_x t$	$v_y = v_{y0} + a_y t$
$x = x_0 + v_{x0}t + \dfrac{1}{2}a_x t^2$	$y = y_0 + v_{y0}t + \dfrac{1}{2}a_y t^2$
$v_x^2 = v_{x0}^2 + 2a_x(x - x_0)$	$v_y^2 = v_{y0}^2 + 2a_y(y - y_0)$

In vector notations, the first two equations are:

$$\vec{v} = \vec{v}_0 + \vec{a}t \qquad [\,\vec{a} = \text{constant}\,] \qquad (3\text{–}13a)$$

$$\vec{r} = \vec{r}_0 + \vec{v}_0 t + \frac{1}{2}\vec{a}t^2 \qquad [\,\vec{a} = \text{constant}\,] \qquad (3\text{–}13b)$$

3–7 Projectile Motion

A **projectile** is an object (such as a batted baseball) that has been projected by some means and moves under the influence of gravity only. Projectile motion is an example of a two-dimensional motion.

A projectile motion can be understood by analyzing its horizontal and vertical components of motion separately since the horizontal and vertical components are independent of each other. Because the motion in the vertical direction is not affected by the horizontal motion, *an object projected horizontally reaches the ground at the same time as an object dropped vertically from the same height.*

For a projectile projected horizontally from a certain height, its x-component of velocity, v_x, remains constant whereas its y-component of velocity, v_y, increases as it falls because of the earth's gravity, as shown below. In this case, $a_x = 0$, and $a_y = -g$.

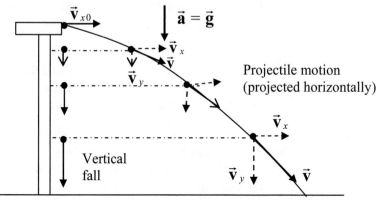

Projectile motion (projected horizontally)

For a projectile projected at an upward angle, the vertical component v_y decreases from its initial value v_{y0} as it goes up, becomes zero at the maximum height, and then increases in magnitude as it falls towards the ground, while v_x remains constant, as shown below.

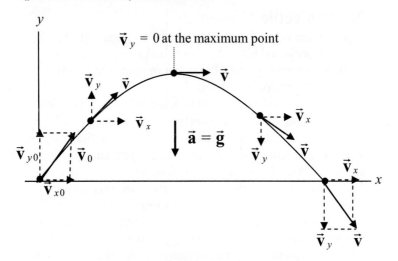

Projectile motion exhibits striking symmetries:

- Because of symmetry in the path of a projectile, the time a projectile takes to reach its highest point is half of the total time of flight.

- The initial y component of velocity and the final y component of velocity when the projectile lands are equal in magnitude but oppositely directed.

- At any given height, the speed of the projectile is the same on the way up as on the way down.

Symmetry considerations are useful for solving problems.

3–8 Solving Problems Involving Projectile Motion

To study motion in two dimensions, we apply the kinematic equations (listed in Table 3–1) separately for the horizontal (x) and vertical (y) components of motion.

TABLE 3–2 Kinematic Equations for Projectile Motion ($a_x = 0$, $a_y = -g = -9.80 \ \text{m/s}^2$)

Horizontal Motion	**Vertical Motion**
($a_x = 0$, v_x = constant)	($a_y = -g$ = constant)
$v_x = v_{x0}$	$v_y = v_{y0} - g\,t$
$x = x_0 + v_{x0}t$	$y = y_0 + v_{y0}t - \dfrac{1}{2}g\,t^2$
	$v_y^2 = v_{y0}^2 - 2g(y - y_0)$

For a projectile motion acceleration in the y direction, $a_y = -g = -9.80 \ \text{m/s}^2$ (assuming upward direction is positive). Since gravity causes no acceleration in the x direction, $a_x = 0$. The equations are listed in Table 3–2.

A projectile that is launched at an angle θ_0, with respect to the horizontal has $v_{x0} = v_0 \cos \theta_0$ and $v_{y0} = v_0 \sin \theta_0$.

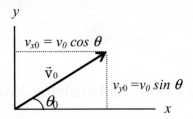

The maximum horizontal distance that a projectile travels before it lands is called the range, R, of the projectile. To determine range, we first determine the total time of flight by setting $y = 0$ in the expression for y. It can be shown that the total time of flight is

$$t = \left(\frac{2v_0}{g}\right) \sin \theta_0$$

Substituting this time into the x equation of motion, we get:

$$R = \frac{v_0^2 \sin 2\theta_0}{g}$$

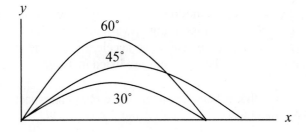

- Range is a maximum when $\sin 2\theta_0 = 1$. That is $\theta_0 = 45°$.

- Also, there are two angles (such as, 30° and 60°) for which sin $2\theta_0$ and, hence, range R is the same, as shown.

Projectile Motion Is Parabolic

For a projectile motion:

$$x = v_{x0}t$$

$$y = v_{y0}\,t - \frac{1}{2}gt^2$$

Eliminating t from these equations, we get,

$$y = \left(\frac{v_{y0}}{v_{x0}}\right)x - \left(\frac{g}{2v_{x0}^2}\right)x^2 = Ax - Bx^2. \quad (3\text{–}14)$$

This is an equation of a parabola. Thus, the path of a projectile is a parabola.

3–9 Relative Velocity

When an object is in motion in a reference frame that is also in motion, the motion of the object will generally appear different to an observer at rest outside the reference frame versus an observer at rest in the reference frame. For example, the motion of a boat inside moving water in a river will appear different to an observer on the shore versus an observer in the river moving with the same velocity as the velocity of water. If $\vec{V}_{BW}$ and $\vec{V}_{WS}$ are the velocity of the **B**oat with respect to the **W**ater and velocity of the **W**ater with respect to the **S**hore, then velocity of the **B**oat with respect to the **S**hore, $\vec{V}_{BS}$, is given by

$$\vec{V}_{BS} = \vec{V}_{BW} + \vec{V}_{WS}.$$ (3–15)

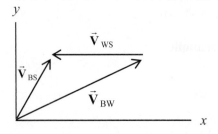

For any two objects or reference frames, A and B, the velocity of A relative to B has the same magnitude, but opposite direction, as the velocity of B relative to A. That is, reversing the subscript reverses the velocity

$$\vec{V}_{BA} = -\vec{V}_{AB}$$ (3–16)

Tips for Solving the Problems

1. To find the components of a vector $\vec{V}$ use $V_x = V \cos \theta$ and $V_y = V \sin \theta$. Remember that θ is the angle made by the vector with the positive x axis.

2. Given the components of a vector, find the vector by using the Pythagorean theorem (for its magnitude) and the inverse tangent of the ratio of its y and x components (for its direction).

3. To add two vectors by components, simply add the x components of the individual vectors to find the x component of the resultant vector ($V_x = V_{1x} + V_{2x}$). Add the y components of the individual vectors to find the y component of the resultant vector ($V_y = V_{1y} + V_{2y}$). The

result is conveniently expressed in unit vector notations as $\vec{V} = V_x \hat{i} + V_y \hat{j}$. If needed, use the Pythagorean theorem and the inverse tangent of the ratio of the y and x components of the resultant to find the magnitude and direction of the resultant, respectively (problems such as those in Sections 3–2 to 3–5).

4. Two-dimensional motion can be described as *two* one-dimensional motions independent of each other. Motion in each direction can be studied using the equations listed in Table 3–1.

5. For projectile motion, there is acceleration $a_y = -g = -9.80 \text{ m/s}^2$ in the y direction. There is no acceleration in the x direction. Thus, $v_{x0} = v_x = v_0$.

6. For a projectile launched horizontally, $v_{x0} = v_0$ and $v_{y0} = 0$. For a projectile launched with an arbitrary angle θ, its initial x and y components of velocity are $v_{x0} = v_0 \cos \theta_0$ and $v_{y0} = v_0 \sin \theta_0$, respectively. Make these substitutions to solve problems such as those in Sections 3–7 and 3–8.

7. At the maximum height, velocity is in the x direction, that is, $v_y = 0$. Use this fact to determine the *maximum height*. When a projectile reaches the ground, $y = 0$. Use this fact to determine its *time of flight*. Find the *range* by multiplying the time of flight by the speed in the x direction.

8. Take the advantage of symmetry when solving projectile motion problems. Because of symmetry in projectile motion, the time it takes to reach the maximum height is half of the total time of flight.

CHAPTER 4
Dynamics: Newton's Laws of Motion

This chapter describes Newton's three laws of motion, which describe the dynamics of linear motion. It also introduces the concepts of weight, tension, and normal force.

Important Concepts

>Force
>Newton's first law of motion
>Inertia and mass
>Inertial frame of reference
>Newton's second and third laws of motion
>Weight
>Normal force
>Free-body diagram

4–1 Force

Force is simply a push or pull. For example, when you push a grocery cart, you exert a force on it; an apple falls toward the earth because of the force of gravity.

A force has a *magnitude* and a *direction*. Force is, thus, a vector quantity and is represented by an arrow. The direction of the arrow is the direction of the force and the length of the arrow is proportional to the magnitude of the force.

4–2 Newton's First Law of Motion

Newton's first law of motion states that *everybody continues in its state of rest or of uniform speed in a straight line unless acted on by a nonzero net force.*

Newton's first law is known as the **law of inertia. Inertia** is the natural tendency of an object to maintain its state of rest or of motion with a constant velocity.

Inertial Reference Frame
A frame of reference in which the law of inertia holds is called an **inertial frame of reference**. A frame of reference that moves with a constant velocity relative to an inertial frame of reference is also inertial. A frame of reference that accelerates relative to an inertial frame is called a **noninertial** frame. Earth is considered to be an inertial frame, since the acceleration of the surface of the Earth due to its rotational and orbital motions is very small.

4–3 Mass
The **mass** of an object is a measure of the *quantity of matter* in the object. Mass is also a *measure of inertia of a body*, or, in other words, mass is a measure of how much an object resists changes in its motion.

Mass is a property of a body, while weight is a force, the force of gravity acting on it. If we take an object on the moon, its mass will be the same but its weight will be about one-sixth of its weight on the earth, since the force of gravity on the moon is weaker.

4–4 Newton's Second Law of Motion
Newton's second law of motion provides a relationship between the cause (force) and the effect (acceleration) of a motion. It states that *the acceleration of an object is directly proportional to the net force acting on it and is inversely proportional to its mass. The direction of acceleration is in the direction of the net force acting on*

the object. If $\sum \vec{F}$ is the *net force* acting on an object of mass m, the acceleration $\vec{a}$ of the object is given by

$$\sum \vec{F} = m\,\vec{a} \qquad\qquad (4\text{--}1a)$$

Since $\vec{F} = F_x\,\vec{i} + F_y\,\vec{j} + F_z\,\vec{k}$, in terms of components:

$$\Sigma F_x = ma_x \quad \Sigma F_y = ma_y \quad \Sigma F_z = ma_z \qquad (4\text{--}1b)$$

From Equation 4–1, we can define *force as an action capable of accelerating an object.*

Force is measured in **newtons** (N), from Equation 4–1, $1\ \text{N} = 1\ \text{kg} \cdot \text{m/s}^2$. In cgs units, the unit of force is *dyne*. 1 dyne $= 1\ \text{g} \cdot \text{cm/s}^2 = 10^{-5}\ \text{N}$. In the British system, the unit of force is *pound* (lb). $1\ \text{lb} = 1\ \text{slug} \cdot \text{ft/s}^2$. $1\ \text{lb} = 4.45\ \text{N}$.

Precise Definition of Mass

From the second law, if the same net force acts on two objects of masses m_1 and m_2, the ratio of their masses can be defined as:

$$\frac{m_2}{m_1} = \frac{a_1}{a_2}$$

4–5 Newton's Third Law of Motion

Whenever one object exerts a force on a second object, the second object exerts an equal and opposite force on the first object. This is **Newton's third law**. This law is also stated as "to every action there is an equal and opposite reaction."

For example, a person walks by pushing with his feet against the ground. If the force exerted by a person on the ground is $\vec{\mathbf{F}}_{PG}$ and the force exerted by the ground on the person is $\vec{\mathbf{F}}_{GP}$, then

$$\vec{\mathbf{F}}_{GP} = -\vec{\mathbf{F}}_{PG}. \qquad (4\text{--}2)$$

- Because of the third law, when an ice skater pushes against a railing, the railing pushes the skater back, causing her to move away.

- When a rocket is launched, the rocket exerts a strong force downward on the escaping gases and the rocket moves upward because of the equal and opposite reaction force by the gases on the rocket.

- When a person throws a packet from a boat, the boat moves backward because of the reaction force on the person and the boat by the packet.

Note: Action–reaction forces act on two *different* objects; hence they do *not* cancel. Action–reaction forces generally produce different accelerations, since the masses of the objects are likely to be different.

4–6 Weight—the Force of Gravity; and the Normal Force

The **weight** of an object is the amount of force of gravity exerted by Earth on the object. Since the downward acceleration due to gravity near the surface of the earth is $\vec{\mathbf{g}}$, from Newton's second law (Equation 4–1), the force of gravity $\vec{\mathbf{F}}_{G}$ is given by

$$\vec{\mathbf{F}}_{G} = m\vec{\mathbf{g}} \qquad (4\text{--}3)$$

$$= -mg\,\vec{\mathbf{j}} \quad \text{(assuming } y\text{-axis is upwards)}$$

The SI unit of weight is the newton, N. Since $g = 9.8 \text{ m/s}^2$, the weight of a 1.00-kg mass on Earth is $1.00 \text{ kg} \times 9.8 \text{ m/s}^2 = 9.8 \text{ N}$.

When an object rests on a table, the table provides a force on the object in a direction perpendicular to its surface. This is called the **normal force**, $\vec{\mathbf{F}}_N$. For an object on a horizontal surface (such as on a table), the normal force is vertical. The force exerted by the table is also called **contact force**, since it occurs when two objects are in contact. The origin of the normal force is the interaction between atoms in a solid that act to maintain its shape.

In most cases, the normal force is equal to the weight of the object; however, the normal force can be larger or smaller than the weight of the object.

4–7 Solving Problems with Newton's Laws: Free-Body Diagrams

The **net force** exerted on an object is the *vector sum* of the individual forces acting on it. When solving problems involving force and Newton's second law, it is convenient to draw a **free-body diagram**, showing *all external forces* acting on the object.

In constructing a **free-body diagram** or **force diagram** for solving problems:

- Isolate the object of interest,

- Sketch the forces,

- Choose a convenient coordinate system, and

- Resolve the forces into components.

Then, apply Newton's second law for each component.

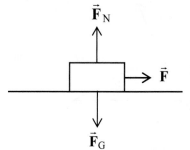

Free-body diagram of an object at rest on a smooth horizontal surface when an external force acts on it to the right

When more than one force acts, the x and y components of the net or total force in that direction can be determined as:

$$\Sigma F_x = F_{1,x} + F_{2,x} + \cdots$$

$$\Sigma F_y = F_{1,y} + F_{2,y} + \cdots$$

Tension in a Flexible Cord

A common way to exert force on an object is to pull it with a flexible cord or a string. The term **tension** is used to describe the force transmitted through a string and it is denoted by $\vec{F}_T$. If the cord has a negligible mass, the force exerted at one end is transmitted undiminished to each adjacent piece of the cord along its entire length.

Inclines

For an inclined surface, it is convenient to choose the x and y axes parallel and perpendicular to the inclined surface, respectively. For an inclined plane, the normal force is still at right angles to the surface, but not in the vertical direction.

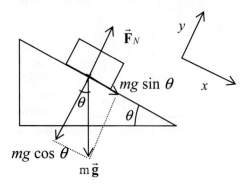

If an object goes down with an acceleration a_x on a frictionless inclined surface, then from the second law:

$\Sigma F_x = mg \sin \theta = ma_x$

$\Sigma F_y = F_N - mg \cos \theta = ma_y = 0$

Solving, we get

$F_N = mg \cos \theta$, and $\quad a_x = g \sin \theta$.

4–8 Problem Solving—A General Approach

A basic part of physics is how to solve problems effectively. The problem-solving strategy is discussed here emphasizing Newton's laws, but the strategy can be applied also for other problems.

1. Read the problem carefully.

2. Draw a diagram of the situation. When involving forces, draw a *free-body diagram*

showing *all* the forces acting on the object(s) of interest.

3. Choose a convenient *xy* coordinate system. Resolve force vectors along the axes. Then apply $\sum \vec{F} = m\vec{a}$ separately for each component.

4. Identify the unknowns. In general, determine appropriate equations that relate to the problem.

5. Try to solve the problem approximately and find out if it is doable and reasonable.

6. Solve the equations and carry out the calculation and report the result in scientific notation to the correct number of significant figures.

7. Keep track of units that may serve as a check on your solution.

8. Check if the result is reasonable.

Tips for Solving the Problems

1. Begin solving a problem with a sketch. Isolate each object in the problem and draw a free-body diagram for each object of interest, showing all external forces with their directions.

2. Establish a convenient coordinate system. Find $\sum \vec{F}$ for each component. Apply Newton's second law of motion separately for each component.

3. Action–reaction forces always act on two *different* bodies, and they *never cancel* each other.

4. Remember that the normal force is always normal to the surface. The force of gravity always acts vertically downward.

5. There is no acceleration in the normal direction. The sum of the normal components of the forces must be zero.

6. For an object moving on a flat surface, the normal force and the weight are equal and opposite. For an object on an inclined plane, their magnitudes are different.

7. If the forces are neither parallel nor perpendicular to the actual or impending motion, resolve the forces into x and y components. Find the net force ΣF_x and ΣF_y in the x and y directions.

8. If you need to resolve forces in the x and y directions for inclined surfaces, choose x and y axes parallel and perpendicular to the inclined surface, respectively.

9. Pulleys are used to change the direction of tension. If the mass of a string is negligible, the tension in the string has the same magnitude throughout its length.

10. If an object is in equilibrium (that is, at rest or moving at a constant velocity), the net acceleration $\vec{\mathbf{a}} = 0$. This means that $\Sigma F_x = 0$ and $\Sigma F_y = 0$. Use these equations to solve the problems for the objects in equilibrium.

11. For problems involving systems moving with a constant acceleration, apply Newton's second law separately for the motion of each object.

CHAPTER 5
Using Newton's Laws: Friction, Circular Motion, Drag Forces

This chapter describes applications of Newton's laws of motion, including friction, and also dynamics of circular motion.

Important Concepts
> Kinetic friction
> Coefficient of kinetic friction
> Static friction
> Coefficient of static friction
> Centripetal acceleration and centripetal force
> Tangential acceleration
> Drag force
> Terminal velocity

5–1 Applications of Newton's Laws Involving Friction

Friction is a resistive force that opposes the relative motion between two surfaces of contact. On the atomic level, it is electromagnetic in nature and acts between atoms and molecules of the surfaces in contact. From a macroscopic point of view, friction is caused by the random irregularities of the two surfaces. When an object rolls across a surface, there is also friction, called *rolling friction*, which is usually much less than when a body slides across a surface. There are two types of friction—kinetic friction and static friction.

Kinetic friction is the friction that acts opposite to the direction of body's velocity, when an object is in motion along a surface. Force of kinetic friction depends on the nature of the surfaces of contact, but is independent of the area of contact between the surfaces and independent also of the relative speed of motion. The force of kinetic friction, $F_{fr,}$ between two surfaces is proportional to the magnitude of the normal force, F_N, between the two surfaces, and can be expressed as

$$F_{fr} = \mu_k F_N$$

where the constant μ_k is called the *coefficient of kinetic friction*.

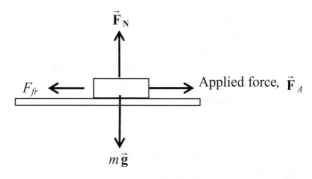

Static friction tends to keep the two surfaces in contact from moving relative to each other. Static friction is stronger than kinetic friction. As the applied force increases, the force of static friction increases. The force of static friction is always equal but oppositely directed to the applied force until the body is about to move, in which case the force of static friction attains its maximum value. Thus, the force of static friction can have any value between zero and a maximum value, which is given by F_{fr} (max) $= \mu_s F_N$ where μ_s is called the *coefficient of static friction*. Thus, we can write $F_{fr} \leq \mu_s F_N$.

Ramps and Inclines

For an object of mass m moving down at a constant velocity on an inclined plane of angle θ, from Newton's second law:

$$\Sigma F_x = mg \sin \theta - \mu_k F_N = ma_x = 0$$

$$\Sigma F_y = F_N - mg \cos \theta = ma_y = 0$$

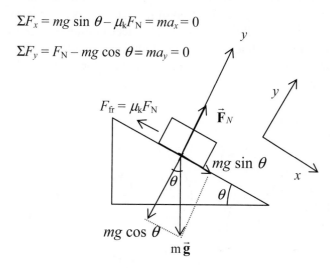

Solving the preceding equations, we get $\mu_k = \tan \theta$. Similarly, when an object is about to slide on the inclined surface, the coefficient of static friction can be shown to be $\mu_s = \tan \theta$.

If the object goes down with an acceleration a_x on the inclined surface, then the equations are:

$$\Sigma F_x = mg \sin \theta - \mu_k F_N = ma_x$$

$$\Sigma F_y = F_N - mg \cos \theta = ma_y = 0$$

Solving, we get

$$F_N = mg \cos \theta, \text{ and } a_x = g (\sin \theta - \mu_k \cos \theta).$$

5–2 Uniform Circular Motion—Kinematics

An object moving in a circle at a constant speed v is said to execute uniform circular motion. Although speed is constant, the velocity $\vec{v}$ of the object continually changes due to its change of direction of motion, and, as a result, the object continually accelerates. This acceleration is called the **centripetal acceleration** or radial acceleration since the acceleration is directed always toward the center of the circle.

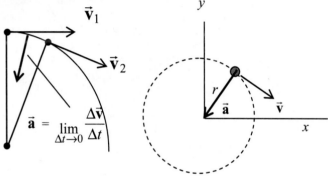

For an object of mass m rotating with a constant speed v in a circular path of radius r, the magnitude of the centripetal acceleration is given by

$$a_R = \frac{v^2}{r} \tag{5–1}$$

Note that the acceleration vector points toward the center of the circle but the velocity vector points toward the direction of motion, which is tangential to the circle. Thus, the velocity and acceleration vectors are always perpendicular to each other.

The frequency f of a circular motion is defined as the number of revolutions per second. The period T of the motion is the time required to complete one revolution. f and T are related by

$$T = \frac{1}{f}.$$ (5–2)

By definition, speed v = distance/time = $2\pi r/T$.

5–3 Dynamics of Uniform Circular Motion

According to Newton's second law, whenever there is an acceleration of an object, there must be a net force acting on the object. The net force needed to cause centripetal acceleration is called centripetal force.

The centripetal force is given by

$$\Sigma F_R = ma_R = m\frac{v^2}{r}$$ (5–3)

The centripetal force is directed toward the center of the circle.

Centripetal force is not a new kind of force; it is a new name for a force directed toward the center of a circle, and can be produced in many different ways. For a ball tied to a string that is rotating in a circle about a person's head, the tension in the string provides the centripetal force. For a car turning a corner, the force of static friction between tires and the road is the car's centripetal force. For an orbiting satellite and for the Moon orbiting the Earth, the force of gravity is the centripetal force. For a car turning on a smooth banked road, the normal force is the centripetal force.

5–4 Highway Curves: Banked and Unbanked

A common example of centripetal force occurs when an automobile makes a circular turn on a road. A centripetal force is needed to make a turn.

On a flat road the force is supplied by the static friction between the tires and the pavement.
Thus, the maximum speed safe turn is given by

$$\mu_s F_N = m \frac{v^2}{r}.$$

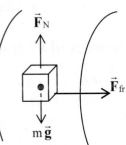

If the friction is not great enough, as under wet or icy conditions, sufficient centripetal force cannot be applied and the car will skid out of a circular path into a nearly straightline path.

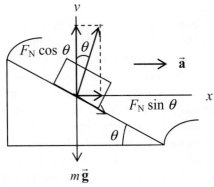

The banking of curves reduces the chance of skidding because the normal force of the inclined road that acts perpendicular to the inclined surface will have a component toward the center of the circle. If friction is neglected, the maximum speed for a safe turn on a banked road with banking angle θ is given by

$$F_N \sin \theta = m \frac{v^2}{r}.$$

Also, in this case,

$$F_N \cos \theta = mg.$$

The banking angle θ from the preceding two equations is given by $\tan \theta = v^2/rg$.

*5–5 Nonuniform Circular Motion

If the net force on an object moving in a circular path is not directed toward the center of the circle, the force has two components. The component of the force directed toward the center F_R causes a centripetal acceleration,

$$a_R = \frac{v^2}{r}.$$

The component of the force F_{tan} that acts to increase or decrease the speed gives rise to a component of the acceleration tangent to the circle, a_{tan}.

$$a_{tan} = \frac{dv}{dt} \qquad (5\text{--}4)$$

In vector notation

$$\vec{a} = \vec{a}_{\tan} + \vec{a}_R. \qquad (5\text{–}5)$$

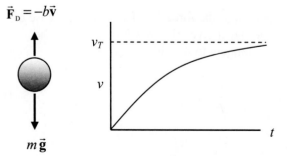

$$a = \sqrt{a_{\tan}^2 + a_R^2}.$$

5–6 Velocity-Dependent Forces: Drag and Terminal Velocity

When an object moves through a fluid (such a liquid or air), the fluid offers resistance to the motion of the object. The resistive force or **drag force** depends on the velocity of the object. For small objects moving at very low speeds, the drag force, F_D, can be expressed as:

$$F_D = -bv \qquad (5\text{–}6)$$

where b is a constant that depends on the viscosity of the fluid and on the size and shape of the object, and v is the velocity.

Thus, $\Sigma F = ma = m\,\dfrac{dv}{dt} = mg - bv \qquad (5\text{–}7)$

The velocity initially increases, but at a slower rate as the object falls until the drage force $-bv$ becomes equal and opposite to the force of gravity, mg. The object then falls with a constant velocity called the **terminal velocity**, v_T, which is given by the equation:

$$mg - bv = 0. \qquad (5\text{–}8)$$

This gives

$$v_T = \frac{mg}{b}. \qquad (5\text{–}9)$$

Solving Equation 5–7, we get

$$v = v_T\,(1 - e^{\,(b/m)\,t}).$$

The graph of v vs. t is shown.

Tips for Solving the Problems

1. Find the normal force in order to calculate the force of friction. If the object is at rest, calculate the force of static friction. If the object is moving with a constant velocity, calculate the force of kinetic friction.

2. If the forces are neither parallel nor perpendicular to the actual or impending motion, resolve the forces into x and y components. Find the net force ΣF_x and ΣF_y in the x and y directions.

3. If you need to resolve forces in the x and y directions for inclined surfaces, choose x and y axes parallel and perpendicular to the inclined surface, respectively.

4. If an object is in equilibrium (that is, at rest or moving at a constant velocity), the net acceleration $\vec{a} = 0$. This means that $\Sigma F_x = 0$, and $\Sigma F_y = 0$. Use these equations to solve the problems for the objects in equilibrium.

5. Note that centripetal acceleration is v^2/r, whereas tangential acceleration is dv/dt.

6. To solve problems involving centripetal force, determine the net force toward the center and set that equal to mv^2/r.

7. To calculate terminal velocity in the presence of drag forces, set the net force to zero in the direction of motion and solve for v. For $F_D = -bv$, $v = v_T \left(1 - e^{(b/m)\,t}\right)$ where $v_T = mg/b$.

CHAPTER 6
Gravitation and Newton's Synthesis

This chapter describes the dynamics of uniform motion of an object in a circle and the force of gravity.

Important Concepts

> Newton's law of universal gravitation
> Satellites
> Weightlessness
> Geosynchronous satellite
> Kepler's laws
> Gravitational field
> Fundamental forces
> Inertial mass and gravitational mass
> Principle of equivalence
> Black hole

6–1 Newton's Law of Universal Gravitation

Besides developing three laws of motion, Newton studied the motion of planets and the moon. Since all objects falling toward the earth accelerate, earth must exert a force on them, which Newton called the force of gravity. The force of gravity acts also on the moon. Newton calculated the magnitude of the gravitational force that the earth exerts on the moon compared to the gravitational force on objects at the earth surface. He found that, in terms of the acceleration due to gravity at the earth's surface, this ratio is equivalent to $a_R = g/3600$ where a_R is the acceleration of the moon toward the earth. That is, a_R is diminished by a factor of 3600, which is proportional to the square of the distance between the earth and the moon. Thus, Newton concluded that the gravitational force exerted by an object decreases as the square of the distance r from the earth's center. Newton also realized that the force of gravity on an object depends on the mass of the object.

Every object in the universe attracts every other object with a force, called the force of gravitation. According to Newton's **law of universal gravitation**, the force of attraction between two objects (particles) of mass m_1 and m_2 is given by

$$F = G \frac{m_1 m_2}{r^2} \qquad (6\text{--}1)$$

where r is the distance between the masses, and G is a constant called the *universal gravitational constant*.

The force of gravity is an action–reaction force. That is, the force on m_1 is equal in magnitude but in opposite direction to the force exerted on m_2.

In 1798, Cavendish experimentally measured the value of the universal gravitational constant, G. Its accepted value is $G = 6.67 \times 10^{-11} \text{ N} \cdot \text{m}^2/\text{kg}^2$. The force of gravity between two ordinary objects on the Earth is negligibly small, since G is very small. The force of gravity by the Earth on a macroscopic object on or near the surface of the Earth is noticeable, however.

Newton's law of gravity applies to point objects. For finite objects, in general, the force of gravity is calculated by dividing the object into a collection of small mass elements, then using superposition and methods of calculus to determine the net gravity force. For an object with a uniform spherical shape, the final result is simple. The net force exerted by a sphere on a point mass m is the same as if all the mass of the sphere were concentrated at the center of the sphere.

6–2 Vector Form of Newton's Law of Universal Gravitation

In vector form the law of gravitation is written as

$$\vec{\mathbf{F}}_{12} = -G\frac{m_1 m_2}{r_{21}^2}\hat{\mathbf{r}}_{21} \qquad (6\text{--}2)$$

where $\vec{\mathbf{F}}_{12}$ is the vector force on mass m_1 exerted by mass m_2, $\hat{\mathbf{r}}_{21}$ is a unit vector that is directed from m_2 to m_1, and r_{21} is the distance between m_1 and m_2.

$$\vec{\mathbf{F}}_{21} = -\vec{\mathbf{F}}_{12},$$

where $\vec{\mathbf{F}}_{21}$ is the force acting on mass m_2 exerted by the mass m_1.

When more than one mass are present, the net force on mass m_1,

$$\vec{\mathbf{F}}_1 = \vec{\mathbf{F}}_{12} + \vec{\mathbf{F}}_{13} + \vec{\mathbf{F}}_{14} + \dots \qquad (6\text{--}3)$$

6–3 Gravity Near the Earth's Surface; Geophysical Applications

Thus, the gravity force exerted by the earth (mass m_E and radius R_E) on a mass m on the surface of the earth is

$$F = G\frac{mm_E}{R_E^2}$$

This gravitational force is the weight of the object and is simply $F_G = mg$. Comparing the preceding two equations, we get

$$g = \frac{Gm_E}{R_E^2} = 9.80 \text{ m/s}^2. \qquad (6\text{–}4)$$

Before Cavendish did his experiment, the values of g and R_E were known from direct measurements. Knowing G, we can calculate the mass of the Earth:

$$m_E = \frac{gR_E^2}{G} = 5.98 \times 10^{24} \text{ kg.}$$

The value of g given by Equation (6–4) does not give precise values of g at different locations because the earth is not a perfect sphere. The value of g also varies locally because of the presence of irregularities and rocks of different densities. Such variations of g, known as "gravity anomalies" are very small but can be measured by gravimeters that can detect variations in g to 1 part in 10^9.

6–4 Satellites and "Weightlessness"
Satellite Motion

Satellites are objects circling the earth at high speeds. For the motion of a satellite around the earth, the necessary centripetal force is provided by the force of gravity. For a satellite of mass m_{sat} and the earth of mass m_E, the force of gravity between them is

$$F = G\frac{m_{sat}m_E}{r^2}$$

Thus, $\qquad G\dfrac{m_{\text{sat}}m_{\text{E}}}{r^2} = m_{\text{sat}}\dfrac{v^2}{r}$ $\qquad$ (6–5)

Also, $v = \dfrac{2\pi r}{T}$

From the preceding two equations

$$T^2 = \left(\dfrac{4\pi^2}{Gm_{\text{E}}}\right)r^3 \text{ and}$$

$$v = \sqrt{\dfrac{Gm_{\text{E}}}{r}}.$$

Weightlessness

A person inside an artificial satellite feels apparent weightlessness. A similar situation occurs in a freely falling elevator.

Applying Newton's second law to the elevator problem, we find that when the elevator accelerates upward with acceleration a, our apparent weight is

$$w = m(g + a).$$

Thus, when we are in an elevator accelerating upward, we feel heavier. When the elevator accelerates downward with uniform acceleration a,

$$w = m(g - a).$$

Thus, when the elevator accelerates downward, we feel lighter. If the elevator is in free fall, the downward acceleration is g; therefore, $w_a = m(g - g) = 0$. Thus, a person feels apparent "weightlessness" in a freely falling elevator. A satellite is also in a state of continuous free fall. Thus, although force of gravity acts on the satellite, a person inside a satellite, experience apparent weightlessness.

6–5 Kepler's Laws and Newton's Synthesis

Johannes Kepler (1571–1630) formulated three laws called **Kepler's laws of planetary motion** that govern the motion of planets around the Sun.

Kepler's First Law: The path of each planet around the sun is an ellipse with the sun at one focus.

Kepler's Second Law: Each planet moves so that an imaginary line drawn from the sun to the planet sweeps out equal areas in equal intervals of time.

As a result of the second law, a planet moves faster as it approaches the sun and moves more slowly as it goes away from the sun.

Kepler's Third Law: The ratio of the squares of the periods (denoted by T) of any two planets is equal to the ratio of the cubes of their mean distances (denoted by r) from the sun. That is,

$$\left(\frac{T_1}{T_2}\right)^2 = \left(\frac{r_1}{r_2}\right)^3$$

where subscripts 1 and 2 refer to planets 1 and 2, respectively.

Kepler laws can be derived using Newton's laws of motion and the law of universal gravitation. If the orbit of a planet around the Sun (mass M_s) is assumed to be circular of radius r, we get (similar to the motion of a satelliete around the earth)

$$\frac{T^2}{r^3} = \frac{4\pi^2}{GM_s} \qquad (6\text{--}6)$$

Thus, comparing the motion of two planets (subscript 1 refers to the first planet and subscript 2 refers to the second planet) around the Sun

$$\frac{T_1^2}{r_1^3} = \frac{T_2^2}{r_2^3}$$

Or $$\left(\frac{T_1}{T_2}\right)^2 = \left(\frac{r_1}{r_2}\right)^3 \qquad (6\text{--}7)$$

Kepler's third law is also valid for other systems such as a satellite or moon circling the earth or any other planet.

Geosynchronous satellites are satellites having the same period T as the period of rotation of the Earth about its axis ($T = 24$ hours). From the earth they look stationary in the sky. These satellites are used for communications, weather forecasts, and other purposes. From Kepler's third law (the preceding equation) we can determine the distance r of all geosynchronous satellites. It is found that the distance is 42,000 km from the center of the earth or 36,000 km above the earth's surface.

Accurate measurements of the orbits of planets have indicated that they do not precisely follow Kepler's law. Such deviations or perturbations from the elliptic orbits led to the discovery of new planets such as Neptune and Pluto.

*6–6 Gravitational Field

We define **gravitational field, $\vec{g}$**, as the gravitational force per unit mass at any point in space. Thus,

$$\vec{g} = \frac{\vec{F}}{m}. \tag{6–8}$$

The unit of $\vec{g}$ is N/kg.

If the gravitational field is due to a body of mass M, then

$$\vec{g} = -G\frac{M}{r^2}\hat{r}$$

where $\hat{r}$ is a unit vector pointing radially outward from mass M and the minus sign indicates that the field is directed toward mass M.

6–7 Types of Forces in Nature

There are four fundamental forces in nature:
  (1) the gravitational force,
  (2) the electromagnetic force,
  (3) the strong nuclear force, and
  (4) the weak nuclear force.

Electromagnetic and weak nuclear forces have been united and found to be two different manifestations of a single electroweak force. Physicists are looking for grand unified theories (GUT) that will further unify the forces. Ordinary forces, other than gravity, such as pushes, pulls, normal forc, and friction are electromagnetic in origin.

*6–8 Principle of Equivalence; Curvature of Space; Black Holes

Newton's second law $F = ma$, provides the definition of **inertial mass**. The more the inertial mass a body has, the less acceleration it produces. Another kind of mass is **gravitational mass**. Gravitational force is proportional to the product of gravitational masses of two objects (Newton's law of gravitation). No experiment has been able to find any difference between inertial mass and gravitational mass. This is called the **principle of equivalence**. The principle of equivalence can be stated in another way: No observer can determine by experiment if acceleration arises because of a gravitational force or because the reference frame is accelerating.

The principle of equivalence shows that light is deflected due to the gravitational force of a massive body. The general relativity theory predicted that the light from a star would be deflected by 1.75 seconds of arc as it passes near the Sun. Such prediction was first verified during a solar eclipse of the sun in 1919 and then by other subsequent experiments. When a large galaxy or a cluster of galaxies lies between Earth and a more distant galaxy, the intermediate galaxy produces a significant amount of bending of light resulting in multiple images of the distant galaxy. This is called gravitational lensing.

Light travels by the shortest, most direct path between two points. If the light beam follows a curved path (as mentioned above), then the *space must be curved* in the presence of gravity. Thus, according to general relativity, gravity causes a curvature in the four–dimensional space-time continuum.

According to Einstein general theory, objects and light rays move along geodesics in curved space–time. The extreme curvature of space–time is produced by a **black hole**. A black hole is a compact but extremely dense star where gravity is so strong that even light cannot escape.

Tips for Solving the Problems

1. Remember that the force of gravity is *always attractive and is applicable to point objects*. The force of gravity can be obtained from the expression

$$F = G \frac{m_1 m_2}{r^2}.$$

2. If more than one gravitational force is involved in a problem, the net force of gravity is the *vector sum* of the individual forces.

3. A symmetrical object behaves as if all its mass were concentrated at its center. Use this fact to calculate the force of gravity between two extended symmetrical objects.

4. Use Kepler's third law, $T = \left(2\pi / \sqrt{GM_S}\right) r^{3/2}$ to calculate the period of rotation of a planet around the sun. For the period of rotation of a satellite around the earth or any other planet replace M_S by the mass of the Earth or the planet. Once you know the period, T, you

can calculate the speed, v (circumference of the orbit/period).

5. The speed of a satellite can also be calculated directly from the expression $v = \sqrt{G\, m_E / r}$. Note that r represents the distance of the satellite from the center of the earth, which is the radius of the earth plus the height of the satellite. To determine the speed of a planet replace M_E by the M_S in the expression. Tips 4 and 5 will be useful to solve problems in Section 6–5.

CHAPTER 7
Work and Energy

This chapter provides a precise definition of work, calculates work done by constant and variable forces, defines scalar product of vectors, and shows how work is related to the energy of motion, called kinetic energy.

Important Concepts

> Work
> Joule
> Positive work and negative work
> Scalar product or dot product of two vectors
> Kinetic energy
> Work–energy principle

7–1 Work Done by a Constant Force

When we push a shopping cart or pull a suitcase, we do work. The **work** done by a constant force is defined to be the *product of the magnitude of the displacement times the component of the force parallel to the displacement.* That is, work W is

$$W = F_{\parallel} d$$

where $F_{\parallel}$ is the component of $\vec{F}$ parallel to the displacement $\vec{d}$. Thus, if the angle between the force $\vec{F}$ and the displacement $\vec{d}$ is θ then

$$W = Fd \cos \theta \qquad (7\text{–}1)$$

The SI unit of work is newton·meter, $N \cdot m$, or **joule**, J. Work is a scalar quantity.

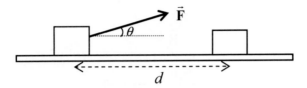

It is clear from Equation 7–1 that

$W = Fd$ when $\theta = 0$,

W is zero when $\theta = \pm 90°$

For example, for the circular motion of the moon around the earth, moon's displacement at any moment is along the direction of its velocity (along the tangent to the circle) whereas the force of gravity on the Moon is along the radius, which is perpendicular to the instantaneous displacement. Thus, the work done by the force of gravity on the Moon is zero.

When more than one force acts on an object, the total work done is the sum of the work done by the individual forces:

$W_{total} = W_1 + W_2 + \dots$

7–2 Scalar Product of Two Vectors

The **scalar product** (also called the **dot product**) of two vectors is defined as the product of the magnitudes of the vectors and the cosine of the angle between them. That is,

$$\vec{A} \cdot \vec{B} = A B \cos \theta. \qquad (7–2)$$

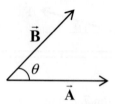

It is clear from the preceding equation that $\vec{A} \cdot \vec{B}$ can be positive, negative, or zero (zero when $\vec{A}$ and $\vec{B}$ are perpendicular, since $\cos 90° = 0$) depending on θ. Also,

$$\vec{A}^2 = \vec{A} \cdot \vec{A} = A A \cos 0° = A^2$$

The work done by a constant force can be written as the scalar product as:

$$W = \vec{F} \cdot \vec{d} = Fd \cos \theta \qquad (7\text{–}3)$$

In unit vector notation,

$$\vec{A} = A_x \hat{\mathbf{i}} + A_y \hat{\mathbf{j}} + A_z \hat{\mathbf{k}}, \quad \vec{B} = B_x \hat{\mathbf{i}} + B_y \hat{\mathbf{j}} + B_z \hat{\mathbf{k}},$$

and

$$\vec{A} \cdot \vec{B} = (A_x \hat{\mathbf{i}} + A_y \hat{\mathbf{j}} + A_z \hat{\mathbf{k}}) \cdot (B_x \hat{\mathbf{i}} + B_y \hat{\mathbf{j}} + B_z \hat{\mathbf{k}}).$$

$\hat{\mathbf{i}} \cdot \hat{\mathbf{j}} = \hat{\mathbf{j}} \cdot \hat{\mathbf{k}} = \hat{\mathbf{k}} \cdot \hat{\mathbf{i}} = 0$ (since θ between each pairs is $90°$)
$\hat{\mathbf{i}} \cdot \hat{\mathbf{i}} = \hat{\mathbf{j}} \cdot \hat{\mathbf{j}} = \hat{\mathbf{k}} \cdot \hat{\mathbf{k}} = 1$ (since the magnitudes of $\hat{\mathbf{i}}, \hat{\mathbf{j}}$, and $\hat{\mathbf{k}}$ are unity)

By direct multiplication of $\vec{A}$ and $\vec{B}$,

$$\vec{\mathbf{A}} \cdot \vec{\mathbf{B}} = A_x B_x + A_y B_y + A_z B_z. \qquad (7\text{--}4)$$

> Note: To determine the dot product, multiply like components and then add.

The dot product is *commutative*: $\vec{\mathbf{A}} \cdot \vec{\mathbf{B}} = \vec{\mathbf{B}} \cdot \vec{\mathbf{A}}$.

The dot product is *distributive*:

$$\vec{\mathbf{A}} \cdot (\vec{\mathbf{B}} + \vec{\mathbf{C}}) = \vec{\mathbf{A}} \cdot \vec{\mathbf{B}} + \vec{\mathbf{A}} \cdot \vec{\mathbf{C}}.$$

7–3 Work Done by a Varying Force

Most forces in nature vary with position. Examples include the force exerted by a spring, which increases with the amount of stretch, and the force of gravity which decreases with distance.

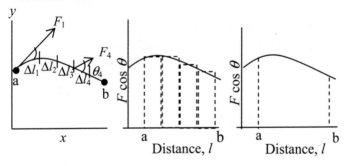

As an object moves from the point *a* to *b*, (as shown in the figure), the whole path is divided into short intervals each of length Δl_1, Δl_2..., etc. during which the force is approximately constant. The work done for segment Δl_1 is $\Delta W = F_1 \cos \theta_1 \, \Delta l_1$, which is the area of the rectangle of width Δl_1 and height $F_1 \cos \theta_1$. The total work done approximately is the sum of the work done for each segment

$$W \approx \sum_i \cos \theta_i \Delta l_i \qquad (7\text{--}5)$$

which is equal to the sum of the areas of the rectangles. If we let each Δl_i approach zero, we obtain the exact value of the work done

$$W = \sum_i F_i \cos \theta_i \, \Delta l_i = \int_a^b F \cos\theta \, dl \qquad (7\text{--}6)$$

which is equal to the area under the $(F \cos \theta)$ versus (l) curve between the points a and b.

In the notation of dot product:

$$W = \int_a^b F\cos\theta \, dl \;=\; \int_a^b \vec{F} \cdot d\vec{l} \qquad (7\text{--}7)$$

Work Done by a Spring Force
The force needed to stretch or compress as spring by x, is $F_p = Fx$. The restoring force on the spring is

$$F_S = -Fx \qquad (7\text{--}8)$$

This is called **Hooke's law**.

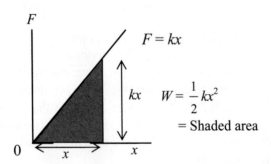

68

In this case, since the force and the displacement are parallel (along x-axis), the work done to stretch a spring by x:

$$W = \int_a^b \vec{F} \cdot d\vec{l} = \int_0^x F_p \, \hat{i} \cdot dx\hat{i} = \int_0^x F_p \, dx$$

$$= \int_0^x kx \, dx = \frac{1}{2} kx^2$$

7–4 Kinetic Energy, and the Work–Energy Principle

Energy is an important concept in physics. It is defined as the "ability to do work." A moving object has the ability to do work and thus has energy. The **kinetic energy** of an object is the energy of its motion.

When positive work is done on an object, its speed increases, and when negative work is done, its speed decreases. The work–energy principle makes the connection between work W_{net} and the change of speed of an object from v_1 to v_2 as expressed by

$$W_{net} = F_{net} \, d = \frac{1}{2} mv_2{}^2 - \frac{1}{2} mv_1{}^2 \qquad (7\text{–}9)$$

For an object of mass m moving with a speed v, its **translational kinetic energy** (K) is defined as

$$K = \frac{1}{2} mv^2 \qquad (7\text{–}10)$$

Thus, $\qquad W_{net} = \Delta K = \dfrac{1}{2} m v_2^2 - \dfrac{1}{2} m v_1^2 \qquad (7\text{–}11)$

Thus, **net work done = the change in its kinetic energy**. This is called **work–energy principle**, which is one of the fundamental results in physics. The work-energy principle is valid even if the force is variable and the motion is in two or three dimensions.

The SI unit of kinetic energy is the joule, same as that of work. However, unlike work, kinetic energy is never negative.

Tips for Solving the Problems

1. Work done by a constant force is $W = Fd \cos \theta$, where θ is the angle between the force and the displacement. The result can be positive, negative, or zero depending on θ. Make sure that the value of F is in newtons and d is in meters to have W in joules. If $\theta = 0$, $W = Fd$. The expression for W will be useful for solving problems in Section 7–1.

2. The dot product $\vec{A} \cdot \vec{B} = A B \cos \theta = A_x B_x + A_y B_y + A_z B_z$. Use this relation to determine the dot product of two vectors or the angle between two vectors.

3. When you determine the angle θ between two vectors $\vec{A}$ and $\vec{B}$ using the relation $\vec{A} \cdot \vec{B} = A B \cos \theta$, always use the knowledge of quadrant to chose the correct angle because you will get two answers for the angle from this relation.

4. For a variable force, the work done is $W = \int_a^b F\cos\theta \, dl$ $= \int_a^b \vec{F} \cdot d\vec{l}$. This integral represents the area between the $F\cos\theta$ curve and the distance l (x axis) between the points a and b. This concept can be used to calculate the work done to stretch (or compress) a spring, or work done by other variable forces such as in Section 7–3.

5. Many problems in this chapter (such as those in Section 7–4) can be solved using the work–energy principle: $W_{\text{total}} = \Delta K = \frac{1}{2} m v_2^2 - \frac{1}{2} m v_i^2$. Use the principle to calculate change in kinetic energy or change in speed as a result of work done. When using this principle, make sure that W_{total} is the *total work* with the *proper sign*.

6. When computing change in kinetic energy, remember that $v_2^2 - v_1^2$ is *not* equal to $(v_2 - v_1)^2$.

CHAPTER 8
Conservation of Energy

This chapter provides a precise definition of work, calculates work done by constant and variable forces, defines scalar product of vectors, and shows how work is related to the energy of motion, called kinetic energy.

Important Concepts

        Conservative and nonconservative forces
        Gravitational potential energy
        Elastic potential energy
        Mechanical energy and its conservation
        Conservation of energy
        Escape velocity
        Power
        Stable and unstable equilibrium

8–1 Conservative and Nonconservative Forces

Forces for which *the work done between two points does not depend on the path* are called **conservative forces**.

The work done by the force of gravity in moving an object of mass m from a height y_1 to a height y_2 is

$$W_G = -\int_{y_1}^{y_2} mg\, dy = -mg(y_2 - y_1) \qquad (8\text{–}1)$$

Since $(y_2 - y_1)$ is the vertical height h, W depends only on the vertical height but not on the particular path taken. Thus, the force of gravity is a conservative force. Elastic force of a spring is also a conservative force.

Also, a force is conservative *if the net work done by the force on an object moving around any closed path is zero*. This is evident from the preceding definition since, as the work done is independent of the path taken, moving an object between two points 1 and 2 and then from 2 to 1 (via a different path such as A and B on the figure on the right below) yields the same work in both situations but with opposite signs.

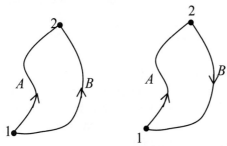

W is the same for   *W* is zero for the closed loop
paths *A* and *B*

Friction is a **nonconservative force**. Work done against friction does depend on path—longer the path the more is the work done against friction.

8–2 Potential Energy

Potential energy (PE) is the energy stored in an object because of its position or configuration. Potential energy can be defined only for a conservative force.

Gravitational Potential Energy

A heavy brick held high in the air has a potential energy because of its height relative to the earth's surface. If it is released, it will fall to the ground because of the force of gravity, and it can do work, such as on a stake, driving it into the ground. The energy associated with force of gravity is called *gravitational potential energy*.

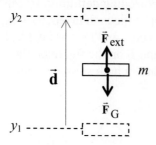

In order to lift an object of mass m vertically an upward force equal to its weight mg must be supplied. When the mass is lifted from a height y_1 to a new height y_2, the work done by the external force is equal to the change in gravitational potential energy, ΔU.

$$W_{ext} = \vec{\mathbf{F}}_{ext} \cdot \vec{\mathbf{d}} = mg\,(y_2 - y_1). \text{ Thus,}$$

$$\Delta U = U_2 - U_1 = mg\,(y_2 - y_1) \qquad (8\text{--}2)$$

Choosing $U = 0$ when $y = 0$, the gravitational potential energy at a height y is

$$U = mgy. \qquad (8\text{--}3)$$

The work done by the force of gravity

$$W_G = \vec{\mathbf{F}}_G \cdot \vec{\mathbf{d}} = mg\,(y_2 - y_1)\cos 180°$$

$$= - mg\,(y_2 - y_1) = -\Delta U.$$

The choice of zero potential energy is arbitrary. However, *change* in potential energy is the same regardless of the choice of the reference position.

Potential Energy in General

The change in gravitational potential energy

$$\Delta U = -W_G = -\int_1^2 \mathbf{F}_G \cdot d\vec{l}$$

There are other kinds of potential energy. In general, for a conservative force $\vec{F}$,

$$\Delta U = U_2 - U_1 = -W = -\int_1^2 \vec{F} \cdot d\vec{l} \qquad (8\text{--}4)$$

Elastic Potential Energy

For small deformations, the force to stretch or compress an ideal spring by an amount x from its natural length is $F_P = kx$. The *restoring force* is given by, $F_S = -kx$. Thus,

$$\Delta U = U(x) - U(0) = -\int_1^2 \vec{F} \cdot d\vec{l} = -\int_0^x \left(-kx\right)\, dx = \frac{1}{2} kx^2$$

If we choose PE = 0 when $x = 0$, the *elastic potential energy* of the spring when it is stretched by a small amount x is

$$U(x) = \frac{1}{2} kx^2 \qquad (8\text{--}5)$$

Note:

1. A potential energy is always associated with a conservative force and $\Delta U = -W = -\int_1^2 \vec{F} \cdot d\vec{l}$.

2. The choice of where $U = 0$ is arbitrary. ΔU is same regardless of the choice.

3. In one-dimensional case,

$$U(x) = -\int F(x)dx. \tag{8-6}$$

This gives, $F(x) = -\dfrac{dU(x)}{dx}$. $\tag{8-7}$

8–3 Mechanical Energy and Its Conservation

The **mechanical energy**, E, of an object is the sum of its potential and kinetic energies: $E = \text{KE} + \text{PE}$.

If only conservative forces are acting on a system, the total mechanical energy of an object is conserved throughout its motion. This is called the **principle of conservation of mechanical energy**.

From the work–energy principle,

$$W_{\text{net}} = \Delta K$$

$$\Delta U_{\text{total}} = -\int_{1}^{2} \vec{\mathbf{F}}_{\text{net}} \cdot d\vec{\boldsymbol{l}} = -W_{\text{net}} \tag{8-8}$$

Combining, $\Delta K + \Delta U = 0$ $\tag{8-9}$

$$K_2 + U_2 = K_1 + U_1 \tag{8-10}$$

or, $\quad E_2 = E_1 = \text{constant}$

or, $\quad \dfrac{1}{2}mv_1{}^2 + U_1 = \dfrac{1}{2}mv_2{}^2 + U_2$ $\tag{8-11}$

8–4 Problem Solving Using Conservation of Mechanical Energy

For an object moving under the influence of gravity, the

total mechanical energy $= \frac{1}{2}mv^2 + mgy$

where y is the height of the object above the ground at a given instant and v is its speed at that instant. If the subscripts 1 and 2 represent two different points along the path of the object, then

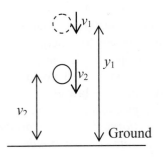

$$\frac{1}{2}mv_1^2 + mgy_1 = \frac{1}{2}mv_2^2 + mgy_2 \qquad (8\text{–}12)$$

For example, a roller coaster car moving without friction illustrates the conservation of mechanical energy. As the car goes up and down, its gravitational potential energy is converted to kinetic energy and vice versa, but the total mechanical energy at any instant remains the same.

Equation (8–12) is very useful for solving problems for objects moving without friction under the action of gravity.

Conservation of mechanical energy holds for the motion of a mass m attached to a spring. In this case,

$$\frac{1}{2}mv_1^2 + \frac{1}{2}kx_1^2 = \frac{1}{2}mv_2^2 + \frac{1}{2}kx_2^2 \qquad (8\text{–}13)$$

8–5 The Law of Conservation of Energy

Besides mechanical energy (kinetic and potential energies) there are other forms of energy. These include: electric energy, nuclear energy, thermal energy, and the chemical energy stored in fuels and food. Energy can be converted from one form to others. For example, a stone held high in air has potential energy; when it is released, its potential energy is converted to its kinetic energy. At the base of a dam, the kinetic energy of water is transferred to turbine blades and then transformed into the electric energy.

One of the important results of physics is that whenever energy is transferred or transformed, no energy is gained or lost. This is expressed in the **law of conservation of energy** as: *Energy cannot be created nor be destroyed. It can only be transformed from one form to others. The total energy of an isolated system remains constant.* Thus,

$$\Delta K + \Delta U + \text{(change in all other forms of energy)} = 0 \qquad (8\text{–}14)$$

8–6 Energy Conservation with Dissipative Forces: Solving Problems

When nonconservative forces such as friction are present, the mechanical energy is not constant. Frictional forces reduce the total mechanical energy; that is why they are also called **dissipative forces**. In this case,

$$\Delta K + \Delta U + F_{fr}\, l = 0$$

where l is the actual distance traveled by the object from point 1 to point 2. This gives,

$$\frac{1}{2}mv_1^2 + mgy_1 = \frac{1}{2}mv_2^2 + mgy_2 + F_{\text{fr}}\,l \quad (8\text{–}15)$$

Thus, initial energy = final energy (including thermal energy).

Work-Energy versus Energy Conservation
Conservation of energy is more general than the work–energy principle. If you choose a particle or rigid object on which external forces do work as your system, then apply the work–energy principle. On the other hand, if you choose a system on which external forces do not work, apply conservation of energy.

8–7 Gravitational Potential Energy and Escape Velocity

From Newton's law of gravitation, the force of gravity on an object of mass m at a distance r from the center of the earth (mass M_E, radius r_E):

$$\vec{\mathbf{F}} = -G\frac{mM_E}{r^2}\hat{\mathbf{r}} \quad (r > r_E).$$

The work done by the gravitational force to move the object from point 1 to point 2:

$$W = \int_1^2 \vec{\mathbf{F}} \cdot d\vec{\mathbf{l}}$$

$$= G\frac{mM_E}{r_2} - G\frac{mM_E}{r_1}$$

$$\Delta U = U_2 - U_1 = -W = -G\frac{mM_E}{r_2} + G\frac{mM_E}{r_1} \qquad (8\text{-}16)$$

The work done by the gravitational force to move the object from point 1 to point 2:

$$W = \int_1^2 \vec{\mathbf{F}} \cdot d\vec{\mathbf{l}} = G\frac{mM_E}{r_2} - G\frac{mM_E}{r_1}$$

Thus, $\Delta U = U_2 - U_1 = -W = -G\dfrac{mM_E}{r_2} + G\dfrac{mM_E}{r_1}$

or $\qquad U(r) = -G\dfrac{mM_E}{r} + C \text{ (constant)}$

If we choose $U = 0$ at $r = \infty$, then $C = 0$. Thus,

$$U(r) = -G\frac{mM_E}{r}. \qquad (8\text{-}17)$$

The plot of $U(r)$ as a function of r is shown.

From the conservation of mechanical energy, as the object moves from r_1 to r_2,

$$\frac{1}{2}mv_1^2 - G\frac{mM_E}{r_1} = \frac{1}{2}mv_2^2 - G\frac{mM_E}{r_2} \qquad (8\text{-}18)$$

The minimum velocity with which an object on the earth is projected so that it never returns to the earth is called the **escape velocity**. In this case the object needs to reach $r_2 = \infty$ with merely $v_2 = 0$. Thus from Equation (8–19)

$$\frac{1}{2} m v_{esc}^2 - G\frac{mM_E}{r_E} = 0 + 0$$

$$V_{esc} = \sqrt{\frac{2GM_E}{r_E}} = 1.12 \times 10^4 \text{ m/s} \qquad (8\text{–}19)$$

8–8 Power

Average **power** is defined as the *rate at which work is done* or the *rate at which energy is transformed.* If work W is done in time t, the average power $\overline{P}$ is given by

$$\overline{P} = \frac{W}{t} \qquad (8\text{–}20a)$$

The *instantaneous power* is

$$P = \frac{dW}{dt} \qquad (8\text{–}20b)$$

or $\qquad P = \frac{dE}{dt} \qquad$ (E is energy) $\qquad (8\text{–}20c)$

The SI unit of power is J/s, which is defined as a **watt**, W. Another common unit of power is **horsepower**, hp. 1 hp = 746 W.

For an object moving with a velocity $\vec{v}$

$$P = \frac{dW}{dt} = \vec{F} \cdot \frac{d\vec{l}}{dt} = \vec{F} \cdot \vec{v} \qquad (8\text{–}21)$$

*8–9 Potential Energy Diagrams; Stable and Unstable Equilibrium

For an object moving under the action of conservative forces, $E = K + U$ = constant. $K = E - U(x)$, and $U(x) \leq E$.

For the potential energy diagram as shown, $E = E_0$ (the minimum value of E) at $x = x_0$ where the object is at rest. An object with energy E_1 can oscillate between x_1 and x_2, which are known as the **turning points** of the motion. If the object has energy E_2, there are four turning points and the object can move only in one of the potential "valleys" depending on where it is initially.

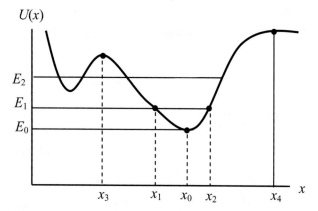

The force on the object is $F = -dU/dx$, which is the negative slope of the U vs x curve. At $x = x_0$, the slope and, hence, $F = 0$. The object is said to be in **equilibrium**.

- If the object at rest at $x = x_0$ is slightly moved left or right, a nonzero force will act on the object to bring it back to x_0. An object that has a tendency to return toward the equilibrium position when displaced slightly is said to be at a point of **stable equilibrium**.

- For a point such as x_3 where the object is at equilibrium and the potential curve has a maximum, if the object is displaced slightly a force will pull the object away from equilibrium position. Points like x_3 are points of **unstable equilibrium**.

- For points such as x_4, U is constant and F is zero over some distance. The object is in equilibrium and it does not have any tendency to move if it is displaced slightly. The object is in **neutral equilibrium** in this region.

Tips for Solving the Problems

1. To calculate gravitational potential energy, choose a horizontal level for which $y = 0$, then use $U = mgy$.

2. Remember that gravity is a conservative force and that the work done in a gravitational field, $W_G = -\Delta U$.

3. The principle of conservation of mechanical energy ($E_2 = E_1$) can be used when conservative forces (such as friction) are absent. The principle can be used to solve many problems in this chapter, such as those in Sections 8–3 and 8–4.

4. The potential energy stored in a spring $U(x) = \frac{1}{2} kx^2$. Apply conservation of mechanical energy in a system involving a spring if no conservative forces are present.

5. Use the law of conservation of energy such as expressed by Equations 8–14 and 8–15 when nonconservative forces are present to determine thermal energy produced. The law is useful to solve problems in Sections 8–5 and 8–6.

6. Gravitational energy of an object of mass m at a distance r from the center of the earth is $U(r) = -GmM_E/r$. For an object moving under the influence of gravity, from the conservation of mechanical energy, $\frac{1}{2}mv_1^2 - GmM_E/r_1 = \frac{1}{2}mv_2^2 - GmM_E/r_2$. This equation is useful for determining speed of objects such as meteoroids or for determining escape velocity (for which set $v_2 = 0$ and $r_2 = \infty$ in the equation). These equations are useful for solving problems in Section 8–7.

7. To calculate power, use either $\overline{P} = W/t$ or $P = dW/dt$ or dE/dt or $\vec{F} \cdot \vec{v}$ depending on the problem. Use the conversion 1 hp = 746 W where necessary.

CHAPTER 9
Linear Momentum

This chapter introduces linear momentum and describes conservation of linear momentum. Newton's second law of motion is restated using momentum. Two types of collisions are discussed. The concept of center of mass and center of gravity are described.

Important Concepts

> Linear momentum
> Newton's second law and momentum
> Impulse
> Conservation of momentum
> Elastic and inelastic collisions
> Completely inelastic collision
> Center of mass and center of gravity
> Thrust

9–1 Momentum and Its Relation to Force

The **linear momentum** (or simply momentum), $\vec{p}$, of an object is the product of its mass m and the velocity $\vec{v}$. That is,

$$\vec{p} = m\,\vec{v} \qquad\qquad (9-1)$$

Since velocity is a vector, momentum is a vector. The SI unit of momentum is $\text{kg} \cdot \text{m/s}$.

A force is needed to change the momentum of an object, whether it is to increase the momentum, to decrease it, or to change its direction. In terms of momentum, **Newton's second law of motion** states that the **time rate of change of momentum of an object is equal to the net force $\Sigma\,\vec{F}$ applied to an object**.

$$\Sigma \vec{F} = \frac{d\vec{p}}{dt} \qquad (9\text{--}2)$$

This is the original statement of Newton's second law that holds even when the mass of an object changes. When m is a constant, this statement reduces to

$$\Sigma \vec{F} = \frac{d(m\vec{v})}{dt} = m\frac{d\vec{v}}{dt} = m\vec{a} .$$

9–2 Conservation of Momentum

The concept of momentum is important because the vector summation of the momenta of two or more objects in a collision remains constant. If $m_A \vec{v}_A$ and $m_B \vec{v}_B$ are the momenta of the two balls A and B before a collision and if $m_A \vec{v}'_A$ and $m_B \vec{v}'_B$ are their momenta after the collision, then

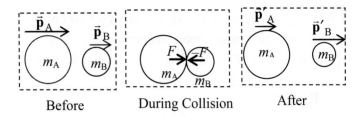

Before During Collision After

$$m_A \vec{v}_A + m_B \vec{v}_B = m_A \vec{v}'_A + m_2 \vec{v}'_B \qquad (9\text{--}3)$$

During the collision, because of action and reaction, the forces on body 2 and body 1 are $\vec{F}$ and $-\vec{F}$, respectively. Thus,

86

$$\vec{\mathbf{F}} = \frac{d\vec{\mathbf{p}}_B}{dt} \text{ , and } -\vec{\mathbf{F}} = \frac{d\vec{\mathbf{p}}_A}{dt}$$

Adding, $\frac{d}{dt}\left(\vec{\mathbf{p}}_A + \vec{\mathbf{p}}_B\right) = 0$, or

$$\vec{\mathbf{p}}_A + \vec{\mathbf{p}}_B = \text{constant.}$$

For a system of objects, if $\vec{\mathbf{P}}$ represents the total momentum of the system,

$$\vec{\mathbf{P}} = m_1 \vec{\mathbf{v}}_1 + m_2 \vec{\mathbf{v}}_2 + \cdots + m_n \vec{\mathbf{v}}_n = \sum \vec{\mathbf{p}}_i . \text{ Thus,}$$

$$\frac{d\vec{\mathbf{P}}}{dt} = \sum \frac{d\vec{\mathbf{p}}_i}{dt} = \Sigma \vec{\mathbf{F}}_i \qquad (9\text{--}4)$$

where $\vec{\mathbf{F}}_i$ represents the force on the ith object. Since *internal forces* occur in pairs and cancel each other (action and reaction), E+quation (9–4) can be written as:

$$\frac{d\vec{\mathbf{P}}}{dt} = \Sigma \vec{\mathbf{F}}_{ext} \qquad (9\text{--}5)$$

where $\Sigma \vec{\mathbf{F}}_{ext}$ represents the total *external forces* acting on the system. Clearly when $\Sigma \vec{\mathbf{F}}_{ext} = 0$, $\vec{\mathbf{P}} =$ constant. This is the law of **conservation of momentum**, which states that the **total momentum remains constant when the net external force on a system is zero**. Such a system is called an **isolated system**.

The law of conservation of momentum is useful when we deal with systems such as collisions and certain types of explosions. *Rocket propulsion*, the recoil of a gun, and throwing a package from a boat, which can be understood from Newton's third law, can also be explained by conservation of momentum. For example, before a rocket is fired, the total momentum of the rocket plus fuel was zero. As the fuel burns, since the total momentum still is zero, the backward momentum of the expelled gas must be balanced by a forward momentum of the rocket. Thus, a rocket can accelerate in the upward direction.

9–3 Collisions and Impulse

Real forces vary with time, as shown below. For example, the force between a ball and a bat rises rapidly to a large value and then falls back to zero. The integral of the force over the time interval during which it acts is called the **impulse, $\vec{\mathbf{J}}$** :

$$\vec{\mathbf{J}} = \int_{t_{\mathrm{i}}}^{t_{\mathrm{f}}} \vec{\mathbf{F}} \, dt \, .$$

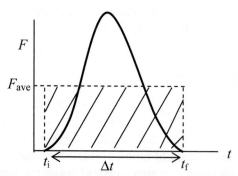

That is, the impulse of a force is equal to the area under the F versus t curve between t_{i} and t_{f}.

From Newton's second law (Equation 9–2), the change in momentum of an object is equal to the impulse on the object.

$$\Delta\vec{P} = \vec{p}_f - \vec{p}_i = \int_{t_i}^{t_f} \vec{F}\, dt = \vec{J} \qquad (9\text{--}6)$$

The concept of impulse is useful when dealing with forces that act during a short interval of time. Physicists usually describe such a situation with the terms average force, $\vec{F}_{av}$, such that the impulse:

$$\vec{J} = \vec{F}_{ave}\, \Delta t = \int_{t_i}^{t_f} \vec{F}\, dt$$

where $\Delta t\ (= t_f - t_i)$ is the duration of the force.

The SI unit of impulse is N·s. Impulse is a vector that points in the same direction as the average force.

9–4 Conservation of Energy and Momentum in Collisions

The total momentum is conserved in the collision of objects. Collisions are classified according to whether total kinetic energy is or is not conserved in the collision. The collision in which the total kinetic energy is conserved is called an **elastic collision**.

If we use subscripts A and B to represent two objects, then from the conservation of kinetic energy applied to the collision of the two objects

$$\frac{1}{2} m_A v_A^2 + \frac{1}{2} m_B v_B^2 = \frac{1}{2} m_A {v'_A}^2 + \frac{1}{2} m_B {v'_B}^2 \qquad (9\text{--}7)$$

where primed quantities represent after the collision and unprimed represent before the collision. Collisions between atoms or molecules are often elastic, but in the collisions of ordinary objects an elastic collision is an ideal situation since at least a small amount of heat and sound is always produced during such a collision.

The collision in which kinetic energy is not conserved is called an **inelastic collision**. The lost kinetic energy in an inelastic collision is converted to other forms of energy and the *total energy* is always conserved.

9–5 Elastic Collisions in One Dimension

We can apply the conservation laws of momentum and kinetic energy to a head-on elastic collision of two objects, so that all the motion is along a line. From the conservation of momentum

$$m_A v_A + m_B v_B = m_A v'_A + m_B v'_B$$

From the conservation of kinetic energy

$$\frac{1}{2} m_A v_A^2 + \frac{1}{2} m_B v_B^2 = \frac{1}{2} m_A v'_A^2 + \frac{1}{2} m_B v'_B^2$$

The preceding two equations can be solved to calculate the velocities after collisions, v'_A and v'_B. Also, from the preceding two equations it can be shown that

$$v_A - v_B = -(v'_A - v'_B) \tag{9–8}$$

That is, the relative speed of the two objects after the collision has the same magnitude but opposite direction as before the collision. Equation (9–8) is useful and often more convenient than conservation of kinetic energy equation because it involves the first power of v.

9–6 Inelastic Collisions

Inelastic collisions are collisions in which total kinetic energy is *not* conserved. Some of the initial kinetic energy is lost in the form of heat, sound, etc.

In the special case of an inelastic collision in which the objects stick together after collision, the collision is called **completely inelastic**. The maximum amount of kinetic energy is lost in a completely inelastic collision.

9–7 Collisions in Two or Three Dimensions

Conservation of momentum and energy can be applied to two- and three-dimensional collisions. Since momentum is a vector, both the x- and y-components of the momentum are conserved in a two-dimensional collision. Let a particle A of mass m_A initially moving with a velocity v_A in the x-direction collide with a particle of mass m_B initially at rest. After collision, the particles move with velocity v'_A and v'_B at angles θ'_A and θ'_B, respectively.

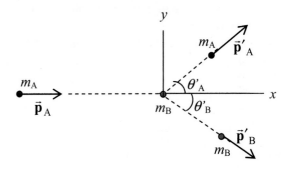

From the conservation of momentum in the x-direction:

$$m_A v_A = m_A v'_A \cos \theta'_A + m_B v'_B \cos \theta'_B \quad (9–9a)$$

Since the total momentum is zero before the collision,

$$0 = m_A v'_A \sin \theta'_A + m_B v'_B \sin \theta'_B. \qquad (9-9b)$$

If the collision is elastic, kinetic energy is also conserved:

$$\frac{1}{2} m_A v_A{}^2 = \frac{1}{2} m_A v'_A{}^2 + \frac{1}{2} m_B v'_B{}^2 \qquad (9-9c)$$

The preceding three equations can be solved for three unknowns.

9–8 Center of Mass (CM)

Observation shows that even if an extended object undergoes both translational and rotational motion, there is one point that moves in the same path that a particle would if subjected to the same net force. This point is called the **center of mass** (CM). For example, the center of mass of a diver moves in the same parabolic path as a projected particle follows under the action of gravity only, even when the diver rotates.

For a system in a plane containing objects of masses m_A, m_B, ... at locations x_A, x_B, ... the x coordinate of the center of mass is

$$x_{CM} = \frac{m_A x_A + m_B x_B + \cdots}{m_A + m_B + \cdots} = \frac{\sum\limits_{i=1}^{n} m_i x_i}{M} \qquad (9-10)$$

In general, the coordinates of the CM in two or three dimensions are:

$$x_{CM} = \frac{\sum m_i x_i}{M}, \; y_{CM} = \frac{\sum m_i y_i}{M}, \; z_{CM} = \frac{\sum m_i z_i}{M}$$

(9–11)

Equation (9–11) can be written in vector form as:

$$\vec{r}_{CM} = \frac{\sum m_i \vec{r}_i}{M}$$

(9–12)

For an extended body made up of a continuous distribution of matter:

$$x_{CM} = \frac{1}{M} \int x \, dm, \; y_{CM} = \frac{1}{M} \int y \, dm, \; z_{CM} = \frac{1}{M} \int z \, dm$$

(9–13)

$$\vec{r}_{CM} = \frac{1}{M} \int \vec{r} \, dm$$

(9–14)

The **center of gravity** (CG) is a point at which the force of gravity or the weight of the body acts. For symmetrically shaped objects such as uniform cylinders and spheres, the CG is located at the geometric center of the object. Since the force of gravity acts at the CG, a body tends to rotate about its pivoting point unless the CG is on a vertical line directly below the pivot, in which case the body remains at rest. There is a conceptual difference between center of mass and center of gravity, although in most common situations they are actually the same point.

9–9 Center of Mass and Translational Motion

The motion of the center of mass of a system is simple compared with the motions of different parts of the system.

Differentiating Equation (9–12) with respect to time we get:

$$M\vec{\mathbf{v}}_{CM} = \sum m_i \, \vec{\mathbf{v}}_i \qquad (9\text{--}15)$$

Differentiating again with respect to time:

$$M\vec{\mathbf{a}}_{CM} = \sum m_i \, \vec{\mathbf{a}}_i = \Sigma \vec{\mathbf{F}}_i \qquad (9\text{--}16)$$

Since internal forces cancel each other, we are left only with the external forces on the right hand side of Equation (9–16):

$$M\vec{\mathbf{a}}_{CM} = \Sigma \vec{\mathbf{F}}_{ext} \qquad (9\text{--}17)$$

This is **Newton's second law** for a system of objects. This shows that the system moves as if its whole mass is concentrated at its CM and all external forces acted at the CM.

From Equation (9–15)

$$\vec{\mathbf{P}} = \sum \vec{\mathbf{p}}_i = \sum m_i \, \vec{\mathbf{v}}_i = M\vec{\mathbf{v}}_{CM} \qquad (9\text{--}18)$$

Differentiating with respect to time, we get:

$$\frac{d\vec{\mathbf{P}}}{dt} = M\vec{\mathbf{a}}_{CM} = \Sigma \vec{\mathbf{F}}_{ext.}$$

*9–10 Systems of Variable Mass; Rocket Propulsion

Examples of systems of variable mass include motion of: rockets (which propel themselves forward by the ejection of burnt gases) and conveyor belts onto which material (gravel, packaged goods) is dropped.

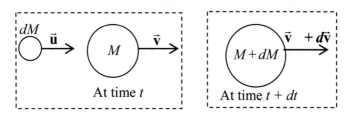

In general, at time t, a mass dM (moving with velocity $\vec{u}$) is about to be added to a system of mass M (moving at a velocity $\vec{v}$). At time $t + dt$, the mass dM has beeen added to M and the velocity of the whole system is $\vec{v} + d\vec{v}$. Change in momentum,

$$d\vec{P} = (M + dM)(\vec{v} + d\vec{v}) - (M\vec{v} + dM\vec{u}).$$

Thus, from Newton's second law, after simplification:

$$\Sigma \vec{F}_{ext} = \frac{d\vec{P}}{dt} = M\frac{d\vec{v}}{dt} - (\vec{u} - \vec{v})\frac{dM}{dt} \qquad (9\text{–}19a)$$

or
$$M\frac{d\vec{v}}{dt} = \Sigma \vec{F}_{ext} + \vec{v}_{rel}\frac{dM}{dt} \qquad (9\text{–}19b)$$

where $\vec{v}_{rel} = \vec{u} - \vec{v}$. The second term on the right represents the rate at which momentum is being transferred to (or from) the mass M because of the mass that is added to (or leaves) it. This is interpreted as the force exerted on the mass M due to the addition (or ejection) of mass. For rocket, this is called the *thrust*.

Tips for Solving the Problems

1. Begin with a sketch. Remember that velocity and momentum are *vector* quantities. Take the direction of motion to the right as positive and stick with it for the whole problem.

2. The total momentum of a system of objects is the vector sum of the individual momenta of all the objects.

3. Determine whether the net external force is zero in the problem. If the net external force is zero, you can apply conservation of momentum ($m_A \vec{\mathbf{V}}_A + m_B \vec{\mathbf{V}}_B = m_A \vec{\mathbf{V}}'_A + m_B \vec{\mathbf{V}}'_B$) to solve problems such as those in Section 9–2.

4. Note that impulse, $\vec{\mathbf{J}} = \vec{\mathbf{F}}_{av} \Delta t = \int_{t_i}^{t_f} \vec{\mathbf{F}}\, dt$ = area under the F vs t curve. Also, change in momentum,

    $\vec{\mathbf{p}}_f - \vec{\mathbf{p}}_i = \int_{t_i}^{t_f} \vec{\mathbf{F}}\, dt = \vec{\mathbf{J}}$. These relations will be useful to solve for problems in Section 9–3.

5. Apply the *momentum conservation principle* to solve any collision problem. If the collision is elastic, *total*

initial KE = total final KE, and as a result $v_A - v_B = -(v'_A - v'_B)$. Use the preceding relation together with momentum conservation when you study an elastic collision in one dimension.

6. For two-dimensional collisions (such as problems in Section 9–7) apply momentum conservation separately for the x and y components.

7. Use Equations 9–11 through 9–14 to find the location of the center of mass. Use Equations 9–15 and 9–16 to determine the velocity and acceleration of the center of mass.

8. For systems of variable mass use Equation 9–19 to solve for unknown(s) in the problem. Note that dM/dt can be positive or negative depending on whether the mass of the object increases or decreases.

CHAPTER 10
Rotational Motion

In this chapter various aspects of rotational motion are described for rigid bodies for which the distance between two points remains the same. Analogies between rotational and linear motions are emphasized.

Important Concepts
Angular position
Radian
Average angular velocity
Instantaneous angular velocity
Average angular acceleration
Instantaneous angular acceleration
Frequency and period
Rolling motion
Torque
Moment of inertia
Rotational kinetic energy
Rolling

10–1 Angular Quantities
To describe rotational motion it is necessary to define angular quantities that are analogous to position, velocity, and acceleration for linear motion.

The angular position is described by an angle, θ, measured from an arbitrary reference line, such as the x-axis. The reference line begins at the axis of rotation and simply defines $\theta = 0$. The SI unit of θ is the radian (rad), which is dimensionless.

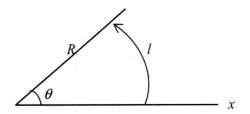

A **radian** is defined as the angle for which the arc length, s, is equal to the radius, R, of the circle. This means that the angle θ in radians is given by:

$$\theta = \frac{l}{R} \qquad (10\text{--}1)$$

or $l = R\theta$

For one complete revolution, $l = 2\pi R$; thus $\theta = 2\pi$ rad = 360°.

1 rad = 57.3°

For a rotating object, the *angular displacement*, $\Delta\theta = \theta_2 - \theta_1$ where θ_1 and θ_2 are its initial and final angular positions, respectively. The **average angular velocity, ϖ**, is the time rate of change of angular displacement:

$$\varpi = \frac{\Delta\theta}{\Delta t} \qquad (10\text{--}2a)$$

The **instantaneous angular velocity** (or simply called angular velocity) is the limit of ϖ as the time interval Δt tends to zero:

$$\omega = \lim_{\Delta t \to 0} \frac{\Delta\theta}{\Delta t} = \frac{d\theta}{dt} \qquad (10\text{--}2b)$$

The SI unit for angular velocity is rad/s.

When the angular velocity of a rotating object changes, the object experiences an angular acceleration. The **average angular acceleration** is the time rate of change of angular velocity and is given by

$$\overline{\alpha} = \frac{\Delta\omega}{\Delta t} \qquad (10\text{--}3a)$$

The **instantaneous angular acceleration** is

$$\alpha = \lim_{\Delta t \to 0} \frac{\Delta\omega}{\Delta t} = \frac{d\omega}{dt} \qquad (10\text{--}3b)$$

The SI unit of angular acceleration is rad/s^2.

Differentiating Equation 10–1 with respect to time, we get linear velocity

$$v = R\frac{d\theta}{dt} = R\,\omega \qquad (10\text{--}4)$$

Also, linear acceleration, $a_{\text{tan}} = \dfrac{dv}{dt} = R\dfrac{d\omega}{dt} = R\,\alpha \quad (10\text{--}5)$

From Equations (10–4) and (10–5), although ω is the same for all points in the rotating object, v and a_{tan} are greater when R is greater. The centripetal acceleration,

$$a_{\text{R}} = \frac{v^2}{R} = \frac{\left(\omega R\right)^2}{R} = \omega^2 R \qquad (10\text{--}6)$$

The total linear acceleration is the vector sum of the tangential and centripetal accelerations in cases where both are present. That is, $\vec{a} = \vec{a}_{tan} + \vec{a}_R$.

The following table shows the analogies between the linear and the angular quantities:

Linear	Type	Rotational	Relation
x	displacement	θ	$x = R\theta$
v	velocity	ω	$v = R\omega$
a_{tan}	acceleration	α	$a_{tan} = R\alpha$

The angular frequency ω is related to the linear **frequency** f, which is the number of revolutions per second, by the relation

$$\omega = 2\pi f \qquad (10\text{–}7)$$

The time required to complete one revolution is called the **period**, T, and it is related to the frequency, f, by,

$$T = \frac{1}{f} \qquad (10\text{–}8)$$

The SI unit of T is second, s. and the unit of f is revolution per second (rev/s) or, hertz (Hz).

10–2 Vector Nature of Angular Quantities

Rotational quantities such as angular velocity, ω, and angular acceleration, α, are vectors and have directions along the axis of rotation. The direction of these vectors is given by the **right-hand rule**:

Curl the fingers of the right hand in the direction of rotation. The thumb points in the direction of the rotational quantity in question.

10–3 Constant Angular Acceleration

Rotational kinematics describes rotational motion in the same manner as kinematics describes a linear motion. We assume constant acceleration (a or α) in both the cases. The kinematic equations in linear and rotational motions are summarized here:

Angular	Linear	
$\omega = \omega_0 + \alpha t$	$v = v_0 + at$	(10–9a)
$\theta = \omega_0 t + \dfrac{1}{2}\alpha t^2$	$x = v_0 t + \dfrac{1}{2}at^2$	(10–9b)
$\omega^2 = \omega_0^2 + 2\alpha\theta$	$v^2 = v_0^2 + 2ax$	(10–9c)
$\varpi = \dfrac{1}{2}(\omega_0 + \omega)$	$\bar{v} = \dfrac{1}{2}(v_0 + v)$	(10–9d)

In problems involving rotation motion, the angular equations are applied in the same manner as the linear equations are applied for a linear motion.

10–4 Torque

The ability of a force to rotate an object is measured by a quantity called **torque**, τ. Torque takes into account both the magnitude of the force, F, and its perpendicular distance from the axis of rotation to the line along which the force acts. The distance is called **lever arm** or **moment arm** of the force.

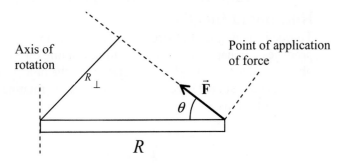

Torque is defined in terms of the moment arm, $R_\perp$ (= R sin θ) as:

$$\tau = R_\perp F \qquad (10\text{–}10a)$$

This can also be written as,

$$\tau = R F_\perp \qquad (10\text{–}10b)$$

For a force applied at a distance R from the axis of rotation and at an angle θ with respect to the radial direction, the torque is

$$\tau = R F \sin \theta \qquad (10\text{–}10c)$$

Torque is a vector quantity. The SI unit of torque is m · N.

When more than one torque acts on a body, the net torque is the vector sum of the individual torques.

Sign convention for torque:

- τ is *positive* if it causes a *counterclockwise* angular acceleration.

- τ is *negative* if it causes a *clockwise* angular acceleration.

10–5 Rotational Dynamics; Torque and Rotational Inertia

Torque is analogous to force. Force acting on an object produces a linear acceleration and torque acting on an object produces an angular acceleration on the object. Let us consider a simple case of particle of mass m rotating in a circle of radius R.

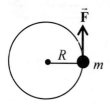

In this case, $a_{tan} = R\alpha$. Therefore, from Newton's second law:

$$F = ma = mR\alpha.$$

Therefore, the torque on the particle

$$\tau = RF = mR^2 a \qquad (10\text{-}11)$$

The quantity mR^2 represents the *rotational inertia* of the particle and is called its *moment of inertia.*

For a rotating rigid object, the net torque on the object is the summation of the individual torques acting on its different particles and is given by

$$\Sigma\tau = \Sigma(mR^2)\alpha \qquad (10\text{–}12)$$

The quantity ΣmR^2 is called the moment of inertia I of the object.

$$I = \Sigma mR^2 = m_1 R_1^2 + m_2 R_2^2 + \dots \qquad (10\text{–}13)$$

The relationship (10–12) is called Newton's second law for rotational motion, which is,

$$\Sigma \tau = I\alpha. \qquad (10\text{–}14)$$

This is the rotational equivalent of Newton's second law. This shows that the moment of inertia I plays the same role for rotational motion that mass m does for translational motion. The larger the I, the more resistant an object is to change its rotational motion.

Equation (10–14) is also valid for an object that rotates about an axis fixed in direction but may accelerate as long as I and a are determined about the center of mass:

$$(\Sigma \tau)_{CM} = I_{CM}\alpha_{CM} \qquad . \qquad (10\text{–}15)$$

Clearly, I not only depends on the mass but also how the mass is distributed with respect to the axis of rotation. For ordinary objects, mass is distributed continuously and the use of calculus is needed to determine the moment of inertia. The SI unit of I is $kg \cdot m^2$.

10–6 Solving Problems in Rotational Dynamics

Use SI units for α and I. Follow these steps for solving a problem in rotational dynamics:
1. Draw a sketch.
2. Draw a free-body diagram for the body under consideration.
3. Determine the torques about the axes of rotation and provide correct signs to each torque.
4. Apply Newton's second law of rotation, $\Sigma \tau = I\alpha$.
5. Also apply $\Sigma F = ma$, if needed.

6. Solve the resulting equation(s) for unknown(s).

7. Verify if the result makes sense.

10–7 Determining Moments of Inertia

By Experiment

Moment of inertia can be determined from the relation $I = \Sigma\tau/\alpha$, by measuring the net torque $\Sigma\tau$ and angular acceleraton α.

Using Calculus

For bodies of simple geometric shapes having a continuous distribution of mass, moment of inertia can be determined by performing the integral

$$I = \int R^2 dm \qquad (10\text{–}16)$$

where dm is mass of an infinitesiemal particle of the body whose distance from the axis is R.

The Parallel-Axis Theorem

According to the **parallel-axis theorem**, the moment of inertia I about a parallel axis is

$$I = I_{CM} + Mh^2 \qquad (10\text{–}17)$$

Parallel axis

where I_{CM} is the moment of inertia about an axis passing through the center of mass and h is the distance between the axes.

*The Perpendicular-Axis Theorem

According to the **perpendicular-axis theorem**, the sum of the moments of inertia of a plane body about any two perpendicular axes in the plane of the body (say, I_x and I_y

for a body in xy plane) is equal to the moment of inertia about an axis passing through their points of intersection perpendicular to the plane of the object (I_z). That is,

$$I_z = I_x + I_y \qquad (10\text{--}18)$$

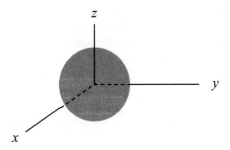

10–8 Rotational Kinetic Energy

An object rotating about an axis has rotational kinetic energy. For a rigid object the total kinetic energy is the sum of the kinetic energies of all its particles. If R represents the distance of any one particle from the axis of rotation, $v = R\omega$. Therefore, for a rigid object,

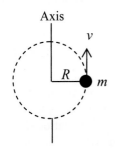

107

$$K = \Sigma(\frac{1}{2}mv^2) = \Sigma\frac{1}{2}m(R\omega)^2 = \frac{1}{2}(\Sigma mR^2)\omega^2$$

Thus, $K = \frac{1}{2}I\omega^2$ (10–19)

Torque acting through an angular displacement does work in rotational motion, just as force acting through a distance does work for linear motion The work done by a torque τ acting on a rigid body rotating about a fixed axis is given by the integral

$$W = \int \tau \, d\theta$$ (10–20)

The power, $P = \dfrac{dW}{dt} = \tau\dfrac{d\theta}{dt} = \tau\omega$ (10–21)

The work–energy principle for a body rotating about a fixed axis is:

$$W = \frac{1}{2}m\omega_2{}^2 - \frac{1}{2}m\omega_1{}^2$$ (10–22)

where ω_1 and ω_2 are initial and final angular velocities of the rotating body, respectively.

10–9 Rotational Plus Translational Motion; Rolling

A rolling object it rotates while its center of mass (CM) undergoes translational motion. In this case the total kinetic energy is summation of translational and rotational kinetic energies. Thus, for an object of mass M, kinetic energy

$$K_{\text{tot}} = \frac{1}{2} I_{CM} \omega^2 + \frac{1}{2} M v_{CM}^2 \qquad (10\text{--}23)$$

where v_{CM} is the velocity of the center of mass and I_{CM} is the moment of inertia about an axis through the center of mass.

*10–10 Why Does a Rolling Sphere Slow Down?

For a rolling sphere the frictional force provides a torque that acts to increase angular acceleration and thus to increase the velocity of the sphere. But, since the objects are deformable to some extent, the sphere flattens slightly and the level surface also acquires a slight depression where the two are in contact. The torque associated with the normal force that the table exerts over the area of contact slows down the sphere.

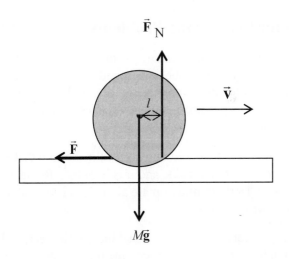

Note the following analogies between linear motion and rotational motion:

Linear	Rotation	Connection
x	θ	$x = R\theta$
v	ω	$v = R\omega$
a	α	$a_{\tan} = R\alpha$
m	I	$I = \Sigma mR^2$
F	τ	$\tau = RF\sin\theta$
$K = \dfrac{1}{2}mv^2$	$\dfrac{1}{2}I\omega^2$	
$W = Fd$	$W = \tau\theta$	
$\Sigma F = ma$	$\Sigma\tau = I\alpha$	

Tips for Solving the Problems

1. Rotational motion is analogous to linear motion. Note the analogy and get the corresponding equation or law for rotational motion replacing the variables of linear motion by the corresponding variables for rotational motion.

2. To use $l = R\theta$, $v = R\omega$, and $a = R\alpha$, make sure that θ is in radians, ω is in rad/s, and α is in rad/s^2. Remember π rad = 180°. θ, ω, and α are all *positives for counter-clockwise rotations*.

3. Use the equations for rotational motion (in Section 10–2) in the same way you used kinematic equations for

linear motion in Chapter 2. In this case you will use rotational variables instead of linear ones.

4. Torque can be obtained using the expression $\tau = RF \sin \theta$. If $\theta = 90°$, $\tau = RF$. Use the expression for torque to solve problems in Section 10–4.

5. Torque is a vector quantity. The magnitude of torque can be positive, negative, or zero. When adding two or more torques, add them with their proper signs. Remember that the SI unit of torque is m·N (it is not Joule).

6. Always draw the sketch of the problem and the free-body diagram. When calculating a torque, think about the point about which the torque will be calculated. Remember, you can avoid a force that you are not interested in by taking torque about a point through which the force passes.

7. Problems relating to torque and angular acceleration can be solved using $\Sigma \tau = I \alpha$.

8. Problems relating to moment of inertia, such as those in Sections 10–6 to 10–8, can be solved using $I = \Sigma mR^2$ or $I = \int R^2 dm$. To determine moment of inertia about a parallel axis use $I = I_{CM} + Mh^2$. Use the perpendicular axis theorem $I_z = I_x + I_y$ as appropriate.

9. Problems relating to kinetic energy, work done W by a torque τ, and rotational power (Section 10–10) can be solved by using the relations: $K = \frac{1}{2} I \omega^2$, $W = \int \tau \, d\theta$, and $P = dW/dt = \tau \omega$, respectively. Note the

work–energy principle is similar and can be obtained by replacing m by I and v by ω in the principle that applies to linear motion.

10. For rolling motion, $K = K_{\text{CM}} + K_{\text{rot}}$.

CHAPTER 11
Angular Momentum; General Rotation

This chapter describes angular momentum, vector cross product, general rotation using vector cross product notation, law of conservation of angular momentum, and motion of a top.

11–1 Angular Momentum—Object Rotating About a Fixed Axis

Angular momentum is a rotational analog of linear momentum. An object of mass m moving with a speed v in a straight line has a linear momentum, $p = mv$. An object of mass m moving with an angular speed ω along a circular path of radius r has an **angular momentum**, L, defined as

$$L = I\omega \tag{11–1}$$

where I is the moment of inertia of the object. The SI unit of angular momentum is $kg \cdot m^2/s$.

Newton's second law for rotational motion can be expressed in terms of rate of change of angular momentum as:

$$\Sigma \tau = I\alpha = \frac{\Delta L}{\Delta t} \tag{11–2}$$

This is the rotational analog of $\Sigma F = \dfrac{\Delta P}{\Delta t}$.

Conservation of Angular Momentum

Angular momentum is important in physics because, from Equation (11–2), if $\Sigma \tau = 0$, $\Delta L = 0$; that is, angular momentum is then conserved. This means

$$I\omega \text{ (final)} = I_0\omega_0 \text{ (intial)}$$

Therefore, the total angular momentum of the system is conserved, if the net external torque acting on a system is zero. This is called the **law of conservation of angular momentum**.

Conservation of angular momentum explains why a collapsing star spins faster, why a planet moves faster in its orbit as it approaches the Sun, and why a skater spins faster when she pulls her arms inward.

Directional Nature of Angular Momentum

$$\vec{L} = I\ \vec{\omega}$$

The direction of $\vec{L}$ is the same as the direction of $\vec{\omega}$, which is given by the right-hand rule.

The vector nature of angular momentum can be used to explain a number of phenomena. For example, if a person, standing at rest on a circular platform capable of rotating without friction about an axis through its center, starts walking along the edge of the platform, the platform rotates in the opposite direction. Initial angular momentum of the system is zero. As the person rotates, say counterclockwise, he creates an upward angular momentum which is balanced by the downward angular momentum due to the clockwise rotation of the platform so that the total angular momentum is still zero.

11–2 Vector Cross Product; Torque as a Vector

The **vector product** (also called **cross product**) of two vectors $\vec{A}$ and $\vec{B}$ is given by a vector $\vec{C} = \vec{A} \times \vec{B}$. $\vec{C}$ is perpendicular to both $\vec{A}$ and $\vec{B}$, and its magnitude is given by

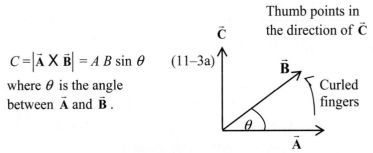

$$C = |\vec{A} \times \vec{B}| = A B \sin\theta \qquad (11\text{–}3a)$$

where θ is the angle between $\vec{A}$ and $\vec{B}$.

The direction of $\vec{C}$ is given by the right-hand rule, which states that if the fingers of the right hand are curled in a way that will rotate $\vec{A}$ through the smaller angle into coincidence with $\vec{B}$, then the thumb will point in the direction of $\vec{C}$.

In unit vector notation,

$$\vec{A} = A_x \hat{i} + A_y \hat{j} + A_z \hat{k}, \quad \vec{B} = B_x \hat{i} + B_y \hat{j} + B_z \hat{k}.$$

$$\hat{i} \times \hat{i} = \hat{j} \times \hat{j} = \hat{k} \times \hat{k} = 0$$

$$\hat{i} \times \hat{j} = \hat{k}, \quad \hat{j} \times \hat{k} = \hat{i}, \quad \hat{k} \times \hat{i} = \hat{j}$$

$$\hat{j} \times \hat{i} = -\hat{k}, \quad \hat{k} \times \hat{j} = -\hat{i}, \quad \hat{i} \times \hat{k} = -\hat{j}$$

(since $\hat{i}$, $\hat{j}$, and $\hat{k}$ are mutually perpendicular)

115

Thus,

$$\vec{A} \times \vec{B} = (A_x\, \hat{i} + A_y\, \hat{j} + A_z\, \vec{k}) \times (B_x\, \hat{i} + B_y\, \hat{j} + B_z\, \hat{k})$$

$$= (A_yB_z - A_zB_y)\, \hat{i} + (A_zB_x - A_xB_z)\, \hat{j} + (A_xB_y - A_yB_x)\, \hat{k}$$
(11–3b)

$$= \begin{vmatrix} \hat{i} & \hat{j} & \hat{k} \\ A_x & A_y & A_z \\ B_x & B_y & B_z \end{vmatrix}.$$
(11–3c)

To determine $\vec{A} \times \vec{B}$, expand the determinant whose first row is $\hat{i}$, $\hat{j}$, $\hat{k}$, whose second row is the components of $\vec{A}$, and whose third row is the components of $\vec{B}$.

To get the first term in 11–3b, draw a horizontal line through the first row and a vertical line through the first column. Cross multiply the remaining four terms, obtaining $A_yB_z - A_zB_y$, and then multiply by $\hat{i}$.

$$\begin{vmatrix} \hat{i} & \hat{j} & \hat{k} \\ A_x & A_y & A_z \\ B_x & B_y & B_z \end{vmatrix}$$

To get the second term, draw a horizontal line through the first row and a vertical line through the second column. Cross multiply the remaining four terms, obtaining $A_xB_z - A_zB_x$, and then multiply by $-\hat{j}$.

$$\begin{vmatrix} \hat{i} & \hat{j} & \hat{k} \\ A_x & A_y & A_z \\ B_x & B_y & B_z \end{vmatrix}$$

To get the third term, draw a horizontal line through the first row and a vertical line through the third column. Cross multiply the remaining four terms, obtaining $A_xB_y - A_yB_x$, and then multiply by $\hat{k}$.

$$\begin{vmatrix} \hat{i} & \hat{j} & \hat{k} \\ A_x & A_y & A_z \\ B_x & B_y & B_z \end{vmatrix}$$

From the definition of cross product,

$$\vec{A} \times \vec{A} = 0 \qquad (\theta \text{ is zero}) \tag{11--4a}$$

$$\vec{A} \times \vec{B} = - \vec{B} \times \vec{A} \tag{11--4b}$$

$$\vec{A} \times (\vec{B} + \vec{C}) = (\vec{A} \times \vec{B}) + (\vec{A} \times \vec{C}) \tag{11--4c}$$

$$\frac{d}{dt}\left(\vec{A} \times \vec{B} \right) = \frac{d\vec{A}}{dt} \times \vec{B} + \vec{A} \times \frac{d\vec{B}}{dt} \times \vec{B} \tag{11--4d}$$

The Torque Vector
Torque is an example of a physical quantity that can be expressed as a cross product:

$$\vec{\tau} = \vec{r} \times \vec{F}, \tag{11--5}$$

where $\vec{r}$ is the position vector of a point on an object and $\vec{F}$ is the force acting on the object at the point tending to rotate the object.

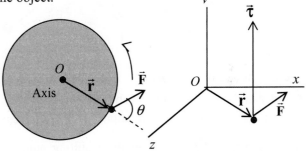

This definition is consistent with the magnitude of the torque as $rF \sin\theta$ and its direction as given by the right hand rule. Note that this definition involves $\vec{r}$, and, $\vec{\tau}$ is being calculated about a point O.

For a system of particles, the total torque $\vec{\boldsymbol{\tau}} = \sum \vec{\mathbf{r}}_i \times \vec{\mathbf{F}}_i$.

11–3 Angular Momentum of a Particle

Angular momentum is the rotational analog of linear momentum. Suppose a particle of mass m has momentum $\vec{\mathbf{p}}$ and position vector $\vec{\mathbf{r}}$ with respect to the origin O in some chosen inertial reference frame. Then, the angular momentum of the particle about point O is given by:

$$\vec{l} = \vec{\mathbf{r}} \times \vec{\mathbf{p}}. \tag{11–6}$$

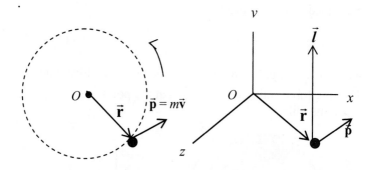

The magnitude of $\vec{l}$ is $rp \sin\theta$. It is perpendicular to both $\vec{\mathbf{r}}$ and $\vec{\mathbf{p}}$ and its direction is given by the right-hand rule.

Taking derivative of Equation 11–6 with respect to time and using $\sum \vec{\mathbf{F}} = \dfrac{d\vec{\mathbf{p}}}{dt}$ in the resulting equation, we find that

$$\sum \vec{\boldsymbol{\tau}} = \frac{d\vec{l}}{dt} \tag{11–7}$$

This is the rotational equivalent of Newton's second law.

11–4 Angular Momentum and Torque for a System of Particles; General Motion

Relation Between Angular Momentum and Torque
For a system of n particles, total angular momentum

$$\vec{L} = \sum_{i=1}^{n} \vec{l}_i \qquad (11-8)$$

and the resultant torque

$$\vec{\tau}_{net} = \Sigma \vec{\tau}_i$$

Because of action and reaction (which cancel each other), the sum of all internal torques adds to zero. Thus,

$$\vec{\tau}_{net} = \Sigma \vec{\tau}_i = \Sigma \vec{\tau}_{ext}$$

Taking derivative of Equation (11–8) with respect to time and using Equation (11–7) for each particle, we get:

$$\frac{d\vec{L}}{dt} = \Sigma \vec{\tau}_{ext} \qquad (11-9a)$$

Equation (11–9a) is valid when $\vec{L}$ and $\vec{\tau}$ are calculated (i) with reference to a point fixed in an inertial frame, and (ii) about a point, which is moving uniformly in an inertial frame. It is also valid when these are calculated about the center of mass even if accelerating:

$$\frac{d\vec{L}_{CM}}{dt} = \sum \vec{\tau}_{CM} \qquad (11\text{--}9b)$$

11–5 Angular Momentum and Torque for a Rigid Body

Let us consider a rigid body rotating about an axis that has a fixed direction in space. The total of angular momentum of the body

$$\vec{L} = \sum_i \vec{l}_i = \sum_i \vec{r}_i \times \vec{p}_i$$

It can be shown from the preceding equation that the component of the total angular momentum along the rotation axis is given by:

$$L_\omega = I\,\omega \qquad (11\text{--}10)$$

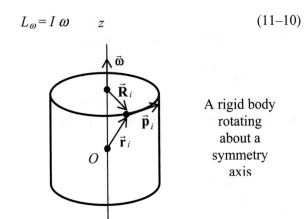

A rigid body rotating about a symmetry axis

where $I = \sum m_i R i^2$, is the moment of inertia of the body about the rotation axis; R_i being the perpendicular distance of mass m_i from the axis of rotation. The result does not depend on the choice of the origin O as long as it is on the axis of rotation.

If a body rotates about a symmetry axis through the CM, then components of $\vec{l}_i$ parallel to the axis add together, but the components perpendicular to the axis cancel, since for each point in one side of the symmetry axis there is a corresponding point on the opposite side. Thus, L_ω is the only component of $\vec{L}$ in this case and we can write

$$\vec{L} = I\vec{\omega} \qquad (11\text{–}11)$$

From Equation (11–9), for a rigid body the component along the rotation axis is:

$$\sum \tau_{axis} = \frac{dL_\omega}{dt} = \frac{d(I\omega)}{dt} = I\alpha.$$

*Rotational Imbalance

In a system where $\vec{L}$ and $\vec{\omega}$ are not parallel, the system will have rotational imbalance.

11–6 Conservation of Angular Momentum

Newton's second law for rotational motion is:

$$\frac{d\vec{L}}{dt} = \sum \vec{\tau}$$

where $\sum \vec{\tau}$ is the net external torque and $\vec{L}$ is the total angular momentum. If $\sum \vec{\tau} = 0$, then

$$\frac{d\vec{L}}{dt} = 0 \text{ and } \vec{L} = \text{constant} \qquad (11\text{–}12)$$

Thus, *the total angular momentum of a system remains constant if the net torque acting on the system is zero*. This is called the **law of conservation of angular momentum**.

*11–7 The Spinning Top and Gyroscope

A spinning top and a gyroscope are interesting examples where Newton's second law for rotational motion (Equation 11–9) can be applied. As a top spins, its axis of rotation also rotates, sweeping out a cone about the vertical. This type of motion is called **precession** and the rate at which the rotation axis moves about the vertical (z) axis is called the angular velocity of precession, Ω.

When the top is spinning with angular velocity ω about its symmetry axis, it has an angular momentum $\vec{L}$ directed along its axis, which would remain constant if no external torque is applied. But the slightest tip to one side results in a net torque $\vec{\tau}_{net} = \vec{r} \times M\vec{g}$ about O, where $\vec{r}$ is the position vector of the CM of the top and M is its mass. This causes a change in $\vec{L}$, and the change $d\vec{L}$ in time dt is $\vec{\tau}_{net}\, dt$, which is perpendicular to $\vec{L}$ and horizontal. Since $d\vec{L}$ is perpendicular to $\vec{L}$, only the direction of $\vec{L}$ changes. The vector $\vec{L}$ and the top's axis move together in a horizontal circle. $\vec{\tau}_{net}$ and $d\vec{L}$ also rotate to remain horizontal and perpendicular to $\vec{L}$.

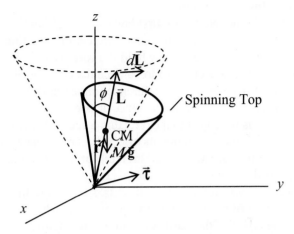

It can be shown that the angular velocity of precession is given by:

$$\Omega = \frac{\tau}{L \sin \phi} \qquad (11\text{–}13\text{a})$$

$$= \frac{Mgr}{L} \qquad (11\text{–}13\text{b})$$

*11–8 Rotating Frames of Reference; Inertial Forces

Inertial and Noninertial Reference Frames
An **inertial reference frame** is the frame in which Newton's first law of motion holds. A reference frame, such as a rotating platform, is called a **noninertial reference frame** if Newton's first law does not hold in the frame. In an inertial frame, Newton's second and third laws also hold. In a noninertial frame, Newton's second law also does not hold.

Fictitious (Inertial Forces)

In a rotating frame, Newton's first law (or second law) does not hold. However, if an object is at rest in a rotating frame, we can still make the law hold if a force (called centrifugal force) acts outward to balance the centripetal force (as seen from an inertial reference frame) so that the net force is zero. It is a **fictitious force** or **pseudoforce** because there is no object that exerts this force and, when viewed from an inertial frame, its effect does not exist at all. Pseudo forces are also called **inertial forces** since they arise only if the frame is not an inertial one.

Since the Earth rotates about its axis, strictly speaking, Newton's laws of motion are not valid on the Earth. Usually the effect of the Earth's rotation is negligible although it does influence, for example, the movement of large air masses and ocean currents.

*11–9 The Coriolis Effect

In a frame that rotates with a constant angular speed ω, another pseudoforce, called the Coriolis force, appears to act on an object to deflect it sideways if the object is moving relative to the rotating frame. For example, a woman at A throws a ball with a horizontal velocity $\vec{v}$ radially outward to a man at B near the edge of the platform. Observed from an inertial frame of reference, since v_B is greater than v_A, when the ball reaches the outer edge of the platform it passes a point that the man at B has already gone. Thus, the ball passes behind him. Observed from the rotating frame of reference, both A and B are at rest, but the ball deflects to the right as shown and passes behind B. This effect acts sideways (perpendicular to $\vec{v}$),

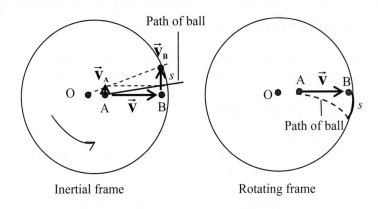

Inertial frame   Rotating frame

and is called a **Coriolis acceleration**. This occurs due to the Coriolis force, which is a fictitious inertial force. That is, when viewed from a rotating frame we can describe the motion using Newton's second law, if we add a "pseudoforce" term representing this Coriolis effect.

It can be shown that the sideways displacement

$$s = \omega v t^2 \qquad (11-14)$$

where t represents time the ball takes to move from A to B. The Coriolis acceleration

$$a_{\text{Cor}} = 2\omega v. \qquad (11-15)$$

Because of the Coriolis effect, the winds are directed to the right (instead of rushing directly into a region of low pressure) in the Northern hemisphere since the Earth rotates from west to east. Because of the same effect, cyclones rotate counterclockwise in the Northern Hemisphere and clockwise in the Southern Hemisphere.

Tips for Solving the Problems

1. The angular momentum is conserved when the net torque is zero (such as in a rotating collision). That is, $I\omega$(final)$= I_0\omega_0$ (intial). This relation will be useful for solving problems in Section 11–1.

2. Note that the cross product of two vectors is a third vector, which is perpendicular to the plane containing the first two vectors.

3. To determine the cross product of two vectors $\vec{A}$ and $\vec{B}$ or to determine torque ($\vec{\tau} = \vec{r} \times \vec{F}$) and angular momentum ($\vec{l} = \vec{r} \times \vec{p}$) vectors using cross product, first set up the determinant that represents the cross product, such as given by Equation 11–3c, and then expand the determinant.

4. When you determine the angle θ between two vectors $\vec{A}$ and $\vec{B}$ using the relation $C == A B \sin \theta$, always use the knowledge of quadrant to choose the correct angle because you will get two answers for the angle from this relation.

CHAPTER 12
Static Equilibrium; Elasticity and Fracture

This chapter describes the two conditions of equilibrium, stability and balance, and elastic properties of substances.

Important Concepts

       Conditions of equilibrium
       Stable, unstable, and neutral equilibrium
       Hooke's law
       Elastic limit
       Ultimate strength
       Elastic modulus or Young's modulus
       Stress
       Strain
       Shear modulus
       Bulk modulus

An object at rest on a table is in *equilibrium* (since the net force is zero) under the action of two equal and opposite forces: the force of gravity and normal force. The study of **statics** deals with the calculation of forces acting on or within the structures that are in equilibrium. This study is important for engineering, medicine, and physical therapy.

12–1 The Conditions for Equilibrium

The First Condition for Equilibrium
Vector sum of the external forces on the object must be zero: $\Sigma \vec{\mathbf{F}} = 0$.

In terms of components

$$\Sigma F_x = 0, \ \Sigma F_y = 0, \ \Sigma F_z = 0. \qquad (12\text{--}1)$$

The Second Condition for Equilibrium

Vector sum of the external torques acting on the object must be zero:

$$\Sigma \vec{\tau} = 0. \qquad (12\text{--}2)$$

In terms of components: $\Sigma \tau_x = 0$, $\Sigma \tau_y = 0$, $\Sigma \tau_z = 0$.
Condition (12–1) means that there is no linear acceleration, and condition (12–2) means that there is no angular acceleration of the object.

12–2 Solving Statics Problems

The following steps are in general helpful to solve statics problems.

1. Draw a free-body diagram showing all the forces with directions.
2. Choose a convenient coordinate system and resolve the forces into their components.
3. Write down the equilibrium equations:

   $\Sigma F_x = 0$, $\Sigma F_y = 0$, and $\Sigma \tau = 0$.

4. For the $\Sigma \tau = 0$ equation, choose an axis perpendicular to the *xy* plane and assign the proper sign to individual torques (+ for counter-clockwise or – for clockwise). You can reduce the number of unknowns by choosing the axis passing through one of the unknown forces.
5. Solve these equations for unknowns.

Note that for an extended object the force of gravity acts through its center of gravity.

12–3 Stability and Balance

For an object in equilibrium $\Sigma \vec{F} = 0$ and $\Sigma \vec{\tau} = 0$.

- If an object in equilibrium returns to its original position when slightly disturbed, the object is in **stable equilibrium**. For example, a ball suspended freely from a string is in stable equilibrium.

- If the object, after it is slightly disturbed from equilibrium, moves even farther from its equilibrium position, the equilibrium is said to be **unstable equilibrium**. A pencil standing on its point is in unstable equilibrium.

- If the object, after it is slightly distributed, remains in its new position, the object is said to be in **neutral equilibrium**. A sphere resting on horizontal tabletop is an example of neutral equilibrium.

Maintaining a stable equilibrium is important in many situations such as in the designing of structures. In general, *a body whose CG is above its base of support will be stable if a vertical line projected downward from the CG falls within the base of support.* Humans are less stable than four-legged animals which have a larger base of support because of their four legs and because their center of gravity is lower.

12–4 Elasticity and Elastic Moduli; Stress and Strain

The shape or dimension of an object can be changed (usually only slightly) when it is acted on by a force.

If a force is applied to an object (such as a metal wire whose one end is fixed and a load is applied at its other end) it increases its length. For small elongation, ΔL, the applied force can be written as

$$F = k \, \Delta L \qquad (12\text{–}3)$$

where k is a proportionality constant. This relation is known as the **Hooke's law**.

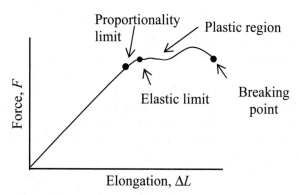

From the graph of applied force versus elongation it is seen that:

- The graph is linear up to the **proportional limit**.

- The object returns to its normal length when the applied force is removed up to a point on the curve called **elastic limit**.
- If the object is stretched beyond the elastic limit, the object enters into the *plastic region*.
- The maximum elongation is reached at the *breaking point*. The maximum strength that can be applied before the material breaks is called the **ultimate strength** of the material.

Young's Modulus

The elongation ΔL is related to the applied force F, original length L_0, area of cross-section A by the relation

$$\Delta L = \frac{1}{E}\frac{F}{A}L_0 \qquad (12\text{–}4)$$

The constant E is a characteristic of the material called the **elastic modulus** or **Young's modulus**.

Stress and Strain

Stress is defined as the force per unit area applied to an object. **Strain** is the resulting fractional change in length or volume or change in shape. From Equation (12–4),

$$\frac{F}{A} = E\frac{\Delta L}{L_0} \qquad (12\text{–}5)$$

$$E = \frac{F/A}{\Delta L/L_0} = \frac{\text{stress}}{\text{strain}}.$$

Tension, Compression, and Shear Stress

A stress is said to be **tensile stress** if it tends to elongate the material and the material is said to be under *tension*. If the material is compressed, the stress is called the **compressive stress**.

Equal and opposite forces applied to two surfaces of a material result in shear deformation (that is, change in shape) and the corresponding stress is called the **shear stress**. The shear strain ΔL is related to the shear force F by

$$\Delta L = \frac{1}{G}\frac{F}{A}L_0 \qquad (12\text{–}6)$$

where A is the area parallel to the applied force F and ΔL is perpendicular to L_0. The constant G is called the **shear modulus** of the material.

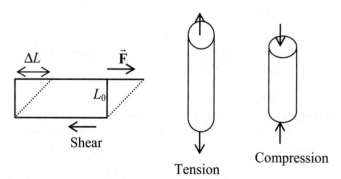

*Volume Change—Bulk Modulus

When the pressure on a solid increases, its volume decreases.

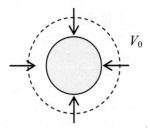

The pressure difference, ΔP, is related to the change in volume, ΔV, and initial volume, V_0, by

$$\frac{\Delta V}{V_0} = -\frac{1}{B}\Delta P$$

$$B = \frac{\Delta P}{\Delta V / V_0} \qquad (12\text{--}7)$$

The constant B is called the **bulk modulus.** The minus sign signifies that when pressure increases, volume decreases, and vice versa. The constants E, G, and B have the units N/m^2.

12–5 Fracture

When the applied stress is very high, an object fractures. The maximum stess that a body can withstand is called ultimate strength. To maintain a *safety factor*, the applied stress should not exceed one-tenth to one-third of the ultimate stress.

Concrete is reasonably strong under compression but extremely weak under tension. Thus, it can be used as vertical columns placed under compression but as a beam.

*12–6 Trusses and Bridges

A **truss** is a framework of rods or struts joined together at their ends by pins or rivets, arranged as rectangles. The place where the struts are joined by a pin is called a *joint*.

Each strut of a struss is assumed to be under tension or compression. Two equal and opposite forces must act along the same line since the net torque is zero. For real struts (having masses) the forces at the joints do not act precisely along the strut.

The tension or compression in each of the struts can be determined by drawing (i) the free-body diagram of the strut as a whole, (ii) drawing a free-body diagram for each of the joints, one by one, and setting $\Sigma \vec{F} = 0$ for each joint.

*12–7 Spanning a Space: Arches and Domes

The introduction of the semicircular **arch** was a technological innovation. For a round or "true" arch, the wedge-shaped stones experience stress that is mainly compressive. A round shaped arch can span a very wide space. A considerable buttressing on the sides is needed to support the horizontal component of the forces.

The pointed arch is used to support heavy loads (such as the tower of a cathedral). The horizontal buttressing force required for a pointed arch is less because the arch is higher. The steeper the arch, the less the horizontal component of the force is, and hence more nearly vertical is the force exerted at the base of the arch. A **dome** is basically an arch rotated about a vertical axis that spans a three-dimensional space as opposed to an arch that spans a two-dimensional space.

Tips for Solving the Problems

1. For problems on the equilibrium of objects, such as those in Sections 12–1 and 12–2, apply the equilibrium equations: $\Sigma F_x = 0$, $\Sigma F_y = 0$, and $\Sigma \tau = 0$.

2. Torque can be obtained using the expression $\tau = rF \sin \theta$. If $\theta = 90°$, $\tau = rF$. The magnitude of torque can be positive, negative, or zero. When adding more two or more torques, add them with their proper signs.

3. Always draw the sketch of the problem and the free-body diagram. When calculating a torque, think about the point about which the torque will be calculated. Remember, you can avoid a force that you are not interested in by taking torque about a point through which the force passes.

4. Area conversions from square centimeters to square meters are common in problems on elasticity. Also, be sure to distinguish between diameter and radius when calculating the area or volume.

5. Stress = force (F)/area (A) or pressure (ΔP) and strain = $\Delta L/L_0$ or $\Delta V/V_0$, depending on the problem. Young's modulus (E), shear modulus (G), and bulk modulus (B) are the ratios of stress and strain for change of length, shape, and volume, respectively. For shear modulus problems remember that the area, A, is the area of the surface parallel to the force.

6. Stress has the unit N/m^2, but strain is dimensionless.

7. To determine the tension or compression in each of the struts of a truss, draw (i) the free-body diagram of the

truss as a whole, (ii) a free-body diagram for each of the joints, one by one, and set $\Sigma \vec{F} = 0$ for each joint.

CHAPTER 13
Fluids

This chapter describes density, pressure, buoyancy, and Archimedes' principle as applied to fluids. The fundamental principles of fluid flow, such as the equation of continuity and Bernoulli's equation, as well as viscosity and surface tension are also presented.

Important Concepts

Solid, liquid, and gas
Density and specific gravity
Pressure
Pascal
Atmosphere
Bar
Gauge pressure
Variation of pressure with depth
Pascal's principle
Barometer
Buoyancy
Archimedes principle
Flotation and submersion
Equation of continuity
Bernoulli's equation
Torricelli's theorem
Viscosity
Coefficient of viscosity
Poiseuille's equation
Surface tension

13–1 Phases of Matter

The three common **phases** of matters are solids, liquids, and gases.

- A **solid** has a fixed size and shape.

- A **liquid** does not have a fixed shape—it takes the shape of the container. Solids and liquids are not readily compressible.

- A **gas** has neither a fixed shape nor a fixed volume. It can be readily compressed. Liquids and gases are collectively called **fluids** since they have the ability to flow.

13–2 Density and Specific Gravity

The **density** ρ of a substance of mass m and volume V is defined as

$$\rho = \frac{m}{V} \qquad (13\text{–}1)$$

The denser a material, the more mass it has in a given volume. The SI unit of density is kg/m^3. The density of water at $4.0^{\circ}C$ is $1000 \ kg/m^3$.

The **specific gravity** of a substance is the ratio of the density of that substance to the density of water at $4.0^{\circ}C$. For example, the specific gravity of lead is 11.3.

13–3 Pressure in Fluids

The force, F, exerted by a fluid perpendicular to a surface divided by the area of the surface gives the **pressure**, P, of the fluid:

$$P = \frac{F}{A} \qquad (13\text{–}2)$$

The SI unit of pressure is N/m², which is called **pascal**.
1 pascal (Pa) = 1 N/m².

The pressure exerted by a fluid
acts equally in all directions and is
perpendicular to any surface.

Fluid pressure

The pressure of a fluid in static equilibrium increases
with depth. This increase of pressure with depth arises
because of the increasing weight of the fluid with depth.
At a depth h, the fluid pressure P is given by

$$P = \rho g h \qquad (13\text{–}3)$$

where ρ is the density of the fluid.

The rate of change of pressure with height, y, within
the fluid can be shown to be

$$\frac{dP}{dy} = -\rho g \qquad (13\text{–}4)$$

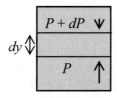

It follows that $P_2 - P_1 = - \int_{y_1}^{y_2} \rho g \; dy$, (13–5)

where P_1 and P_2 are the pressures of the fluid at heights y_1 and y_2, respectively.

If the density ρ is constant for the fluid,

$$P_2 - P_1 = -\rho g (y_2 - y_1)$$ (13–6a)

If y_2 the position on the top of the surface, then P_2 represents the atmospheric pressure P_0. Thus, the pressure at a depth h is:

$$P = P_0 + \rho g h .$$ (13–6b)

13–4 Atmospheric Pressure and Gauge Pressure

Atmospheric Pressure

The atmospheric pressure arises because of the weight of the air above us. At sea level its value (called the **atmosphere** or atm) is:

$\qquad$ 1 atm = $1.013 \times 10^5 \, \text{N/m}^2$ = 101.3 kPa

A common unit of atmospheric pressure is the **bar**:

$\qquad$ 1 bar = $1.00 \times 10^5 \, \text{N/m}^2$.

In British units, pressure is measured in pounds per square inch and its value is 14.7 lb/in^2.

Gauge Pressure

The **gauge pressure** P_G (such as determined by a tire gauge) is the difference between the absolute pressure P and atmospheric pressure P_A: $P_G = P - P_A$

$\qquad$ Note that if a tire gauge registers 220 kPa, the actual pressure within the tire is 220 kPa + 101 kPa = 321 kPa.

13–5 Pascal's Principle

According to **Pascal's principle**, *if an external pressure is applied to a confined fluid, the pressure at every point within the fluid increases by that amount.*

A hydraulic lift works on Pascal's principle. The lift consists of two cylinders (fitted with pistons) of cross-sectional areas A_{out} and A_{in} ($A_{out} > A_{in}$) connected by a tube and filled with a liquid. A force F_{in} applied to piston 1 increases the pressure, which is transmitted through the fluid. As a result, the force F_{out} on piston 2 can be obtained from the equation

$$\frac{F_{out}}{F_{in}} = \frac{A_{out}}{A_{in}}$$

Since A_{out} is greater than A_{in}, force is magnified in a hydraulic lift. The ratio F_{out}/F_{in} is called the **mechanical advantage** of the hydraulic lift.

13–6 Measurement of Pressure; Gauges and the Barometer

Pressure can be measured using several devices. An open tube *manometer* that is used to measure pressure is a U-shaped tube partially filled with a liquid, usually mercury.

The pressure P being measured is related to the difference in height Δh of the two levels of the liquid and is given by

$$P = P_0 + \rho g \Delta h$$

where P_0 is the atmospheric pressure and ρ is the density of the fluid. Sometimes pressures are specified as the difference in height h, such as in so many "mm of mercury". The unit mm of mercury is called **torr**. In an aneroid pressure gauge, a pointer is linked to the flexible ends of an evacuated thin metal chamber. In an electronic pressure gauge, the pressure is applied to a thin metal diaphragm whose resulting distortion is detected electronically.

Atmospheric pressure is often measured by a **barometer**, wherein a glass tube is completely filled with mercury and then inverted into the bowl of mercury. The tube is long enough so that the level of mercury drops, creating a vacuum at the top of the tube. The height of the mercury column indicates the atmospheric pressure, which is about 76 cm of mercury. Thus, the atmospheric pressure is

$$P = \rho g h = (13.6 \times 10^3 \text{ kg/m}^3)\,(9.8 \text{ m/s}^2)\,(0.760 \text{ m})$$

$$= 1.013 \times 10^5 \text{ N/m}^2 = 1.00 \text{ atm}.$$

13–7 Buoyancy and Archimedes' Principle

When an object is immersed in a fluid (water, for example), the upward pressure exerted by the fluid on the lower surface of the object is more than the downward pressure on the top surface, since fluid pressure increases with its depth. As a result, there is a net upward force on any object immersed in a fluid. This upward force is called **buoyant force**.

It can be shown that, if an object of volume V is immersed in a fluid of density ρ_F, the buoyant force F_B is

$$F_B = \rho_F Vg = m_F g = \text{weight of the displaced fluid.}$$

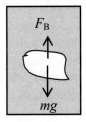

This result is called **Archimedes' principle,** which states that: *the buoyant force on a body immersed in a fluid is equal to the weight of the fluid displaced by that object.*

Archimedes' principle applies also to objects that float, such as wood in water. *An object floats on a fluid if its density is less than that of the fluid.* When floating, an object is in equilibrium so that upward buoyant force on the object is equal to the weight of the object. If the upward buoyant force, even when an object is completely submerged, is less than the weight of the object, the object will sink.

The condition that the weight of a floating object and the buoyant force are equal can be used to determine how much of a floating object is submerged when it is floating. If the volume of the displaced fluid (which is same as the submerged volume of the object) is V_{disp} for a solid of volume V_o and density ρ_o, and if the object is floating in a fluid of density ρ_F, then it can be shown that

$$\frac{V_{displ}}{V_o} = \frac{\rho_o}{\rho_F}$$

13–8 Fluids in Motion; Flow Rate and the Equation of Continuity

Fluid dynamics is the study of fluids in motion. If the fluid is water, the study is called **hydrodynamics**. There are two types of fluid flow:

- If the flow of a fluid is smooth so that neighboring layers of the fluid slide by each other smoothly, the flow is called **streamline** or **laminar flow**. In streamline flow, each particle follows a smooth path and these paths do not intersect.

- If the speed of the fluid is high, the flow is called **turbulent flow**, which is characterized by erratic, small, whirlpool-like circles, called eddies.

The internal friction that is present in a fluid motion is called the **viscosity**.

In order that matter is conserved in the flow of a fluid, the mass **flow rate** (that is, the mass of fluid passing a point) at one point must be the same as the flow rate passing another point. Thus, for a fluid moving from point 1 to point 2,

$$\rho_1 A_1 v_1 = \rho_2 A_2 v_2 \qquad (13\text{–}7a)$$

This is called the **equation of continuity**.

Most liquids are practically incompressible. Thus, the equation of continuity for the flow of an incompressible liquid is

$$A_1 v_1 = A_2 v_2 \qquad (\rho = \text{constant}) \qquad (13\text{–}7b)$$

This equation can be applied to the motion of blood in the body. It is clear from Equation (13–7) that the speed of a fluid changes, and that the smaller the cross-sectional area, the faster the speed of the fluid.

13–9 Bernoulli's Equation

The relationship between the pressure of a fluid, its speed, and its height is known as **Bernoulli's equation**.

The speed of a fluid changes as it moves from point 1 to point 2, as was seen in the equation of continuity. The pressure of a fluid also changes as it moves from point 1 to point 2.

If both the speed (v) and the height (y) of a fluid change during its motion from point 1 to point 2, then from the work–energy theorem for the motion of fluid it can be shown that the pressure and speed are related by

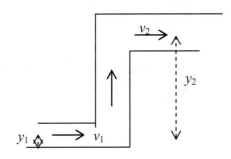

$$P_1 + \rho g y_1 + \frac{1}{2}\rho v_1^{\,2} = P_2 + \rho g y_2 + \frac{1}{2}\rho v_2^{\,2} \qquad (13\text{–}8)$$

This is Bernoulli's equation, according to which

$$P + \frac{1}{2} \rho v^2 + \rho gy = \text{constant}$$

This relation can be thought of as energy conservation since it is based on the work–energy principle.

13–10 Applications of Bernoulli's Principle: Torricelli, Airplanes, Baseballs, and TIA

Bernoulli's equation can be used in a wide range of applications.

The speed of a fluid as it flows through a hole is the same as if the fluid had freely fallen a distance $(y_2 - y_1)$ equal to the distance of the hole from the top of the fluid:

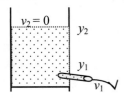

$$v_1 = \sqrt{2g(y_1 - y_2)} \qquad (13\text{–}9)$$

This relation is known as **Torricelli's theorem**.
From Bernoulli's equation when there is no change in height,

$$P_1 + \frac{1}{2} \rho v_1^2 = P_2 + \frac{1}{2} \rho v_2^2 \qquad (13\text{–}10)$$

The preceding equation shows that when the pressure is low, the speed is high and vice versa.

Airplane Wings and Dynamic Lift
The shape of an airplane wing is such that air flows more rapidly over the top surface than over the lower surface. Because the air speed is greater above the wing than below

it, the pressure above the wing is less than the pressure below it (Bernoulli's principle). Thus, there is net upward force on the wing, called **dynamic lift**.

Sailboats
A sailboat can move against the wind because of the Bernoulli's effect by setting the sail at an angle.

Baseball Curve
A spinning baseball drags a thin layer of air around it. On one side of the ball (which is the right side if the spin is counterclockwise), the boundary air layer tends to slow down the oncoming air. As a result, the air speed on the other side of the ball is higher and hence the pressure on the other side is lower. This results in a net force on this side, causing the ball to curve.

Lack of Blood to the Brain—TIA
The rear of the head and shoulders has arteries that lead to the brain and to the arms. High blood velocity past constriction in the left subclavian artery causes low pressure in the left vertebral artery, in which a downward blood flow can occur, resulting in a transient ischemic attack (TIA).

Other Applications
A **venture tube** is a pipe with a narrow constriction (the throat). The fluid speeds up as it passes through the throat, so the pressure is lower. A *venturimeter* is used to determine the speed of fluids such as the blood velocity in arteries.

*13–11 Viscosity
Viscosity refers to the internal friction of a fluid as one layer moves relative to another. Different fluids possess

different amounts of viscosity. Air has low viscosity, water is more viscous than air, and fluids like honey are highly viscous.

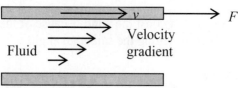

Moving plate

Fluid

Velocity gradient

Stationary plate

A thin layer of liquid is placed between the two plates, one of which is stationary, while the other plate is forced to move at constant velocity v, as shown. The force, F, required to move the moving plate is equal to the area of fluid in contact with each plate, A, and to the speed, v, and is inversely proportional to the separation, l, of the plates. Thus,

$$F = \eta A \frac{v}{l} \qquad (13\text{--}11)$$

where the proportionality constant η is called the *coefficient of viscosity*. The SI unit of η is N·s/m^2 = Pa·s. A common unit of η is the *poise*. 1 poise = dyne·s/cm^2 = 0.1N·m/s^2. The viscosity of liquids decreases with temperature.

*13–12 Flow in Tubes: Poiseuille's Equation, Blood Flow

For a real fluid, a pressure difference between the ends of a tube is necessary for the steady flow of the fluid. The rate of flow of a fluid in a cylindrical tube depends on the viscosity of the fluid, the pressure difference, and the dimensions of the tube. The volume rate of flow Q (that is, volume of fluid flowing past a given point per unit time) is given by *Poiseuille's equation*:

$$Q = \frac{\pi R^4 (P_1 - P_2)}{8 \eta L} \qquad (13\text{–}12)$$

where R is the inside radius of the tube, L its length, and $P_1 - P_2$ is the pressure difference between the ends. Because of the fourth power dependence of R, a small change in radius causes a large change in volume flow rate. The R^4 dependence has an important effect in blood flow in the human body. The body controls the flow of blood by means of tiny bands of muscle surrounding the arteries. As the muscles contract, the radius of an artery decreases and because of R^4 dependence, the flow rate is greatly decreased. Also, the radius can be decreased because of hardening of arteries and cholesterol buildup. When this happens, a large pressure difference must be maintained (for example, by a factor of 2^4 if the radius is reduced by a factor of 2) to maintain the same flow rate. The heart works harder in these situations but still cannot maintain the same flow rate. Therefore, when the blood pressure is high, the flow rate is reduced.

*13–13 Surface Tension and Capillarity

The surface of water and other fluids behaves as if it is under tension like a stretched elastic membrane. This effect is called **surface tension**. More specifically, surface tension γ is defined as the force F per unit length L that acts across any line in a surface tending to pull the surface closed, that is,

$$\gamma = \frac{F}{L} \qquad (13\text{–}13)$$

Surface tension arises because the molecules in a fluid exert attractive forces on one another. For a molecule well inside a fluid the net attractive force is zero because it experiences forces in all directions exerted by other fluid molecules. For a molecule on the surface, the net force by other fluid molecules is in a direction away from the surface.

Because of surface tension, a force is needed and work is done to bring molecules from the interior to the surface. This work increases the potential energy of the molecules and is called *surface energy*. It can be shown that surface tension γ is also equal to the work done per unit increase in surface area.

Because of its surface tension, a fluid tends to pull inward on its surface, which results in a surface of minimum area. This explains why small water drops are always spherical. Because of surface tension, insects can walk on water and a steel needle can float on the surface. Because of surface tension, in tubes with very small diameters, liquids are observed to rise (such as a glass tube in water) or fall (such as a glass tube in mercury) relative to the level of the surrounding liquid. This phenomenon is called **capillarity**.

*13–14 Pumps and the Heart

A *vacuum pump* reduces the pressure of air in a given container. A *force pump* is intended to increase the pressure to push a fluid through a pipe or lift a fluid (such as water from a vessel). In one kind of pump the intake valve opens and air (or the fluid) fills the empty space when the piston moves to the left. When the piston moves to the right, the outlet valve opens and fluid is forced out. In a *centrifugal pump*, its rotating blades force fluids through the outlet pipe and it can be used as a cirulating pump to circulate a fluid around a closed path.

The human heart is essentially a circulating pump where there are two separate paths for blood flow. In the longer path, blood flows to the body via the arteries, brings oxygen to the body tissues, picks up carbon dioxide, and goes back to the heart via veins. In the second path, the blood is then pumped to the lungs where carbon dioxide is released and oxygen is taken up. The oxygen-laden blood goes back to the heart and then is pumped again to the body.

Tips for Solving the Problems

1. Inside a fluid, the pressure difference between two points is $P = \rho gh$. That is, the pressure P_2 of a fluid at a point 2 at a depth h below a point 1 (where pressure is P_1), is $P_2 = P_1 + \rho gh$. Do not forget to add atmospheric pressure in situations where the fluid is in an open container.

2. Pascal's principle provides mechanical advantage, which is the ratio A_{out}/A_{in}. Make sure you can apply the principle to static fluid situations.

3. Make sure you understand Archimedes' principle. Think about *buoyant force* and how it is related to the *weight of the displaced fluid*. Once you are comfortable with these concepts you should be able to solve the problems in Sections 13–7.

4. There are two basic equations in fluid dynamics: the *equation of continuity* and *Bernoulli's principle*. Torricelli's theorem, which is derived from Bernoulli's

equation, gives the speed of a fluid as it flows through a hole. Read the problem carefully to determine which of these concepts is involved for problems in Sections 13–8 to 13–10.

5. If the problem involves a change in the speed of a fluid as the area through which its flows changes, apply the equation of continuity

6. When using Bernoulli's equation, *draw a diagram.* Locate the region where the pressure, elevation, and speed are known. Locate the region where the unknown quantity needs to be determined.

CHAPTER 14
Oscillations

This chapter describes simple harmonic motion.
Relationships between uniform circular motion and simple
harmonic motion, motion of a mass attached to a spring,
motion of different pendulums, and damped and forced
oscillations are discussed.

Important Concepts

Periodic motion
Period, frequency, and amplitude
Simple harmonic motion
Restoring force
Simple pendulum
Physical pendulum
Torsion pendulum
Damped harmonic motion
Forced oscillation
Resonance
Quality factor

14–1 Oscillation of a Spring

A motion in which an object continuously repeats its path
is called a **periodic** motion.

Assume a spring is mounted horizontally and the mass
m slides without friction on a horizontal surface. When the
length of the spring is the same as its natural length, the
force on the mass is zero, and the position of the mass is
called the **equilibrium position**. For small displacements
of the mass, the spring exerts a restoring force, F, that is
proportional to the displacement, x, of the mass from its
equilibrium position. That is,

$$F = -kx, \qquad\qquad (14–1)$$

This is called Hooke's law. The proportionality constant k is called the "spring constant." Because of this restoring force F, the mass m executes a periodic motion between, say, $x = A$ and $x = -A$.

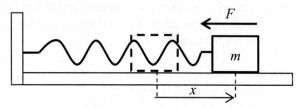

In order to stretch the spring by x, one needs to exert a force at least of amount $F = +kx$. From Equation (14–1), the force F, and hence the acceleration of the mass m, is *not* a constant but depends on x.

In a periodic motion:

- The maximum distance the mass moves from the equilibrium position is called the **amplitude**.

- The time required to complete one cycle of repetitive motion is called the **period**, T. The SI unit of T is seconds/cycle = s.

- The inverse of period is the number of cycles or oscillations per unit time and is called the **frequency**, f.

$$f = \frac{1}{T} \quad \text{and} \quad T = \frac{1}{f} \qquad (14–2)$$

The SI unit of f is cycles/second = $1/s$ = hertz (Hz).

14–2 Simple Harmonic Motion

Any periodic motion, where restoring force F is equal to $-kx$, is called a **simple harmonic motion** (SHM) and the system is often called a **simple harmonic oscillator** (SHO). Many natural vibrations can be well approximated by SHM.

For the simple harmonic motion of a mass m attached to a spring of spring constant k (such as shown in Section 14–1), from Newton's second law:

$$ma = \Sigma F$$

$$m\frac{d^2x}{dt^2} = -kx$$

$$m\frac{d^2x}{dt^2} + kx = 0 \qquad (14\text{--}3)$$

This is the *equation of motion* of the SHO. The general solution of the equation shows that x varies sinusoidally with t as given by:

$$x = A \cos(\omega t + \phi) \qquad (14\text{--}4)$$

where the constant A represents the amplitude of motion and the constant ϕ represents the initial phase angle. It can be shown that

$$\omega^2 = \frac{k}{m} \qquad (14\text{--}5)$$

- If the mass is started at its maximum displacement and is released from rest, $x = A$ at $t = 0$. In this case from Equation (14–4), we get $\phi = 0$, and $x = A \cos \omega t$.

- If at $t = 0$, $x = 0$, we get $\phi = \pm \pi/2$, and in this case, $x = A \cos(\omega t \pm \pi/2) = \pm A \sin \omega t$.

Since the oscillating mass repeats its motion after a time equal to its period T and since a sine or cosine function repeats itself after every 2π radians,

$$\omega T = 2\pi$$
$$\omega = \frac{2\pi}{T} = = 2\pi f$$

ω is thus called the **angular frequency**, and Equation (14–4) becomes

$$x = A \cos\left(\frac{2\pi}{T} t + \phi\right) \qquad (14\text{–}6a)$$

$$= A \cos(2\pi f t + \phi) \qquad (14\text{–}6b)$$

Variation of x with t is shown in the following sketch for $\phi = 0$ and $\phi = -\pi/2$ (for which $v > 0$ at $t = 0$).

From Equation (14–5)

$$f = \frac{1}{2\pi} \sqrt{\frac{k}{m}} \qquad (14\text{–}7a)$$

$$T = 2\pi \sqrt{\frac{m}{k}} \qquad (14\text{--}7b)$$

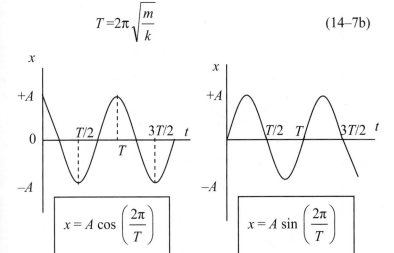

$$x = A \cos\left(\frac{2\pi}{T}\right)$$

$$x = A \sin\left(\frac{2\pi}{T}\right)$$

The frequency f, as expressed by Equation (11–7a), is called the **natural frequency** of the SHO. From Equation (14–4):

$$\text{Velocity, } v = \frac{dx}{dt} = -\omega A \sin(\omega t + \phi) \qquad (14\text{--}8)$$

$$\text{Acceleration, } a = \frac{dv}{dt} = -\omega^2 A \cos(\omega t + \phi) \qquad (14\text{--}9)$$

Clearly, $v_{max} = \omega A$, and, $a_{max} = \omega^2 A$.

It is important to note that in simple harmonic motion both velocity and acceleration continually change. The graph of x vs. t, y vs. t, and a vs. t is shown for a complete cycle. Note that when the magnitude of the velocity is a maximum, the acceleration is zero, and vice versa.

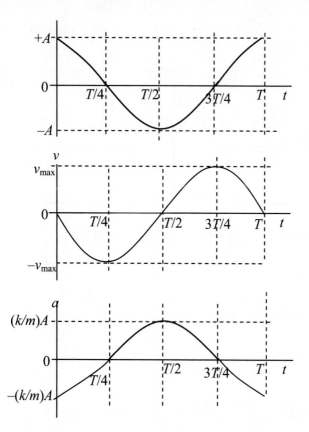

14–3 Energy in the Simple Harmonic Oscillator

To stretch or compress a spring, work is done, which is stored as the elastic potential energy of the spring given by, $U = \dfrac{1}{2} kx^2$. The total energy E of the spring–mass system:

$$E = K + U = \frac{1}{2} mv^2 + \frac{1}{2} kx^2 = \text{constant}$$

At the extreme points ($x = A$ and $-A$), the velocity $v = 0$.
Thus,

$$E = \frac{1}{2} m(0)^2 + \frac{1}{2} kA^2 = \frac{1}{2} kA^2 \qquad (14\text{--}10a)$$

At the equilibrium position, $x = 0$ and total energy is kinetic. Thus,

$$E = \frac{1}{2} mv_{max}^2 + \frac{1}{2} k(0)^2 = \frac{1}{2} mv_{max}^2 \qquad (14\text{--}10b)$$

Thus, $\qquad \frac{1}{2} mv^2 + \frac{1}{2} kx^2 = \frac{1}{2} kA^2 = \frac{1}{2} mv_{max}^2 \qquad (14\text{--}10c)$

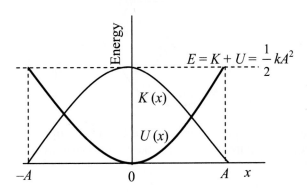

From Equation (14–10c) :

$$v = \pm \sqrt{\frac{k}{m}\left(A^2 - x^2\right)} \qquad (14\text{--}11a)$$

$$= \pm v_{max} \sqrt{1 - \frac{x^2}{A^2}} \qquad (14\text{–}11b)$$

Graphs of U and K versus x are shown. (Note $E = K + U = $ constant for any point x.)

14–4 Simple Harmonic Motion Related to Uniform Circular Motion

There exists a close relationship between simple harmonic motion and uniform circular motion. Let us consider a mass m revolving counterclockwise in a circle of radius A with a speed v_M on the top of a table (say, xy plane).

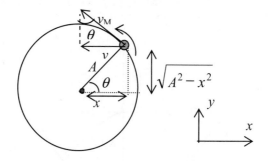

In this case, from the similarity of the two triangles of the preceding figure, we find that the x component of v_M is given by

$$v = v_M \sin \theta = v_M \sqrt{1 - \frac{x^2}{A^2}}$$

which is the same as Equation (14–11b) where $v_M = v_{max}$. Also $x = A \cos \theta = A \cos \omega t$ [or, $A \cos (\omega t + \phi)$ if at $t = 0$, $\theta = \phi$] The angular velocity ω is given by:

$$\omega = \frac{v_M}{A} = \frac{2\pi A/T}{A} = \frac{2\pi}{T} = 2\pi f$$

Thus the projection of the circular motion along x-axis is SHM. Similarly, the projection of the motion on the y-axis is also SHM and, thus, a uniform circular motion is equivalent to two simple harmonic motions operating at right angles.

14–5 The Simple Pendulum

A **simple pendulum** consists of a mass m suspended by a light cord of length L. For oscillations of small amplitude, the motion of a simple pendulum is simple harmonic.

The displacement of the pendulum along the arc is given by $x = L\theta$, where θ is the angle the pendulum makes with the vertical.

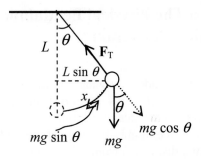

For small amplitudes of oscillation, the restoring force:

$$F = -mg \sin \theta = -mg\theta = -\frac{mg}{L}x.$$

Comparing with $F = -kx$ for the motion of a mass on a spring, we find that the restoring force acting on the pendulum has the same form if we let $k = mg/L$. Consequently,

the angular frequency, $\omega = \sqrt{\dfrac{mg/L}{m}} = \sqrt{\dfrac{g}{L}}$,   (14–12a)

the frequency, $f = \dfrac{1}{2\pi} \sqrt{\dfrac{g}{L}}$,   (14–12b)

and the period, $T = \dfrac{1}{f} = 2\pi \sqrt{\dfrac{L}{g}}$ (14–12c)

Note that the period, T, is independent of the mass, m, and the amplitude of oscillation, A.

*14–6 The Physical Pendulum and the Torsion Pendulum

Physical Pendulum
Any real extended body that oscillates back and forth (such as a baseball bat or a rod suspended from a point O) constitutes a physical pendulum.

The force of gravity acting at the center of gravity (CG) at a distance h from O provides the torque:

$$\tau = -mgh\sin \theta.$$

From Newton's second law:

$$I\alpha = \Sigma\tau$$

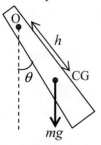

$$I\frac{d^2\theta}{dt^2} + mgh \; \theta = 0 \qquad \text{(for small } \theta, \sin \theta = \theta\text{)}$$

$$\frac{d^2\theta}{dt^2} + \frac{mgh}{I} \; \theta = 0 \qquad (14\text{--}13)$$

This shows the body executes SHM and the period:

$$T = \frac{2\pi}{\omega} = 2\pi \sqrt{\frac{I}{mgh}} \qquad (14\text{--}14)$$

Torsion Pendulum
A disk or a bar suspended from a wire constitutes a torsion pendulum.

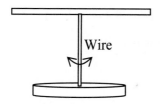

In this case $\tau = -K\theta$ where K is a constant that depends on the properties of the system. Thus the angular frequency:

$$\omega = \sqrt{\frac{K}{I}}.$$

14-7 Damped Harmonic Motion

In a real oscillatory system, the amplitude of oscillation decreases with time due to the resistance of the medium (such as air) and to the internal friction within the oscillatory system. This type of motion is called **damped harmonic motion**.

The damping force can be approximated as:

$$F_{\text{damping}} = -bv = -b\frac{dx}{dt}.$$

From $\Sigma F = ma$, we get, after rearranging the terms:

$$m\frac{d^2x}{dt^2} + b\frac{dx}{dt} + kx = 0 \qquad (14\text{-}15)$$

The solution of this equation is:

$$x = A\,e^{-\gamma t}\cos\omega't \qquad (14\text{-}16)$$

where $\qquad \gamma = \dfrac{b}{2m} \qquad\qquad (14\text{-}17)$

$$\omega' = \sqrt{\frac{k}{m} - \frac{b^2}{4m^2}}. \qquad (14\text{-}18)$$

The frequency, $f = \dfrac{\omega'}{2\pi} = \dfrac{1}{2\pi}\sqrt{\dfrac{k}{m} - \dfrac{b^2}{4m^2}}. \qquad (14\text{-}19)$

Three common cases of damping are shown.

164

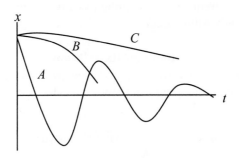

- Curve *A* represents an **underdamped** situation where damping is small ($b^2 < 4mk$) so that ω' is real. The system makes several oscillations before coming to rest.

- Curve *B* represents **critical damping** ($b^2 = 4mk$) where equilibrium is reached within the shortest amount of time. Shock absorbers in an automobile is usually designed to give critical damping.

- The curve *C* represents the **overdamped** situation where the damping is so large ($b^2 >> 4mk$) that the system takes a long time to reach equilibrium.

14–8 Forced Oscillations; Resonance

When a vibrating system is allowed to vibrate on its own, it vibrates at its natural frequency. For example, the natural frequency, f_0, of a mass m attached to a spring of spring constant k is

$$f_0 = \frac{1}{2\pi} \sqrt{\frac{k}{m}}. \qquad (14\text{–}20)$$

When an additional driving force, which has its own particular frequency f, is applied to a vibrating system, **forced vibration** occurs.

*Equation of Motion and Its Solution

If the external force $F_{ext} = F_0 \cos \omega t$, the equation of motion with damping can be written as:

$$m \frac{d^2x}{dt^2} + b \frac{dx}{dt} + kx = F_0 \cos \omega t \qquad (14–21)$$

Its solution is $x = A_0 \sin (\omega t + \phi_0)$, (14–22)

where

$$\text{amplitude, } A_0 = \frac{F_0}{m\sqrt{\left(\omega^2 - \omega_0^2 \right)^2 + b^2\omega^2/m^2}}, \qquad (14–23)$$

$$\omega_0 = \sqrt{k/m}, \text{ and}$$

$$\phi_0 = \tan^{-1} \frac{\omega_0^2 - \omega^2}{b\omega/m}. \qquad (14–24)$$

The figure shows the amplitude of vibration of a system as a function of the external frequency f. When the driving frequency f matches the natural frequency f_0 of vibration of a system, vibration occurs with very large amplitude. This condition is known as **resonance**.

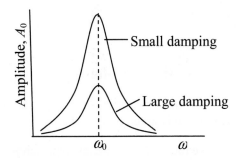

The natural frequency f_0 of a system is called its **resonant frequency**. There are many examples of resonance and resonance plays an important role in many physical systems. For example, some singers can break a wine glass using their voices. This is possible because at resonance the frequency of the voice matches with the natural frequency of the glass. This causes large amplitude of vibration in the glass, causing it to shatter.

Resonance is important in buildings and bridges. Soldiers are asked to break steps when they cross a bridge to avoid any possibilities of collapse of the bridge due to resonance with their footsteps.

*Q value

The height and narrowness of a resonant is expressed by its **quality factor** or ***Q* value**, defined as

$$Q = \frac{m\omega_0}{b} \qquad (14\text{--}25)$$

The Q value is also a measure of the width of the peak. It can be shown that

$$\frac{\Delta\omega}{\omega_0} = \frac{1}{Q} \qquad\qquad (14\text{--}26)$$

Tips for Solving the Problems

1. Remember that simple harmonic motion occurs when the restoring force is given by $F = -kx$.

2. Do not confuse the linear frequency, f (which is in s^{-1} or Hz), with the angular frequency, ω (which is in rad/s).

3. Use the following expressions for position, velocity, and acceleration, respectively: $x = A\cos(\omega t + \phi)$, $v = -a\omega\sin(\omega t + \phi)$, and $a = -\omega^2 A\cos(\omega t + \phi)$, where $\omega = 2\pi/T$ to solve problems illustrating the general features of a simple harmonic motion (such as those in Section 14–1 and 14–2).

4. For simple harmonic motion, the basic facts to keep in mind are the definitions of spring constant ($F = kx$), conservation of energy, and the fact that $\omega = \sqrt{k/m}$ and $U = \frac{1}{2} kx^2$ for a spring, and $\omega = \sqrt{g/l}$ for a pendulum. Most everything else can be quickly derived from these.

5. In problems where a single mass is connected to more than one spring, find the equivalent spring constant of a single spring that can replace the springs in the problem.

6. Energy conservation ($KE + PE$ = constant) is useful for solving several problems in this chapter involving simple harmonic motion such as those in Section 14–3.

7. It is usually convenient to determine total energy from the expression $E = \frac{1}{2} kA^2$ ($=\frac{1}{2} m\omega^2 A^2$) and potential energy from $U = \frac{1}{2} kx^2$ ($=\frac{1}{2} m\omega^2 x^2$). Kinetic energy can then be calculated by subtracting U from E.

CHAPTER 15
Wave Motion

This chapter describes wave motion, wave equation, and different properties including reflection, refraction, interference, standing wave, and diffraction.

Important Concepts

> Wave
> Transverse and longitudinal waves
> Velocity of waves
> Intensity
> Wave equation
> The principle of superposition
> Reflection and refraction
> Constructive and destructive interferences
> Standing waves
> Nodes and antinodes
> Diffraction

A **wave** is a disturbance that propagates from one place to another without any actual transfer of particles of the medium through which it propagates. For example, when you throw a stone in a pool of water, a circular wave is formed in water that moves outward. Waves also travel along a cord if you vibrate one end of the cord back and forth. Waves carry energy.

15–1 Characteristics of Wave Motion

A single wave **pulse** can be formed on a rope by a quick up and down motion of the hand. The source of the pulse is a disturbance and the cohesive forces between adjacent pieces of rope cause the disturbance to travel to the other end. A **continuous** or **periodic wave** can be created by the continuous vibration of a source. For example, a continuously vibrating tuning fork or drum membrane gives rise to continuous sound waves in air. The source of a periodic wave is thus vibration and it is the vibration that propagates outward, constituting the wave.

A wave repeats itself both in space and in time. If the vibration of the source is sinusoidal and the medium is perfectly elastic, the wave will have sinusoidal shape in space and in time.

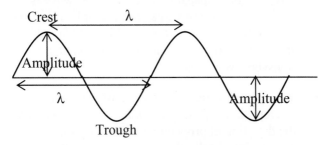

- The maximum height of a crest or depth of trough relative to the normal level is called the **amplitude** of the wave.

- The distance over which a wave repeats is called the **wavelength**, λ, of the wave. It is the distance between two consecutive crests (or troughs) of a wave.

- The *frequency* of a wave is the number of cycles (that is, say, number of crests) that pass a given point in one second. The time required for one wavelength to pass a given point is called the **period**, T.

Since a wave travels a distance λ in time T, its speed is

$$v = \frac{\text{distance}}{\text{time}} = \frac{\lambda}{T} = \lambda f \qquad (15\text{--}1)$$

15–2 Types of Waves; Transverse and Longitudinal

- When the direction of vibration of the individual particles of the medium is *at right angles* to the direction of propagation of a wave, the wave is called a **transverse wave**. For example, if a string or rope tied to a wall at one end is pulled on the free end and moved up and down by hand, a transverse wave results.

- When the direction of vibration of the individual particles of the medium is *in the same direction* as the direction of propagation of a wave, the wave is called a **longitudinal wave**.

Sound is a longitudinal wave. When a source of sound (such as a vibrating drumhead) vibrates, it creates a series of compressions and expansions in the air molecules in front of the source. These compressions and expansions travel horizontally away from the source with the speed of sound. Clearly, as the sound travels it creates pressure variations in the air. Similarly, when one end of a slinky moves back and forth horizontally, it creates a series of compressions and expansions that propagate along the slinky.

In a longitudinal wave the wavelength is the distance between successive compressions or successive rarefactions. The frequency is the number of compressions or refractions that pass a given point in one second.

Velocity of Transverse Waves

Waves propagate with well-defined speeds determined by the characteristics of the material of the medium.

For example, the velocity of a wave on a string with tension F_T and mass per unit length μ is

$$v = \sqrt{\frac{F_T}{\mu}} \qquad (15\text{--}2)$$

Velocity of Longitudinal Waves

In general, for a wave in a medium, the velocity is given by

$$v = \sqrt{\frac{\text{elastic force factor}}{\text{inertia factor}}}.$$

For a longitudinal wave traveling in a solid,

$$v = \sqrt{\frac{E}{\rho}} \qquad (15\text{--}3)$$

where E is the elastic modulus and ρ is the density of the medium. For a longitudinal wave traveling in a liquid or gas,

$$v = \sqrt{\frac{B}{\rho}} \qquad (15\text{--}4)$$

where B is the bulk modulus of the medium.

Other Waves

In an earthquake, both transverse and longitudinal waves are produced. The transverse wave, called S wave (S for

173

shear), travels through the solid outer core of the earth but cannot pass through the inner core of the earth because it is liquid. The longitudinal wave, called the P wave (P for pressure) can pass through both solids and liquids. Earthquakes also generate waves that travel along the surface of the earth, which causes the most damage. A water wave is a combination of transverse and longitudinal waves. In this case, the water particles move roughly in a circular path as the wave propagates in the horizontal direction.

15–3 Energy Transported by Waves

Waves carry energy. As waves travel, the energy is transferred as vibrational energy from particles to particles. The energy E transported in time t through a cross-sectional area S of a medium of density ρ is found to be:

$$E = 2\pi^2 \rho S v t f^2 D_M^2 \qquad (15\text{–}5)$$

(f = wave frequency and D_M = wave amplitude)

Average power, $\overline{P} = \dfrac{E}{t} = 2\pi^2 \rho S v f^2 D_M^2 \qquad (15\text{–}6)$

which is proportional to the square of the amplitude and the square of frequency.

The intensity I of a wave is given by

$$I = \frac{\text{Average power}}{\text{area}} = 2\pi^2 \rho v f^2 D_M^2 \quad (15\text{–}7)$$

Clearly, $I \propto D_M^2$. The SI unit of I is W/m^2.

The intensity decreases with distance because as the distance increases, the power, P, spreads out over a larger area. For a spherical wave that can spread equally in all directions, the intensity at a distance r is

$$I = \frac{\overline{P}}{4\pi r^2}$$

since area of a spherical surface is $4\pi r^2$. That is,

$$I \propto \frac{1}{r^2} \qquad (15\text{–}8a)$$

If I_1 and I_2 are intensities at distances r_1 and r_2, respectively, for a small source then

$$\frac{I_2}{I_1} = \left(\frac{r_1}{r_2}\right)^2 \qquad (15\text{–}8b)$$

Using Equations (15–7) and (15–8b), we get

$$\frac{D_{M2}}{D_{M1}} = \frac{r_1}{r_2}.$$

Note that for a one-dimensional wave (such as a transverse wave on a string) the amplitude and intensity stays nearly constant since area does not change.

15–4 Mathematical Representation of a Traveling Wave

A wave is a function of position x and time t. For a sinusoidal one-dimensional wave traveling along the x-axis, the wave shape at time $t = 0$ (as shown by the solid curve) can be expressed as:

$$D(x) = D_M \sin \frac{2\pi}{\lambda} x \qquad (15\text{–}9)$$

where $D(x)$ is the **displacement** of the wave at position x and D_M is the wave **amplitude**. (Note the function repeats itself for every λ.)

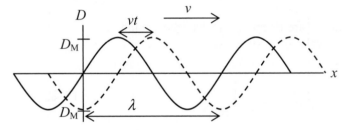

The wave travels a distance vt to the right in time t, and the displacement of the wave at time t (as shown by the dotted curve) can then be expressed as:

$$D(x, t) = D_M \sin\left[\frac{2\pi}{\lambda}(x - vt) \right] \qquad (15\text{–}10a)$$

$$D(x, t) = D_M \sin\left[\frac{2\pi x}{\lambda} - \frac{2\pi t}{T} \right] \qquad (15\text{–}10b)$$

$$(\text{since } v = \lambda/T)$$

This can be written as

$$D(x, t) = D_M \sin (kx - \omega t) \qquad (15\text{--}10c)$$

where $k = \dfrac{2\pi}{\lambda}$ (called the **wave number**) $\qquad (15\text{--}11)$

and $\omega = \dfrac{2\pi}{T}$. Clearly,

$$v = f\lambda = \frac{\omega}{k} \qquad (15\text{--}12)$$

For a wave traveling to the left we get, after replacing v by $-v$,

$$D(x, t) = D_M \sin \left[\frac{2\pi}{\lambda}(x + vt) \right] \qquad (15\text{--}13a)$$

$$D(x, t) = D_M \sin \left[\frac{2\pi x}{\lambda} + \frac{2\pi t}{T} \right] \qquad (15\text{--}13b)$$

$$D(x, t) = D_M \sin (kx + \omega t) \qquad (15\text{--}13c)$$

In general, for a traveling sinusoidal wave (called a harmonic wave)

$$D(x, t) = D_M \sin (kx \pm \omega t + \phi)$$

where ϕ is the initial phase angle.

For a general wave (or wave pulse) of any shape:

- If the wave travels to the right,
  $D(x, t) = D(x - vt).$ $\qquad (15\text{--}14)$

- If the wave travels to the left,
  $D(x, t) = D(x + vt).$ $\qquad (15\text{--}15)$

177

*15–5 The Wave Equation

Many waves satisfy a general equation that is equivalent to Newton's second law of motion for particles. This equation of motion for a wave is called the **wave equation**. For a wave traveling along the x-axis, the **one-dimensional wave equation** is:

$$\frac{\partial^2 D}{\partial x^2} = \frac{1}{v^2} \frac{\partial^2 D}{\partial t^2}. \tag{15–16}$$

The wave equation is a direct consequence of Newton's second law applied to a continuous elastic medium. It can be shown by direct substitution (after taking second derivatives) that Equation (15–10) and (15–13) satisfy the wave equation.

Wave equation is a linear equation: the displacement D appears singly in each term. Thus, if $D_1(x, t)$ and $D_2(x, t)$ are two different solutions, then the linear combination:

$$D_3(x, t) = a\, D_1(x, t) + b\, D_2(x, t)$$

is also a solution (a and b are constants). This is the essence of the *superposition principle*.

15–6 The Principle of Superposition

When two or more waves pass through the same region of space at the same time, the *actual displacement is the vector sum of the individual displacements*. This is called the **principle of superposition**.

Because of the superposition principle, a composite wave can be formed when more than one sinusoidal wave of appropriate amplitudes and frequencies are superimposed at a certain instant in time. It can be shown that any complex wave can be considered to be composed of several (or many) simple sinusoidal waves of different amplitudes, wavelengths, and frequencies. This is called *Fourier's theorem*.

The variation of speed of wave with frequency is called **dispersion** (occurs when restoring force is not precisely proportional to the displacement). As a result of dispersion a complex wave will change shape in such a medium as the component waves will travel at different velocities.

15–7 Reflection and Transmission of Waves

When a wave strikes an obstacle or comes to the end of a medium, it is reflected. For example, a wave pulse traveling along a rope is reflected from its end. If the end of the rope is fixed, the reflected wave is inverted because of the equal and opposite reaction force by the wall on the rope. If the end of the rope is free to move, waves are reflected with no inversion.

In general, when a wave reaches a boundary, a part of the wave is reflected, a part is transmitted, and the remaining part is absorbed, which is converted to heat.

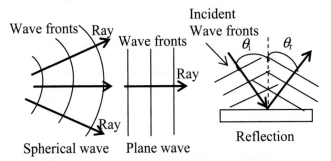

Spherical wave   Plane wave   Reflection

For a two- or three-dimensional wave, the whole width of a wave crest is called **wave fronts**. A line that signifies the direction of motion and perpendicular to the direction of the wave fronts is called a **ray**. Far from the source, the wave fronts are nearly straight or flat, which are called **plane waves**.

According to the law of reflection for two- and three-dimensional plane waves, the angle of incidence (θ_i) is equal to the angle of reflection (θ_r).

15–8 Interference

When two or more waves simultaneously occupy the same location, the *resulting displacement is the algebraic sum of the individual displacement* with proper sign (a crest is considered positive and trough negative).

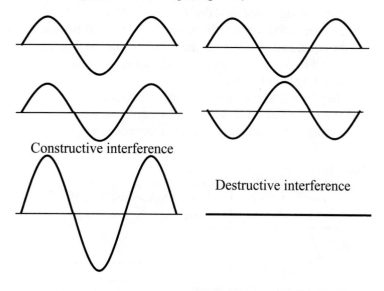

Constructive interference

Destructive interference

- As a result of superposition, when two waves meet **in phase** (crest to crest and trough to trough), the amplitude of the resulting wave is the sum of the amplitudes of the individual waves. This is **constructive interference**.

- When two waves *meet* **out-of-phase** (crest to trough, or, more precisely when the two waves differ by one wavelength), the amplitude of the resulting wave is the difference of the amplitude of the individual waves. This is **destructive interference**.

15–9 Standing Waves; Resonance

When you shake one end of a cord (whose other end is kept fixed) at a right frequency, the two identical waves moving in opposite directions on the cord are superimposed to produce a large amplitude **standing wave**. A standing wave is a pattern or oscillation with time but in a fixed location. Points on a standing wave where there is no displacement are called **nodes**. Halfway between two consecutive nodes there are points where displacement is a maximum. These points are **antinodes**.

The frequencies at which standing waves are produced are called **natural frequencies** or **resonant frequencies**. The lowest frequency at which a standing wave occurs is called the **fundamental frequency** or the **first harmonic**. The other frequencies in general are called **overtones**.

Consider a string stretched between two supports that is plucked like a guitar. The wavelengths of the standing waves bear a simple relationship to the length L of the string. For the fundamental frequency, the length L corresponds to a half wavelength of a wave on the string (one loop). That is, $L = \frac{1}{2} \lambda_1$. The next mode has two loops and is called **second harmonic**. For the second harmonic $L = \lambda_2$. In general, solving for λ,

$$\lambda_n = \frac{2L}{n} \qquad n = 1, 2, 3, \cdots, \qquad (15\text{–}17a)$$

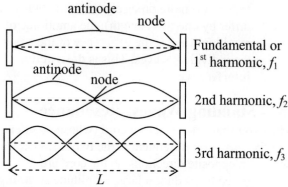

Standing waves in a plucked string

Since, $f = v/\lambda$, The harmonic frequencies are given by

$$f_n = n\frac{v}{2L} = nf_1 \qquad (15\text{–}17b)$$

Standing waves are also produced in other objects that are set into vibration. In general, the resonant frequencies depend on the dimensions of the length.

Mathematical Representation of a Standing Wave
The superposition of two identical waves traveling in the opposite directions (that produce a standing wave) can be written as

$$D(x, t) = D_1(x, t) + D_2(x, t)$$
$$= D_M \sin(kx - \omega t) + D_M \sin(kx + \omega t)$$
$$= 2D_M \sin kx \cos \omega t. \qquad (15\text{–}18)$$

- This shows that particles at any x vibrate in simple harmonic motion (because of cos ωt). The amplitude $2D_M \sin kx$ is a maximum when $kx = \pi/2, 3\pi/2, 5\pi/2,\ldots$ This gives $x = \lambda/4, 3\lambda/4, 5\lambda/4,\ldots$ the position of the antinodes.

- $D(x, t) = 0$ at $x = 0$ and $x = L$. From the second condition, $\sin kL = 0$ or $kL = n\pi$ ($n = 1, 2, 3, \ldots$).

  This gives, $\lambda = \dfrac{2L}{n}$ $\qquad n = 1, 2, 3,\ldots$

*15–10 Refraction

When a wave passes from one medium to another medium wherein its velocity is different, the transmitted wave to the second medium moves in a different direction than the incident wave. This phenomenon is called **refraction**.

If the velocity in the second medium is less than the velocity in the first medium, the angle of refraction θ_r is less than the angle of incidence θ_i. When the velocity changes gradually (such as when a water wave approaches the shore), the wave also refracts slowly and gradually. The quantitative relation between the angles and the velocities v_1 and v_2 in the two

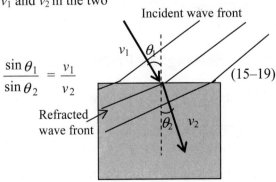

$$\frac{\sin \theta_1}{\sin \theta_2} = \frac{v_1}{v_2} \qquad (15\text{–}19)$$

Earthquake waves refract as they pass through different layers of the earth. Light waves also refract, such as when the pass from air to glass.

*15–11 Diffraction

When a wave encounters an obstacle or passes through a small hole, it bends. This is called the **diffraction** of a wave.

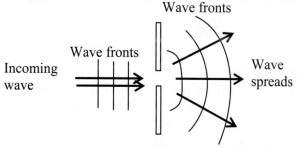

The amount of diffraction depends on the wavelength (λ) of the wave and the size of the obstacle or the opening. If the wavelength is larger than or comparable to the size of the opening (or the obstacle), diffraction is noticeable. The angular spread θ of the wave after it has passed through an opening L (or around an obstacle of width L) is given by θ (radian) $\approx \lambda/L$.

Tips for Solving the Problems

1. Velocity $v = f\lambda$ for any wave; $v = \sqrt{\dfrac{F_T}{\mu}}$ for a transverse wave on a string (with tension F_T and mass per unit length m/L); $v = \sqrt{E/\rho}$ for a longitudinal wave traveling in a solid (where E is the elastic

modulus and ρ is the density of the solid); and $v = \sqrt{B/\rho}$ for a longitudinal wave traveling in a liquid or gas (where B is the bulk modulus of the medium). Remember this when solving problems in Sections 15–1 and 15–2.

2. For a spherical wave ratio of two wave intensities $I_2/I_1 = \left(r_1/r_2\right)^2$, whereas the ratio of amplitudes $D_{M2}/D_{M1} = r_1/r_2$.

3. The velocity of a wave can be determined by either of the equations $v = f\lambda$ or $v = \omega/k$, depending on the problem.

4. The frequencies of vibration for the standing wave on a string of length L fixed at both ends, $f_n = nv/2L = nf_1$, where $n = 1, 2, 3, \cdots$.

5. For standing wave problems you may not need to remember the equation $D(x, t) = 2D_M \sin kx \cos \omega t$. Remember that $D_1(x, t) = D_M \sin (kx - \omega t)$ represents a wave traveling toward the right. Replace the negative sign with positive sign to get an identical wave $D_2(x, t)$ traveling to the left and then add D_1 and D_2 using trigonometric identity to get $D(x, t)$.

6. In general, trigonometric identities involving function–sum relations of sines and cosines are useful to study superposition of two waves.

CHAPTER 16
Sound

This chapter describes sound, sound perception, sound level, Doppler effect of sound, and superposition of sounds giving rise to interference, standing waves, and beats.

Important Concepts

Speed of sound
Loudness and pitch
Audible sound, ultrasonic and infrasonic
Sound intensity and intensity level
Decibel
Node and antinode
Open tube and closed tube
Quality
Constructive and destructive interferences
Loudness level
Fundamental and harmonics
Beats and beat frequency
Doppler effect
Shock wave
Mach number
Sonar

16–1 Characteristics of Sound

Sound is a longitudinal wave that can travel in air. Sound cannot travel in a vacuum. That is why sound cannot be heard by ringing a bell inside an evacuated jar.

The **speed of sound** is different in different media. In air, the speed of sound is 331 m/s at 0°C. The speed increases with temperature and at a temperature $T°C$ its speed in air $v \approx (331 + 0.61\, T)$ m/s. Sound can also travel through other gases as well as in liquids and solids. In general, sound travels faster in solids than in liquids or gases.

Loudness of sound is related to the energy in the sound wave. The **pitch** of a sound depends on the frequency of the sound wave. High-pitched sounds have high frequencies, and low-pitched sounds have low frequencies.

- The **audible** range for humans: 20 Hz to 20,000 Hz.
- **Ultrasonic** sounds: Sound frequencies above 20,000 Hz. Many animals (such as dogs and bats) can hear ultrasonic frequencies. Bats use it for communication or navigation. Ultrasound is used for imaging purposes in medicine.
- **Infrasonic**: Those with frequencies less than 20 Hz.

16–2 Mathematical Representation of Longitudinal Waves

For a longitudinal wave (such as sound) traveling along x-axis,

$$D = D_M \sin (kx - \omega t) \qquad (16\text{–}1)$$

where D is the displacement of the particle at position x and time t.

As the sound travels in air it creates pressure variations (compressions corresponding to higher pressure and expansions corresponding to lower pressure than the normal) in the air. That is why sound and other longitudinal waves are called **pressure waves**. From the analyses of pressure variation, it can be shown that the pressure difference from the normal pressure, ΔP, is:

$$\Delta P = -B \frac{\partial D}{\partial x} \qquad (B = \text{Bulk modulus}) \qquad (16\text{–}2)$$

187

$$= -(BD_{\mathrm{M}}k)\cos(kx - \omega t) \qquad (16\text{--}3)$$

$$= -\Delta P_{\mathrm{M}}\cos(kx - \omega t) \qquad (16\text{--}4)$$

The pressure amplitude, $\Delta P_{\mathrm{M}} = BD_{\mathrm{M}}k = 2\pi\rho v D_{\mathrm{M}}f.\ (16\text{--}5)$

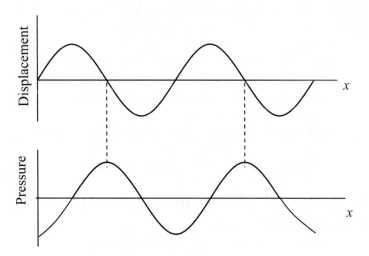

Thus, pressure also varies sinusoidally but is out of phase from the displacement by 90°, as shown.

16–3 Intensity of Sound: Decibels

The loudness of a sound is determined by sound intensity, which is a measure of the amount of sound power that passes through unit area. Sound intensity is measured in W/m^2.

The human ear is incredibly sensitive to sound intensities. It can detect sound intensity as low as 10^{-12} W/m^2 (called the *threshold of hearing*, I_0) and as high as $1\ W/m^2$ without problems. The intensity of sound must be increased by a factor of 10 in order for it to seem twice as loud to us. Since uniform increases in loudness correspond to intensities that increase by multiplicative factors, a logarithmic scale is more often used to express sound loudness.

Sound Level

The **sound level** β of a sound is defined in terms of its intensity I as

$$\beta = 10 \log \left(\frac{I}{I_0} \right) \qquad (16\text{--}6)$$

β is expressed in **decibels** (dB). The threshold of hearing is 0 dB, meaning $I = I_0$.

The smallest change in intensity level that the human ear can detect is 1 to 2 dB. The sound intensity of a number of sources is simply the sum of the individual intensities; but, when two identical sounds are simultaneously present, the intensity level is 3 dB greater than the intensity level of a single sound.

Intensity Related to Amplitude

The intensity I of a wave is proportional to the square of the wave amplitude D_M and is given by equation (15–7).

The Ear's Response

The humar ear is not equally sensitive to all incoming frequencies. At lower intensity levels, the ear is less sensitive to high and low frequencies compared to middle frequencies. The curves representing sounds that seem to have the same loudness are called the curves of equal **loudness level**. The unit of loudness level is called *phons*. The number in phons is numerically equal to the sound level in dB at 1000 Hz. For example, a curve representing 40 phons represents sound of the same loudless as a sound of 40 dB at 1000 Hz. But a sound of 100 Hz must have a sound level much higher (about 62 dB) to sound as loud as a 1000 Hz of only 40 dB.

16–4 Sources of Sound: Vibrating Strings and Air Columns

In musical instruments, the source is set into vibration by striking, plucking, bowing, or blowing. Standing waves are produced and the source vibrates at its natural resonant frequencies.

Stringed Instruments

The pitch is determined by the lowest resonant frequency of the standing wave, called the *fundamental* frequency. For the vibration of a string of length L fixed at both ends, the possible frequencies of vibration are

$$f_n = n\frac{v}{4L} = nf_1, n = 1, 2, 3, \ldots$$

where v is the velocity of the wave through the string. $n = 1$ for the fundamental frequency. Higher frequencies are called **overtones** or **harmonics**. Standing wave patterns for a vibrating string fixed at both ends is shown in Section 15–9.

In string instruments, the strings are mounted on a sounding board (piano) or sounding box (violin, guitar), so that when the string is set to vibration the sounding board or box also vibrates, thus making the sound louder.

Wind Instruments

A standing wave can be formed in the vibration of an air column. For an *open tube* (both ends open), there will be displacement antinodes (or pressure nodes) at both the ends. For the lowest frequency (fundamental), $L = \lambda/2$.

The fundamental frequency, different harmonics, and wavelengths are given by

$$f_n = n\frac{v}{2L} = nf_1 \qquad \text{where } n = 1, 2, 3, \ldots$$

$$\lambda_n = \frac{2L}{n}.$$

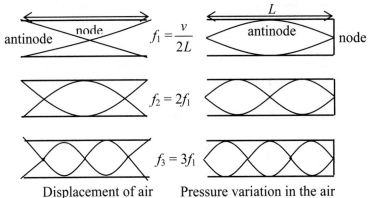

$$f_1 = \frac{v}{2L}$$

$$f_2 = 2f_1$$

$$f_3 = 3f_1$$

antinode   node   antinode   node

Displacement of air   Pressure variation in the air

For a *closed tube* (one end closed), at the closed end there is no displacement of the air; hence, there must be a displacement node (or pressure antinode) at the closed end. At the open end, there is a displacement antinode (or pressure node). For the lowest frequency (fundamental), $L = \lambda/4$. Thus, the fundamental frequency is

$$f_1 = \frac{v}{\lambda} = \frac{v}{4L}$$

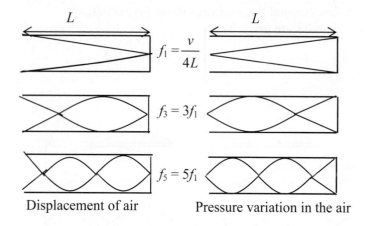

$$f_1 = \frac{v}{4L}$$

$$f_3 = 3f_1$$

$$f_5 = 5f_1$$

Displacement of air   Pressure variation in the air

and the harmonics and wavelengths are given by

$$f_n = n\frac{v}{4L} = nf_1 \text{ where } n = 1, 3, 5, \ldots$$

$$\lambda_n = \frac{4L}{n}.$$

Pipe organs make use of both open and closed pipes. A flute is an open tube and different notes are obtained by changing the length (that is, uncovering holes). In a trumpet, pushing down the valve opens additional length. In all these instruments, higher pitches are obtained by shortening the length.

*16–5 Quality of Sound and Noise; Superposition

A musical sound has three characteristics: loudness, pitch, and **quality**. The terms *timbre* or *tone color* are also used for quality. Because of differences in qualities, the sounds from different instruments (such as from a piano and a flute) are easily distinguishable.

The quality of a sound depends on the presence of overtones or harmonics—their number, values, and relative amplitudes. A sound spectrum of a sound shows the relative amplitudes of different frequencies in the sound. Because of the differences in the sound spectra of two sounds having different qualities, the *waveform* of two sounds are also different.

A noise is a mixture of many frequencies that have no relation to each other. The sound spectrum of a noise is not discrete lines, instead it is a nearly continuous spectrum of frequencies. Loud noise can cause loss of hearing.

16–6 Interference of Sound Waves; Beats

Interference in Space

When two or more waves meet at a point in a medium, they interfere. This occurs for all waves including sound. In the constructive interference of sounds, two identical sounds (such as from two identical loudspeakers) meet compression to compression at a point (such as point *C*) and the resulting sound is louder. In the destructive interference of sounds, the two sound waves meet compression to expansion at a point (such as at point *D*), and the resulting sound is weaker.

The following *two conditions* must be met for interference to occur at a point

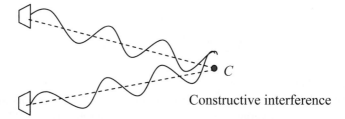

Constructive interference

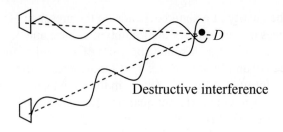

Destructive interference

1. When the path difference from the two sources is
   0, λ, 2λ, 3λ,..., *constructive interference* occurs.

2. When the path difference from the two sources is
   $\lambda/2$, $3\lambda/2$, $5\lambda/2$,..., *destructive interference* occurs.

Beats—Interference in Time

Beats can be thought of as interference in time. When two
sound waves of slightly different frequencies (such as
sounds from two tuning forks of frequencies, say 252 Hz
and 256 Hz) interfere, the resulting sound shows periodic
variations in loudness with time. This phenomenon is
called **beats.** The number of times the sound loudness rises
or falls in one second is called the **beat frequency**.

   If two sources of frequencies f_1 and f_2 are
simultaneously vibrated, the resulting sound at a fixed
point is space, by the principle of superposition, is:

$$D = D_1 + D_2 = D_M \sin 2\pi f_1 t + D_M \sin 2\pi f_2$$

$$= \left[2D_M \cos 2\pi \left(\frac{f_1 - f_2}{2} \right) t \right] \sin 2\pi \left(\frac{f_1 + f_2}{2} \right) t$$

This shows that:
- The resulting sound frequency, f, is the same as the
  average frequency of the two sounds, $(f_1 + f_2)/2$.

- The vibration amplitude changes with time with a frequency of $(f_1 - f_2)/2$. Since two beats occur in each cycle (when cosine function is +1 or −1), the beat frequency is $(f_1 - f_2)$.

Beats are used to compare frequencies. For example, to tune a piano, a source of sound (such as a tuning fork) of the correct frequency is vibrated simultaneously with the piano. The tension in the piano string is then increased or decreased until the beats disappear.

16–7 Doppler Effect

The change in pitch (or frequency) of sound due to relative motion between a source and an observer is called the **Doppler effect**. The observed pitch increases when the source and/or the observer approach each other and decreases when they recede from each other.

Source in Motion

When the source moves toward a stationary observer, the spacing between two consecutive compressions (or rarefactions), or in other words, the wavelength of sound, decreases. As the wavelength decreases, the observed frequency f' increases. The observed frequency f' is given by

$$f' = \frac{f}{\left(1 - \dfrac{v_{source}}{v_{snd}}\right)} \qquad (16\text{–}9a)$$

When the source moves away from the observer,

$$f' = \frac{f}{\left(1 + \dfrac{v_{\text{source}}}{v_{\text{snd}}}\right)} \qquad (16\text{–}9b)$$

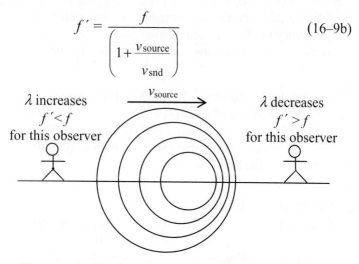

λ increases
$f' < f$
for this observer

v_{source}

λ decreases
$f' > f$
for this observer

Observer in Motion

The frequency of sound is $f = v_{\text{snd}}/\lambda$. If the observer moves toward the source with a speed v_{obs} relative to the stationary source, the speed of sound relative to the observer is $v_{\text{snd}} + v_{\text{obs}}$. As a result, the observed frequency increases. The observed frequency f' is given by

$$f' = \left(1 + \frac{v_{\text{obs}}}{v_{\text{snd}}}\right) f \qquad (16\text{–}10a)$$

If the observer moves away from the source,

$$f' = \left(1 - \frac{v_{\text{obs}}}{v_{\text{snd}}}\right) f \qquad (16\text{–}10b)$$

Equations (16–9) and (16–10) can be combined to show that when both the observer and the source are in motion,

$$f' = f\left(\frac{v_{snd} \pm v_{obs}}{v_{snd} \mp v_{source}}\right) \qquad (16\text{--}11)$$

The upper signs in the numerator and denominator apply to an observer and/or a source *approaching* each other, and the lower signs apply to an observer and/or a source moving away from each other.

The Doppler effect is used to determine the speed of a moving source. In medicine, it is used to determine the speed of blood flow through an artery or in the heart by recording the Doppler shifted frequency of ultrasonic waves reflected from the blood cells.

Doppler Effect for Light

The Doppler effect applies to all waves including light. The effect is used to determine the speed of stars or galaxies by recording the Doppler shifted wavelength of the light emitted by the star or the galaxy. Light coming from a star or galaxy moving away from the earth is **red shifted** (that is, its wavelength increases), and light coming from a star or galaxy approaching the earth is blue shifted (that is, its wavelength decreases).

*16–8 Shock Waves and the Sonic Boom

Any speed greater than the speed of sound is known as the **supersonic speed**. **Mach number** is the ratio of the speed of the object to the speed of sound in the medium.

When an object moves at a supersonic speed, the wavefronts pile up on one another along the sides. The different wave crests overlap one another, forming a large crest or the **shock wave**. When a shock wave such as from an airplane traveling at supersonic speeds passes a listener, it is heard as a loud *sonic boom*. The sonic boom lasts for a fraction of a second and contains high energy that can be enough to break windows and to cause other damage.

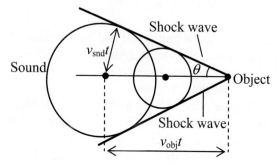

The shock wave contains a cone whose apex is at the source and the angle of this cone is given by

$$\sin \theta = \frac{v_{snd}t}{v_{obj}t} = \frac{v_{snd}}{v_{obj}} \qquad (16\text{–}12)$$

where v_{obj} and v_{snd} are the velocity of the object and the velocity of sound in the medium, respectively.

*16–9 Applications: Sonar, Ultrasound, and Medical Imaging

*Sonar

In a **sonar**, a transmitter sends out a wave pulse through water and a detector receives the reflection. The time interval is accurately measured and the distance of the reflecting object is determined from the speed of the wave in water. Sonar usually uses **ultrasonic** frequencies (that is, frequencies greater than 20 kHz) because thry are inaudible and diffract less (since the wavelength is very short).

*Ultrasound Medical Imaging

Ultrasound is used in medicine to form images, called *sonograms*, which are used for diagnostic purposes. In a pulse–echo technique, a high frequency (1 to 10 MHz) wave pulse is directed into the body and its reflections from boundaries or interfaces between organs etc. are detected. The strength of the reflected pulse depends on the difference in the density of the two materials on either side of the interface and is displayed as a pulse or a dot. Each dot represents a point whose position depends on the time delay and whose brightness depends on the strength of reflection. A two-dimensional image is formed out of the dots from a series of scans.

Tips for Solving the Problems

1. Note that the basic equation of a wave, $v = f\lambda = \omega/k$ is applicable to any wave including sound. Also, for any wave, $f = 1/T$.

2. For sound $D = D_M \sin(kx - \omega t)$ and $\Delta P = -\Delta P_M \cos(kx - \omega t)$. The displacement amplitude D_M and pressure amplitude ΔP_M are related by $\Delta P_M = 2\pi\rho v D_M f$. These expressions will be useful to solve problems in Section 16–2.

3. Make sure you understand the difference between the sound intensity I and the sound level β. They are related by $\beta = 10 \log(I/I_0)$. $I = $ power/area. The unit of I is W/m^2 whereas the unit of β is the decibel (dB). $I_0 = 10^{-12}\ W/m^2$ is the threshold intensity (threshold sound level $\beta = 0$ dB). Doubling sound intensity I increases sound intensity level β by 3 dB. Review

logarithms when solving problems on sound intensity level, such as those in Section 16–3.

4. Standing wave modes for the vibration of an air column and for the vibration of a string have the same basic form. It is not necessary to remember formulas for the modes of vibration. Relate the wavelength to the length L of the air column or the string. Remember that the distance between two consecutive nodes or antinodes is $\lambda/2$, and the distance between a node and the nearest antinode is $\lambda/4$. Using these concepts you should be able to determine the frequencies of the different modes of vibration.

5. For *constructive interference* of sound, the path difference between two sources is an *integral multiple of λ* (that is, an even multiple of $\lambda/2$) whereas for *destructive interference*, the path difference is an *odd multiple of $\lambda/2$*. Use this concept to solve problems on interference, such as those in Section 16–6.

6. Remember that the *beat frequency* of two sounds is the *magnitude* of the difference between the two source frequencies and thus is always positive.

7. In general, trigonometric identities involving function–sum relations of sines and cosines are useful to study superposition of two sounds resulting in standing waves, beats, or interference.

8. To calculate the change in frequency due to the relative motion of the source and the observer, use the expression for the Doppler effect (such as for problems in Section 16–7). If you are confused by the sign of any terms in the expression, remember that

when the source and the observer approach each other the observed frequency increases. If they move away from each other, the observed frequency decreases.

CHAPTER 17
Temperature, Thermal Expansion, and the Ideal Gas Law

This chapter discusses temperature, thermal expansion, and ideal gas laws.

Important Concepts

> Atomic theory of matter
> Temperature
> Celsius, Fahrenheit, and Kelvin scales
> Absolute zero
> Thermal equilibrium
> Zeroth law of thermodynamics
> Constant volume gas thermometer
> Linear expansion
> Coefficient of linear expansion
> Volume expansion
> Coefficient of volume expansion
> Thermal stress
> Boyle's law, Charles's law, and Gay-Lussac's law
> Ideal gas law
> Universal gas constant
> Mole and Avogadro's number
> Ideal gas temperature scale

17–1 Atomic Theory of Matter

The smallest indivisible part of matter was called an **atom** by Democritus. The concept of an atom means that the matter is discontinuous and cannot be subdivided indefinitely. The experimental evidence of the atomic theory came in the eighteenth century from the analysis of chemical reactions.

The **unified atomic mass units** (u) is used to express the relative mass of atoms and molecules.

$$1 \text{ u} = 1.6605 \times 10^{-27} \text{ kg.}$$

The atomic mass of hydrogen is 1.0078 u, which is based on the carbon atom mass having exactly 12.0000 u.

Important evidence of atomic theory is the **Brownian movement**, which is the zigzag motion of tiny particles (pollen grains, for example) suspended in water. Based on his studies of Brownian motion, Einstein determined that the diameter of a typical atom is about 10^{-10} m.

Atoms and molecules exert attractive forces on each other. In a *solid*, the attractive forces are very strong and the atoms vibrate about their mean positions in an array called crystal lattice. In a *liquid*, the force between atoms or molecules is weak and the atoms or molecules move rapidly within the liquid. In a *gas*, the forces are so weak that the molecules move rapidly in random directions.

17–2 Temperature and Thermometers

Temperature is a measurement of the degree of hotness or coldness of objects. Many properties of matter depend on temperature. For example, most substances expand when heated. The electrical resistance of a material changes with temperature, and so on.

Thermometers can measure temperature. Operation of a thermometer depends on a material property that changes with temperature. The most common thermometers consist of a hollow glass tube filled with mercury or alcohol that expands with an increase of temperature. Bimetallic strips that are made of two thin strips of metals having different rates of expansion are used as a thermometer. The shape of a bimetallic strip is very sensitive to temperature changes. As the temperature changes, the strip changes its shape, moving a needle to indicate temperature.

Temperature Scales

According to the **Celsius** scale or **centigrade** scale of temperature, water freezes at 0°C and boils at 100°C. According to the **Fahrenheit** scale, water freezes at 32°F and boils at 212°F.

180 Fahrenheit degrees change in temperature is equivalent to 100 Celsius degrees. That is 1 F° = 5/9 C°. The conversion between the two scales is given by

$$T\,(°C) = \frac{5}{9}\,[T\,(°F) - 32]$$

$$\text{or} \quad T\,(°F) = \frac{9}{5}\,T\,(°C) + 32$$

A constant-volume gas thermometer works on the principle that if the volume of a gas is constant, its pressure is linearly proportional to its temperature. The thermometer consists of a bulb filled with a dilute gas connected by a thin tube to a mercury manometer. The volume of the gas is maintained constant by raising or lowering the right-hand tube of the manometer. This thermometer gives the same result for any working gas in the limit of reducing the gas pressure toward zero (Section 17–10).

17–3 Thermal Equilibrium and Zeroth Law of Thermodynamics

Two objects are said to be in **thermal equilibrium** when they reach the same temperature when in thermal contact. In this situation there is no exchange of energy between them.

According to the **zeroth law of thermodynamics**, *if two systems are in thermal equilibrium with a third system, then they are in thermal equilibrium with each other.* Temperature is the quantity that determines whether two objects will be in thermal equilibrium.

17–4 Thermal Expansion

Most substances expand when heated and contract when cooled; however, water is an exception. Thermal expansion can be linear, area, and volume expansion.

Linear Expansion

When the temperature of a substance of length L_0 increases by ΔT, its length increases by ΔL, where

$$\Delta L = \alpha L_0 \Delta T \qquad (17\text{--}1a)$$

where the constant α is called the *coefficient of linear expansion.* The unit of α is $(C^\circ)^{-1}$. The value of α depends on the nature of the material.

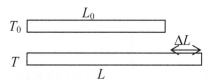

The final length

$$L = L_0 + \Delta L = L_0 (1 + \alpha \Delta T) \qquad (17\text{--}1b)$$

- To provide space for thermal expansion because of temperature changes, bridges as well as rail lines have gaps between different sections.

- Note that a hole in a piece of metal expands when heated just as if it were filled with the surrounding material.

Volume Expansion
When the temperature of an object of volume V_0 increases by ΔT, its volume increases by ΔV, where
$$\Delta V = \beta V_0 \Delta T \qquad (17\text{--}2)$$
The constant β is called the **coefficient of volume expansion**. It can be shown that $\beta \approx 3\alpha$.

Anomalous Behavior of Water Below 4°C
When water is heated from 0°C to 4°C, its volume decreases, but it expands normally with temperature when heated above 4°C (as shown). The density of water is thus a maximum at 4°C.

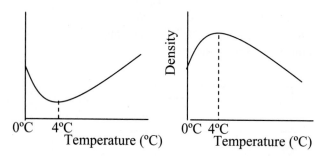

Because of the anomalous behavior of water between 0°C and 4°C during winter although lakes freeze on the surface, the water below stays at 4°C, thus allowing fish and other creatures in the water to survive. Because water expands as it cools between 4°C to 0°C, ice cubes float in water and pipes break in winter when water inside them freezes.

*17-5 Thermal Stresses
In solids where the ends are rigidly fixed, thermal stress develops due to temperature changes.

$$\Delta L = \frac{1}{E}\frac{F}{A}L_0,$$

where E is the elastic modulus and A is the area of cross-section. Also, $\Delta L = \alpha L_0 \Delta T$. From the preceding equations, the stress, $F/A = \alpha E\,\Delta T$.

17–6 The Gas Laws and Absolute Temperature
The **equation of state** of a gas describes how the pressure, P, of the gas depends on its absolute temperature, T, number of molecules, N, and its volume, V.

It is found experimentally that the *volume of a gas is inversely proportional to its pressure when its temperature is kept constant*. This is called **Boyle's law** which can be written as $PV = $ constant (when T is constant). The accompanied plot illustrates Boyle's law.

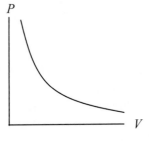

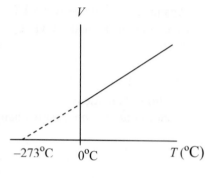

The volume of a gas is directly proportional to the temperature if its pressure is constant. This is **Charles's law**. The graph illustrates Charles's law (where temperature T is plotted along the x-axis and volume V along the y-axis.)

The line when extended meets the tempearature axis at $-273.15°C$. This means that if the gas could be cooled to $-273.15°C$, its volume would be zero. This temperature $-273.15°C$ at which the volume would be zero is called **absolute zero**. Absolute zero forms the basis of the **absolute** or **Kelvin scale**. The relationship between Kelvin and Celsius scale is given by

$$T\,(K) = T\,(°C) + 273.15.$$

According to **Gay-Lussac's law**, *the pressure of a gas is direcly proportional to the absolute temperature when its volume is constant.*

17–7 The Ideal Gas Law
The gas laws of Boyle, Charles, and Gay-Lussac can be combined into a single relation, $PV \propto T$.

A **mole** of a substance is the amount of the substance that contains as many particles as there are atoms in 12 g of carbon-12. In general, the number of moles, n, is:

$$n = \frac{\text{mass (grams)}}{\text{molecular mass (g/mol)}}$$

In terms of number of moles n of the gas, the preceding relation can be written as:

$$PV = nRT, \qquad (17\text{--}3)$$

where R is called the **universal gas constant**. This relation is called the **ideal gas law** or **equation of state for ideal gas**.

$$
\begin{aligned}
R &= 8.315 \text{ J/(mol} \cdot \text{K)} \\
&= 0.0821 \text{ (L} \cdot \text{atm)/(mol} \cdot \text{K)} \\
&= 1.99 \text{ calories/(mol} \cdot \text{K).}
\end{aligned}
$$

An ideal gas is a simplified model of a real gas. In an ideal gas, intermolecular attractions are neglected. A real gas can be approximated as an ideal gas at sufficiently low densities.

17-8 Problem Solving with the Ideal Gas Law

The ideal gas law $PV = nRT$ is useful for solving problems. According to the law, $PV/T = nR = $ constant. If P_1, V_1, and T_1 represent the appropriate variables initially, and P_2, V_2, and T_2 represent the variables finally, then

$$\frac{P_1 V_1}{T_1} = \frac{P_2 V_2}{T_2}.$$

P_1, V_1, T_1

P_2, V_2, T_2

The preceding relation is very useful to solve for an unknown when other variables are given.

17-9 Ideal Gas Law in terms of Molecules: Avogadro's Number

The number of molecules in one mole is called **Avogadro's number**, N_A. This number is $N_A = 6.022 \times 10^{23}$ (molecules/mol). If N is the number of molecules of a gas, $N = nN_A$, then

$$PV = nRT = \frac{N}{N_A} RT = NkT \qquad (17\text{–}4)$$

where $k = R/N_A = 1.38 \times 10^{-23}$ J/K , called **Boltzmann's constant**.

*17-10 Ideal Gas Temperature Scale—a Standard

It is important to have a precisely defined temperature scale so that different temperature measurements can be accurately compared. The standard thermometer for this scale is the constant-volume gas thermometer and the scale is called the ideal gas temperature scale. The temperature is proportional to the pressure in the ideal gas. The two fixed points are: $P = 0$ at $T = 0$ and $P = 4.58$ Torr at the triple point of water (which is $T = 273.16$ K).

The absolute or Kelvin temperature T at any point is then defined using a constant-volume gas thermometer for an ideal gas:

$$T = (273.16 \text{ K}) \left(\frac{P}{P_{\text{tp}}} \right) \qquad (17\text{–}5a)$$

where P and P_{tp} are pressure in the thermometer at temperature T and triple point, respectively. This definition of T with real gas is only approximate because we find a different result for T depending on the type of gas used. However, it is found that as the amount of gas in the thermometer is reduced, the limit $P_{tp} = 0$ always gives the same value of T regardless of the gas used. Thus, the ideal gas temperature scale is defined as:

$$T = (273.16 \text{ K}) \lim_{P_{tp} \to 0} \left(\frac{P}{P_{tp}} \right) \qquad (17\text{--}5b)$$

The scale does not depend on the properties of gases.

Tips for Solving the Problems

1. Remember to include $32°F$ when converting a temperature from Celsius to Fahrenheit, or vice versa, such as in the problems in Section 17–2. A *temperature change* in Celsius degrees is related to a *temperature change* in Fahrenheit degrees, by a factor of 5/9. A change in temperature in Celsius degrees is the same as the change in temperature in kelvins.

2. Remember that $\beta = 3\alpha$ in problems relating to thermal expansion. Usually, α will be given in the problem. In the expression for thermal expansion, the units of L, ΔL (or V, ΔV) need not be SI units so long as the units are the same, because the units cancel.

3. Use the expressions $\Delta L = \alpha L_0 \Delta T$, $L = L_0 (1 + \alpha \Delta T)$ and $\Delta V = 3\alpha V_0 \Delta T$, $V = V_0 (1 + 3\alpha \Delta T)$ to solve problems in Section 17–4.

4. A hole in a material (or an empty volume inside a container) expands at the same rate as if it were made of the material. To find such expansions, use the coefficient of thermal expansion of the material.

5. For an ideal gas $PV = NkT = nRT$. Also note that $P_1V_1/T_1 = P_2V_2/T_2$. These equations will be useful for solving problems in Sections 17–7 to 17–9. Remember that the temperature T in this expression *must be* in kelvins. Also remember that N is the number of *molecules* and n is the number of *moles*. For problems relating to Boyle's law T is constant, so that the ideal-gas equation becomes $PV =$ constant. For problems relating to Charles's law P is constant, so that the ideal gas equation becomes $V/T =$ constant.

CHAPTER 18
Kinetic Theory of Gases

This chapter discusses kinetic theory of gases, real gases and change of phase, van der Waals equation of state, and diffusion.

Important Concepts

Mean speed and rms speed
Most probable speed
Kinetic interpretation of temperature
Velocity distribution
Real gas
Critical point and critical temperature
Phase diagram
Triple point
Evaporation and condensation
Vapor pressure
Humidity
Van der Waals equation of state
Mean free path
Diffusion
Fick's law

18–1 The Ideal Gas Law and the Molecular Interpretation of Temperature

The pressure and temperature of a gas are *macroscopic* quantities and can be measured using a pressure gauge and a thermometer. The kinetic theory of gases relates the *macroscopic* quantities (such as pressure and temperature) to the *microscopic* quantities, such as the velocity of molecules.

The basic postulates of kinetic theory are:

- Gases are composed of a large number of identical pointlike molecules moving randomly in all directions.

- The moleculeas are far apart and they interact only during collisions with others.

- The molecules obey the laws of mechanics.

- The collisions are elastic.

Each collision of a molecule with the walls of a container results in a change in momentum of the molecule. The total change of momentum of the molecules per unit time is the force exerted on the wall. The average force per unit area is the pressure exerted by the gas on the wall.

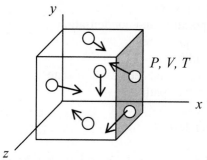

The molecules move with different velocities and the average value of the square of the x-component of velocity is

$$\bar{v}_x^2 = \frac{v_{x1}^2 + v_{x2}^2 + \cdots v_{xN}^2}{N} \qquad (18\text{--}1)$$

The pressure, P, of a gas is related to the volume, V, number of molecules, N, and the mass, m, of a molecule by

$$P = \frac{1}{3} \frac{mN \overline{v^2}}{V} \qquad (18\text{--}2)$$

where $\overline{v^2}$ is the mean square velocity of the molecules.

Equation (18–2) can be written as:

$$PV = \frac{2}{3} N \left(\frac{1}{2} m \overline{v^2} \right) \qquad (18\text{--}3)$$

Thus, the average kinetic energy ($\overline{K}$) of a molecule is

$$\overline{K} = \frac{1}{2} m \overline{v^2} = \frac{3}{2} kT \text{ (since } PV = NKT) \qquad (18\text{--}4)$$

Thus, *the average kinetic energy of a gas is directly proportional to the absolute temperature.* This is the kinetic interpretation of temperature.

The square root of the quantity $\overline{v^2}$ is called the root mean square speed or **rms speed,** v_{rms}. From Equation (18–4), the rms speed of a gas molecule is

$$v_{rms} = \sqrt{\frac{3kT}{m}} \approx 1.73 \sqrt{\frac{kT}{m}} \qquad (18\text{--}5)$$

18–2 Distribution of Molecular Speeds

In a gas at a particular temperature, different molecules will have different speeds. Also, the speed of a molecule changes with time as it collides with other molecules.

Maxwell, with the help of kinetic theory, derived a formula for the most probable distribution of speeds for a gas containing N molecules each mass of m at a temperature T. The distribution is given by:

$$f(v) = 4\pi N \left(\frac{m}{2\pi kT} \right)^{\frac{3}{2}} v^2 e^{-\frac{1}{2}\frac{mv^2}{kT}}. \qquad (18\text{–}6)$$

$f(v)$ is called the **Maxwell distribution of speeds**, which is plotted below.

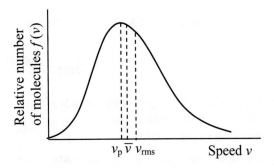

- Maxwell's distribution shows tł ⎵ lity that a given molecule will have ⎵ ⎵ quantity $f(v)\, dv$ represents the number of molecules having speed between v and $v + dv$. Thus, when we integrate over all speeds:

$$\int_0^\infty f(v)dv = N$$

- The speed at which the number of molecules is a maximum is the **most probable speed,** v_p, which can be obtained by calculating $df(v)/dv$ from Equation (18–6) and setting it to zero. The result is:

$$v_p = \sqrt{\frac{2kT}{m}} \approx 1.41\sqrt{\frac{kT}{m}} \qquad (18\text{–}7a)$$

The average speed,

$$\overline{v} = \frac{\int_0^\infty vf(v)dv}{N} = \sqrt{\frac{8kT}{\pi m}} \approx 1.60\sqrt{\frac{kT}{m}} \qquad (18\text{–}7b)$$

Also, $v^2_{rms} = \dfrac{\int_0^\infty v^2 f(v)dv}{N} = \dfrac{3kT}{m}$.

v_p, $\overline{v}$, and v_{rms} of a gas increases with its temperature.

18–3 Real Gases and Changes of Phase

Ideal gas law holds for gases at temperatures far from its liquefaction point and at low pressures. The PV diagram for a real gas is shown below. Curves A, B, C, and D represent the same gas at different temperatures.

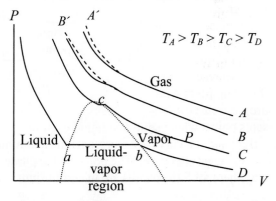

217

Note the following:

- The dashed curve A' represents the behavior of gas as predicted by the ideal gas law, $PV = $ constant; the solid curve A represents the behavior of a real gas at the same temperature.

- Curve D represents the situation where the gas starts becoming liquid at point b as its pressure is increased, and at point a it becomes completely liquid.

- Curve C represents the behavior of the substance at its **critical temperature** (T_c), where the point c is called the **critical point**. A gas can be liquified by increasing pressure only if its temperature is below its critical temperature. A substance in the gaseous state below its critical temperature is called a **vapor**, above the critical temperature is called a **gas**.

The behavior of a susbstance can be expressed also on a PT diagram, called a **phase diagram**. The phase diagram of water is shown where the curve l-v represents where the liquid and vapors are in equilibrium (the curve represents the plot of boiling point with pressure), the curve s-l

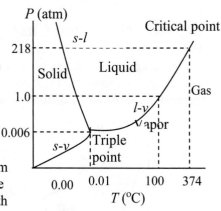

represents points where solid and liquid exist in equilibrium and the curve s-v represents the submimation point versus pressure curve. The point on the phase diagram where all three phases are in equilibrium is called the **triple point**. For water at the triple point, $T = 273.16$ K, and $P = 6.03 \times 10^{-3}$ atm. At this temperature and pressure, ice, water, and water vapor are all in equilibrium.

18–4 Vapor Pressure and Humidity

Evaporation
Evaporation occurs when molecules in the liquid phase attain speeds high enough to escape into the gas phase. Evaporation rate is greater at higher temperature since at higher temperature average molecular energy is higher. Because the fastest molecules escape in evaporation, the average speed of the remaining molecules, and hence, the temperature, is less. Thus, *evaporation is a cooling process*.

Vapor Pressure
In a closed container containing partly a liquid (say, water) and its vapor, a dynamic equilibrium is reached when the number of liquid molecules in evaporation equals the number of vapor molecules returning to the liquid in **condensation**. The space above the liquid is then *saturated*, and the pressure of the vapor when it is saturated is called the **saturated vapor pressure**.

The saturated vapor pressure of a substance depends on the temperature. At higher temperature, equilibrium is reached at a higher pressure, and thus saturated vapor pressure is more at higher temperatures. If the container is large or is not closed, the liquid may evaporate without saturating its vapor.

Boiling
A liquid boils at a temperature when its saturated vapor pressure equals the external pressure. This occurs for water at a pressure of 1 atm at 100°C. The boiling point of a

liquid depends on external pressure. At high elevations, the boiling point of water is less than at sea level.

Partial Pressure and Humidity

In a gas mixture, the total pressure is the sum of the *partial pressures* of each component gas. **Partial pressure** of a gas is the pressure the gas would exert if it were present alone. The **relative humidity** is defined as

$$\text{Relative humidity} = \frac{\text{partial pressure of } H_2O}{\text{saturated vapor pressure of } H_2O} \times 100\%$$

A relative humidity of 40-50% is optimum for comfort.

When the partial pressure of water exceeds the saturated vapor pressure, the air is **supersaturated**. This situation can occur when a temperature decreases and when excess water in air condenses to appear as dew, fog, or rain. When air containing water is cooled, a temperature is reached when the partial pressure of water equals the saturated vapor pressure. This is called the **dew point**.

*18–5 Van der Waals Equation of State

The molecules of an ideal gas are assumed to occupy no volume and they do not exert intermolecular forces (except during collisions). Van der Waals modified the ideal gas equation taking into account (1) the finite size of molecules, and (2) the intermolecular forces.

If b represents the "unavailable volume per mol" due to the finite volume occupied by other molecules, the volume V in the ideal gas law needs to be replaced by $(V - nb)$ (n = number of moles). Thus,

$$P(V - nb) = nRT \qquad (18-8)$$

Because of the intermolecular attractive forces, the molecules at the edge of the gas headed toward a wall of the container are slowed down by a net force pulling them back into the gas. The net force and, hence, the reduced pressure, is proportional to the density of molecules in the layer of gas at the surface and also to the density of the next layer, which exerts the force. Thus, P in Equation (18–8) should be reduced by $a(n/V)^2$ (a = proportionality constant) giving

$$P = \frac{nRT}{V - nb} - a\left(\frac{n}{V}\right)^2$$

$$\left(P + a\frac{n^2}{V^2}\right)(V - nb) = nRT, \qquad (18\text{–}9)$$

which is called the **van der Waals equation of state**.

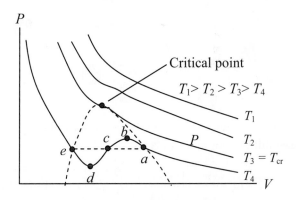

The constants a and b are different for different gases and are obtained by fitting to experimental data for each gas. *PV* diagram, for a van der Waals gas are shown (at different temperatures), which fit experimental data well for most gases. Below critical temperature (T_3) the gas passes through the liquid–vapor region. The points *b* and *d* would seem to be artifacts, since usually we see constant pressure indicated by the horizontal dashed line. However, for very pure supersatured vapor or supercooled liquids, sectons *ab* and *ed*, respectively, have been observed.

*18–6 Mean Free Path

In a gas, molecules move in random directions and they collide with each other as they move. If we were to follow the path of a particular molecule, we would find that it follows a zigzag path, as shown. The distance a molecule travels between two collisions is called a free path, some of which are shorter compared to others since the collisions are random. The average distance between two collisions is called the **mean free path**, l_M.

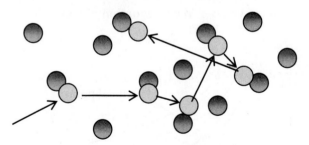

It can be shown that the mean free path l_M of a molecule (assuming other molecules are at rest) is:

$$l_\text{M} = \frac{1}{4\pi r^2 (N/V)} \qquad (18\text{–}10\text{a})$$

where πr^2 is the cross-section of the molecules and N/V is their concentration. Considering other molecules are also moving, the result is:

$$l_{\mathrm{M}} = \frac{1}{4\pi\sqrt{2}\,r^2\left(N/V\right)} \qquad (18\text{--}10\mathrm{b})$$

*18–7 Diffusion

In diffusion, a substance moves from a region of higher concentration to a region of lower concentration. For example, a few drops of food coloring in water spread slowly through the water, eventually becoming uniform.

Diffusion is understood from kinetic theory. The rate of diffusion J (number of molecules or moles or kg per second) through the cross-sectional area A is found to be:

$$J = DA\;\frac{C_1 - C_2}{\Delta x}$$

Region 1: $\overset{\longleftrightarrow}{\Delta x}$ Region 2:
concentration   concentration
$= C_1$ $= C_2$

where $C_1 - C_2/\Delta x$ is called the **concentration gradient** (that is, change in concentration in unit distance), A is the area of cross-section, and the constant D is called the **diffusion constant**. In terms of derivatives:

$$J = DA\;\frac{dC}{dx} \qquad (18\text{--}11)$$

This is called the **diffusion equation** or **Fick's law**.

223

Tips for Solving the Problems

1. Remember that temperature is in *kelvins* in any expression of the kinetic theory of gases involving temperature (such as the expressions for internal energy and rms molecular speed).

2. The expression $1/2\, m\bar{v}^2 = 3/2\, kT$ will be very helpful in solving problems dealing with the kinetic theory of gases such as those in Section 18–1. Most other quantities such as internal energy of a gas and rms molecular speed can be obtained from this expression. This equation together with $PV = NkT$ can be used to calculate pressure.

3. Note that average speed and rms speed can be determined from their definitions (such as Equation 18–1) or from using the formulae: $\bar{v} = \sqrt{8kT/\pi m}$ or $v_{rms} = \sqrt{3kT/m}$, depnding on the situations.

4. When using Van der Waals equation of state, remember that the constants a and b refer to per mole (not per molecule).

CHAPTER 19
Heat and the First Law of Thermodynamics

This chapter introduces heat as energy transfer. Concepts of specific heat, latent heat, and processes of heat transfer (conduction, convection, and radiation) are discussed.

Important Concepts

> Heat
> Calorie and kilocalorie
> Mechanical equivalent of heat
> Internal energy
> Specific heat
> Calorimetry
> Latent heat
> Heat of fusion and heat of vaporization
> The first law of thermodynamics
> Isothermal and adiabatic processes
> Specific heats at constant volume and at constant pressure
> Equipartition of energy
> Conduction, convection, and radiation
> Thermal conductivity and R-value
> Conductors and insulators
> Stefan–Boltzmann equation
> Solar constant

19–1 Heat as Energy Transfer

Heat is a form of energy that can be *transferred from one object to another because of a difference in temperature.*

It was thought that a hot object contained a certain amount of heat fluid called "caloric." Experimental observations, however, support that heat is a form of energy. In an experiment, Joule observed that the gravitational potential energy of a mass used to turn paddles in water in a container raised temperature of the water. By measuring the mechanical work done and the increase in the temperature, Joule found that energy was conserved in this case and that it had been converted from gravitational potential energy to heat energy.

The SI unit of heat is the joule (J). A common unit of heat is the **calorie**, which is the amount of heat needed to raise the temperature of 1 g of water by 1°C. A **kilocalorie** (kcal) is the amount of heat needed to raise the temperature of 1 kg of water by 1°C.

1 cal = 4.186 J. This is called the **mechanical equivalent of heat**. In studies of nutrition a different calorie called the Calorie (C) is used. 1C = 1 kcal.

19-2 Internal Energy

The total energy of all the moleclues of a substance is called its **internal energy**.

Distinguishing Temperature, Heat, and Internal Energy
Temperature is a measurement of the average kinetic energy of individual molecules. Internal energy is the total energy of all molecules. Heat is the transfer of internal energy from one object to another. The direction of the flow of heat depends on the temperature but not on the internal energy of the substance.

Internal Energy of an Ideal Gas
An ideal gas does not have potential energy, since there are no intermolecular interactions. The **internal energy**, E_{int}, of a monoatomic ideal gas is the sum of the kinetic energy of all its atoms:

$$E_{int} = N\left(\frac{1}{2}m\,\overline{v}^2\right) = \frac{3}{2}NkT = \frac{3}{2}nRT \qquad (19\text{--}1)$$

19–3 Specific Heat

When heat flows to a substance its temperature increases. If an amount of heat Q is added to a substance of mass m to raise its temperature by ΔT, then,

$$Q = mc\,\Delta T \qquad\qquad (19\text{--}2)$$

where c is a constant characteristic of the material, called its **specific heat**. Objects with a large c (such as water) require a large amount of heat to increase their temperatures.

The SI unit of c is $J/kg \cdot C^0$.

19–4 Calorimetry—Solving Problems

If there is no loss of heat to the surroundings,

Heat lost (by the hot body) = *Heat gained* (by the cold body).

This is the principle of **calorimetry**.

Insulated containers, called **calorimeters**, usually containing water are used in calorimetric measurements. Calorimeters are used to determine specific heats of substances. A **bomb calorimeter** is used to determine the heat released when a substance burns.

19–5 Latent Heat

During a change of phase, the temperature of a system stays constant but a certain amount of heat energy is transferred. The heat needed to change 1.0 kg of a substance from the solid to the liquid phase is called the **heat of fusion**, L_F. The corresponding heat needed to change from the liquid to vapor phase is called **heat of**

vaporization, L_v. The heat of fusion and the heat of vaporization are also called the **latent heats**, L. The SI unit of L is J/kg.

If Q amount of heat is required to convert a mass m from one phase to another,

$$Q = mL \qquad\qquad (19\text{--}3)$$

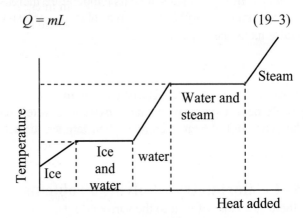

The heat added versus *temperature* curve for an ordinary solid (for example, ice) is shown by the preceding curve.

Evaporation
A solid can change from the liquid to the gas phase at even room temperature. This is called **evaporation**. Evaporation of water is important to control temperature of the body.

Kinetic Theory of Latent Heats
The latent heat, such as in a phase change from solid to liquid, is used not to increase kinetic energy (and, hence, temperature of the substance) but to cause solid molecules to break loose from neighboring molecules and become part of the liquid. Similarly, latent heat is used for molecules held close together in liquid phase to escape into the gas phase.

19–6 The First Law of Thermodynamics

The first law of thermodynamics is a statement of conservation of energy that includes heat.

When heat is added to a system, its internal energy increases. On the other hand, when a system does work, its internal energy decreases. The change in internal energy ΔE_{int} is thus the heat added to the system minus the work done by the system:

$$\Delta E_{int} = Q - W \qquad (19\text{–}4)$$

This is the **first law of thermodynamics**.

Note the usual sign convention:

Q is *positive*	when a system *gains* heat
Q is *negative*	when a system *loses* heat
W is *positive*	if work is done *by* a system
W is *negative*	if work is done *on* a system

*The First Law of Thermodynamics Extended

For a moving system whose kinetic energy (KE) and potential energy (PE) changes, the first law can be written as

$$\Delta\text{KE} + \Delta\text{PE} + \Delta E_{int} = Q - W \qquad (19\text{–}5)$$

19–7 Applying the First Law of Thermodynamics; Calculating the Work

We can analyze simple thermodynamic processes using the first law of thermodynamics.

Isothermal Processs ($\Delta T = 0$)

The temperature, T, remains constant in an **isothermal process**. For this process, $PV = nRT$ = constant.

If a gas that is enclosed in a container fitted in a movable piston gas and in thermal contact with a large heat reservoir is expanded or contracted very slowly so that its temperature does not change during the expansion or compression, the process is an isothermal process.

In this process, $\Delta E_{int} = \dfrac{3}{2} nR \, \Delta T = 0$. Thus, from the first law, $Q = W$.

Adiabatic Process ($Q = 0$)
In an **adiabatic** process there is no heat transfer ($Q = 0$) between the system and its environment.

An adiabatic process occurs when the system is thermally insulated or when the process occurs very quickly so that there is no time for heat to flow. Thus, sudden expansions (such as expansion of gases in an internal combustion engine) or sudden compressions are adiabatic.

Since $Q = 0$, from the first law, $\Delta E_{int} = -W$; that is, the work done on the system is used to increase the internal energy and temperature of the system. This principle is used in a diesel engine to ignite the fuel. After fuel and air are admitted into the cylinder, the piston rapidly compresses the air–fuel mixture. The rise in temperature is sufficient to ignite the fuel.

The *PV* diagram for adiabatic (*AC*) and isothermal (*AB*) processes on an ideal gas is shown at the right.

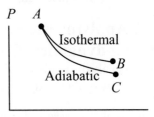

Isobaric and Isovolumetric Processes
In an **isobaric** process, pressure remains the same. Such a process is represented by a horizontal line in a *PV* diagram. In an **isovolumetric** or **isochoric** process,

volume remains the same. Such a process is represented by a vertical line in the PV diagram.

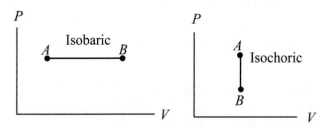

Work Done in Volume Changes

The work done by a gas as its volume changes by an infinitesimal amount dV is

$$dW = P\,dV \qquad (19\text{–}6)$$

For a change in volume from V_A to V_B the work done W is:

$$W = \int dW = \int_{V_A}^{V_B} P\,dV \qquad (19\text{–}7)$$

For an *isothermal process*, we get after substituting $P = nRT/V$ (for an ideal gas):

$$W = \int_{V_A}^{V_B} P \, dV = nRT \int_{V_A}^{V_B} \frac{dV}{V} = nRT \ln \frac{V_B}{V_A}. \qquad (19\text{–}8)$$

= Area under the PV curve AB between V_A and V_B.

For an *isobaric process DB*,

$$W = \int_{V_A}^{V_B} PdV = P_B (V_B - V_A)$$

$$= P \, \Delta V \qquad (19\text{–}9a)$$

$$= nRT_B \left(1 - \frac{V_A}{V_B} \right) \qquad (19\text{–}9b)$$

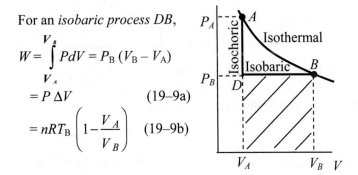

= Area of the shaded rectangle between DB and the volumes V_A and V_B.

Thus, the work done during AB and ADB are different. This result is general. The work done *depends not only on the initial and final states but on the process (path)*. The same is true also for *heat*.

Free Expansion
In a **free expansion,** a gas is allowed to expand adiabatically without doing any work, since it expands from a compartment filled with gas to a vacuum. In this case $Q = W = 0$, and, hence, $\Delta E_{int} = 0$. Thus, *internal energy of a gas remains constant in a free expansion*.

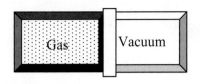

Since for an ideal gas, E_{int} depends only on T, $\Delta T = 0$ in the free expansion of an ideal gas. However, experimentally it was observed that E_{int} of *real gases* drops very slightly in a free expansion.

19–8 Molar Specific Heats for Gases, and the Equipartition of Energy

A gas has two specific heats: specific heat at constant volume (c_V) and specific heat at constant pressure (c_P).

Molar specific heats, C_V and C_P, are defined as the heat required to raise 1 mol of a gas by 1 C° at constant volume and at constant pressure, respectively. Thus,

$$Q = n\,C_V\,\Delta T \qquad\qquad (19\text{–}10a)$$

$$Q = n\,C_P\,\Delta T \qquad\qquad (19\text{–}10b)$$

Now, when heat is added to a gas at constant volume, the added heat is used only to raise its temperature. But when heat is added at constant pressure, the added heat is used not only to raise its temperature by the same amount but also to do some work to keep its temperature constant. As a result, C_P is always greater than C_V and the difference is the amount of work done. Based on this, for an ideal gas, we find that

$$C_P - C_V = R \quad \text{(for one mol)} \qquad (19\text{–}11)$$

For an ideal monoatomic gas, the total internal energy is the total kinetic energy of all the N molecules.

$$E_{int} = N\left(\frac{1}{2}m\,\overline{v^2}\right) = \frac{3}{2}nRT$$

If the heat is added at a constant volume,

$$Q = \Delta E_{int} = \frac{3}{2}nR\Delta T = n\,C_V\,\Delta T \qquad (19\text{--}12)$$

$$C_V = \frac{3}{2}R = 2.98 \text{ cal/mol·K} \qquad (19\text{--}13)$$

$$C_P = \frac{3}{2}R + R = \frac{5}{2}R = 4.97 \text{ cal/mol·K}$$

Equipartition of Energy

According to the **principle of equipartition of energy**, the total energy is divided equally among its various degrees of freedom, and for each degree of freedom the energy is $\frac{1}{2}kT$.

A monoatomic molecule has three independent ways of moving, and, thus, three independent ways of possessing energy. Each independent way a molecule can possess energy is called a **degree of freedom**. Thus, a monoatomic molecule has three translational degrees of freedom. Thus, total energy is

$$E_{int} = 3\,N\left(\frac{1}{2}kT\right) = \frac{3}{2}nRT,$$

which is the same as obtained before, providing $C_V = \frac{3}{2}R$.

A diatomic molecule can rotate in addition to moving linearly. (At low temperature nearly all diatomic molecules have only translational energy, so rotational degrees of freedom are not active.) It can rotate about two axes (as shown) giving two degrees of freedom of rotation, thus providing five degrees of freedom (three translational and two rotational) in total. Therefore, its total internal energy is

$$E_{int} = 5 N \left(\frac{1}{2} kT\right) = \frac{5}{2} nRT.$$

This gives $\quad C_V = \frac{5}{2} R.$

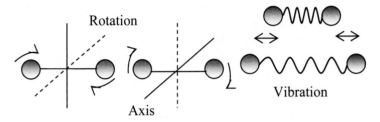

Rotation

Vibration

Axis

At very high temperatures, a diatomic molecule can vibrate as if the two atoms are connected by a spring, as shown. The kinetic energy and the potential energy of vibrational motion each provide one vibrational degree of freedom, making the total degrees of freedom seven. This gives $C_V = \frac{7}{2} R.$ C_V vs. T curve is shown for a diatomic molecule.

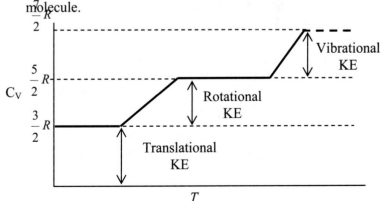

*Solids

Each atom in a crystalline solid can vibrate in x-, y-, and z-directions about its equilibrium position and at high temperatures it has six degrees of freedom (three for kinetic energy and three for potential energy of vibration). Thus, from the equipartition of energy, $C_V = 3R = = 6.0$ cal/mol·K. This is called *Dulong and Petit law*. Solids do not follow Dulong and Petit's law at low temperatures. At low temperatures the specific heat of solids increases as T^3, which can be explained by Debye-T^3 law of specific heat. The variation of molar specific heat for a solid is shown.

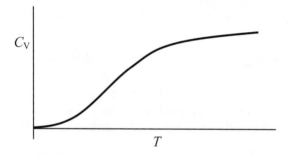

19–9 Adiabatic Expansion of a Gas

The PV curve for an adiabatic process is steeper than for an isothermal process. This means that for the same change in volume the change in pressure will be greater. Hence, temperature must drop in an adiabatic expansion and must rise in an adiabatic compression. Using the first law and the ideal gas equation $PV = nRT$, we find that in an adiabatic process:

$$PV^\gamma = \text{constant} \qquad\qquad (19\text{–}14)$$

where $\gamma = \dfrac{C_P}{C_V}$. (19–15)

19–10 Heat Transfer: Conduction

There are three ways heat can be transferred from one place to another, namely, *conduction*, *convection*, and *radiation*.

Conduction

In heat **conduction** (such as, in heat transfer in a silver spoon when placed in a hot bowl of soup), the atoms and molecules in a hotter section of the material vibrate with greater energy than those in the cooler sections. Energy is transferred by means of collisions from more energetic atoms/molecules to their less energetic neighbors.

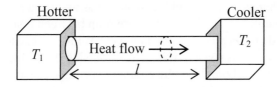

The conducted heat flow Q in time interval t is through a material of length l whose two ends are at temperatures T_1 and T_2 ($T_1 > T_2$) is given by

$$\frac{\Delta Q}{\Delta t} = kA \, \frac{T_1 - T_2}{l}$$ (19–16a)

where A is the cross-sectional area of the material. The constant k is called the **thermal conductivity**, whose value depends on the nature of the material. The SI unit of k is $J/(s \cdot m \cdot K)$.

In the notation of calculus:

$$\frac{dQ}{dt} = -kA \frac{dT}{dx} \qquad (19\text{-}16\text{b})$$

The negative sign means that heat flows in a direction opposite to the temperature gradient, dT/dx.

Substances, such as metals, that conduct heat very rapidly (that is, substances for which k is large) are called **conductors**. Substances, such as wood, that do not conduct heat (that is, k is small) are called **insulators**.

Tile is a better conductor than rug; that is why a tile floor is much colder on the feet than a rug-covered floor even when both are at the same temperature.

A major source of heat loss from a house is by conduction through windows. Usually there are layers of air on either side of the window that acts as an insulator and the major part of the temperature drop between inside and outside takes place across the layer. Wind can cause much greater loss since in this case air outside the window will be constantly replaced by cold air, causing a greater temperature gradient across the window. Two panes of glass separated by an air gap reduce heat loss because of the increased width of air layer.

Since air is a poor conductor of heat, clothes keep us warm by holding air so that it cannot move readily.

R Values for Building Materials
The thermal properties of building materials are specified by R values (or thermal resistance) defined for a given thickness l of material as:

$R = l/k$.

Clearly, R values increase with thickness. In the U.S. R values are given in British units: $\text{ft}^2 \cdot \text{h} \cdot \text{F/Btu}$.

Convection

Convection is a process where heat is transferred by the mass movement of molecules from one place to another. A forced-air furnace is an example of *forced convection* in which hot air is blown by a fan into a room. *Natural convection* also occurs in heating homes (such as by a radiator) in which rising warm air sets up a circulating flow of air that transports heat throughout a room. Hot-air furnaces with registers near the floor also depend on natural convection. In this case, the cold air returns to the furnace so that convective currents circulate throughout the room so that it is uniformly heated.

The inside temperature of the human body is 37°C while the skin temperature in a comfortable environment is about 34°C. The heat is carried by blood by convection to just beneath the surface of the skin. It is then conducted to the surface. Heat is lost from the surface to the environment by convection, evaporation, and radiaton.

Radiation

Radiation is a process of heat transfer in which heat flows in the form of electromagnetic waves such as visible light and infrared radiation. Electromagnetic waves can propagate through a vacuum and, thus, there is no need for a material medium for radiation to occur. The earth receives energy from the sun by radiation.

Each object gives off energy by radiation that is proportional to the fourth power of its Kelvin temperature, T. The radiated energy per unit time or radiated power by an object with surface area A is given by

$$\frac{\Delta Q}{\Delta t} = e\sigma A T^4 \qquad (19\text{--}17)$$

This is known as the **Stefan–Boltzmann law**. The constant σ is a fundamental constant called the **Stefan–Boltzmann constant**. $\sigma = 5.67 \times 10^{-8}$ W/m$^2 \cdot$ K^4. The constant e is called the **emissivity**, which is a dimensionless number between 0 and 1 that indicates how effectively the body radiates energy.

For very black surfaces, emissivity e is close to 1. On the other hand, for shiny surfaces, e close to 0. A *very good emitter (such as a blackbody) is also a very good absorber*. An ideal reflector absorbs no radiation and radiates no energy.

An object absorbs radiation from its surroundings according to the same law. Thus, the net radiated power of an object of temperature T_1 placed in surroundings of temperature T_2 is

$$\frac{\Delta Q}{\Delta t} = e\sigma A(T_1^{4} - T_2^{4}), \qquad (19\text{--}18)$$

Note that:
- If $T_1 > T_2$, net heat flows from the body to the surroundings,
- If $T_1 < T_2$, net heat flows from the surroundings to the body,
- If $T_1 = T_2$, net heat flow is zero.

About 1350 J of radiated energy from the Sun strikes the earth's atmosphere in one second per square meter surface area of the earth at right angles to the Sun's rays. This is called the **solar constant**. On a clear day about 1000 W/m^2 reaches the earth's surface. The rate of energy received by an object of emissivity e and surface area A facing the sun is

$$\frac{\Delta Q}{\Delta t} = (1000 \text{ W/m}^2) \, eA \cos \theta \qquad (19\text{--}19)$$

where θ is the angle between the Sun's ray and the normal to drawn to the surface. Because of the cos θ factor, the sun heats the earth more at noon than at sunrise or sunset, and seasons are formed. At the earth poles, cos θ varies between ½ (in the summer) to 0 (in the winter) so ice can form at the poles.

Thermography is a procedure used for diagnostic purposes that shows the results of the measurements of thermal radiation at different parts of the human body. Places, such as tumors, where metabolic activity is high can be detected in thermograms as regions of higher temperature.

Tips for Solving the Problems

1. In problems involving specific heat and latent heat, remember that *specific heat* is related to the *change in temperature in a given phase*, and *latent heat* is related to *the change of phase at a given temperature*.

2. Heat absorbed or heat given off by a substance is given by $Q = mc\Delta T$ where ΔT can be in C° or K.

3. In calorimetric problems, such as those in Section 19–4, use the equation $Q = mc\Delta T$ to calculate the heat gained by the colder objects and the heat lost by the hotter objects. Use the relation *Total heat gained = Total heat lost* to calculate the unknown in a problem such as the final temperature of mixture, or the specific heat of a substance.

4. Heat added or removed during a phase change is given by $Q = mL$. Make sure that the unit of L is J/kg to have Q in joules (J).

5. Be mindful of the signs of heat Q and work W. If *heat is added*, Q is "+"; otherwise, it is "–". If work is done *by* the system, W is "+"; otherwise, it is "–"

6. First law: $\Delta E_{int} = Q - W$ will be useful for solving problems such as in Sections 19–6 and 19–7.

7. Work done in any process can be calculated by evaluating the integral: $W = \int_{V_A}^{V_B} P \, dV$. Also, *the area under the PV curve is equal to the work done, W.*

8. Remember that $E_{int} = 3/2 \, nRT$ for a monoatomic gas. For a diatomic gas, determine how many degrees of freedom are active and use equipartition of energy to determine internal energy and C_V.

9. For an ideal gas $C_P - C_V = R$ (for one mol). Use this to determine its C_P when C_V is known.

10. For an ideal gas in adiabatic process use: $PV^\gamma =$ constant and $PV = nRT$. Note that $\gamma = 1.40$ for a diatomic gas and $\gamma = 1.67$ for a monoatomic gas.

11. For problems in heat transfer, such as those in Section 19–10, read the problem carefully to find whether the problem is related to conduction or to radiation.

12. For problems of heat flow as a result of conduction use the expression $\Delta Q / \Delta t = kA \left(T_1 - T_2 \right) / l$.

13. For the power radiated by a substance use the Stefan–Boltzmann law of radiation. Remember to include the factor e (emissivity) in the expression if the radiator is not a blackbody. The *net* radiated power when a hot object of temperature T_1 is placed in surroundings of temperature T_2 is $\Delta Q / \Delta t = e\sigma A(T_1^4 - T_2^4)$.

14. Be careful when using the Stefan–Boltzmann law that $(T_1^4 - T_2^4)$ is *not* equal to $(T_1 - T_2)^4$. Also, remember T is in Kelvin.

CHAPTER 20
Second Law of Thermodynamics

This chapter discusses the second and third laws of thermodynamics as well as entropy. Practical applications of the concepts such as the heat engine and refrigerator are also discussed.

Important Concepts

The second law of thermodynamics
Heat engine
Efficiency
Reversible and irreversible processes
Carnot engine
Carnot's theorem
Refrigerator, air conditioner, and heat pump
Coefficient of performance
Entropy
Entropy and disorder
Heat death
Kelvin temperature scale
Third law of thermodynamics

20–1 The Second Law of Thermodynamics—Introduction

The first law of thermodynamics is a statement of conservation of energy. However, out of many possible processes for which energy could be conserved, only some of those that proceed in a certain direction can actually occur in nature.

The **second law of thermodynamics** is a statement about which processes can occur in nature and which cannot. It can be stated in several ways, all of which are equivalent. **Heat flows spontaneously from a hot object to a cold object and never flows spontaneously from a cold object to a hot object**. This is Clausius's statement of the second law.

A more general statement of the law is based partly on the study of heat engines.

20–2 Heat Engines

A **heat engine** is a device that converts thermal energy to mechanical work. An engine operates in a cycle. It receives heat Q_H from a high-temperature reservoir at temperature T_H. It converts a fraction of the received heat to work, W. Finally, it returns the remaining heat, Q_L, to a low-temperature reservoir at temperatue T_L. From energy conservation:

$$Q_H = W + Q_L.$$

Steam Engine and Internal Combusion Engine

In a steam engine, steam is used as the **working substance** and higher temperature is obtained by burning coal, oil, or other fuel to heat the steam. In a reciprocating type steam engine, the heated steam passes through an intake valve, expands against a piston, and when the piston returns to the original position it forces the gases out of an exhaust valve. A steam turbine is essentially a steam engine except the piston is replaced by a rotating turbine. In an internal combustion engine (used in most automobiles), the high temperature is achieved by burning the gasoline–air mixture in the cylinder ignited by spark plug.

Why a ΔT is Needed to Drive a Heat Engine

In an engine, work is done *by* the gas on the piston when it expands. If the working substance (such as steam) is at the same temperature throughout the system, an equal amount of work would be done by the piston *on* the gas. Hence no net work would be done. Thus, net work would be obtained only when there is a difference of temperature.

Efficiency and the Second Law

The efficiency e of an engine is defined as

$$e = \frac{W}{Q_H} \qquad (20\text{–}1a)$$

$$= \frac{Q_H - Q_L}{Q_H}$$

$$= 1 - \frac{Q_L}{Q_H} \qquad (20\text{–}1b)$$

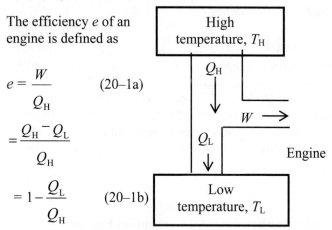

e is greater if Q_L is small. e would be 100 % if Q_L could be reduced to zero, which is not possible as expressed by the Kelvin–Planck statement of the second law of thermodynamics: **No device is possible whose sole effect is to convert a given amount of heat completely to work.**

20–3 Reversible and Irreversible Processes; the Carnot Engine

Reversible and Irreversible Processes
A **reversible process** is a process that is carried out infinitely slowly, so that it can be considered as a series of equilibrium states and the whole process could be done in reverse so that there is no change in magnitude of the work

done or heat exchange. A reversible process can be plotted on a *PV* diagram and the work done in reverse retraces the same path in the diagram.

A process that is not a reversible is called an **irreversible process**. An irreversible process is not done very slowly. The turbulence in the gas and friction could be present and it could not be done precisely in reverse since the heat loss due to friction would not reverse itself, the turbulence could be different, etc. An irreversible process, strictly speaking, cannot be plotted on a *PV* diagram. Although all real processes are irreversible, reversible processes are conceptually important.

Carnot's Engine

In order to study the maximum possible efficiency of a heat engine, Carnot considered an ideal engine (called a Carnot engine) where each process of heat addition or subtraction was done **reversibly**. That is, each process was done very slowly so that the process could be considered to be a series of equilibrium states and the whole process could be done in reverse without any change of the work done or heat exchanged. Real engines are irreversible and have efficiency less than that of a reversible Carnot engine. The **Carnot cycle** consists of the following four operations:

(1) The gas is expanded isothermally, extracting Q_H amount of heat at temperature T_H $(a \rightarrow b)$,

(2) The gas expands adiabatically and its temperature drops to T_L $(b \rightarrow c)$,

(3) The gas is compressed isothermally at temperature T_L $(c \rightarrow d)$ and, finally,

(4) The gas is compressed adiabatically back to its initial state ($d \rightarrow a$).

For a Carnot engine:

$$Q_H = W_{ab} = nRT_H \ln \frac{V_b}{V_a}.$$

$$Q_L = nRT_H \ln \frac{V_c}{V_d}. \text{ (magnitude)}.$$

Also, since bc and ad are adiabatic processes, it can be shown that

$$\frac{V_b}{V_a} = \frac{V_c}{V_d}$$

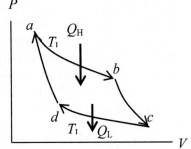

Thus, for a Carnot engine,

$$\frac{Q_L}{Q_H} = \frac{T_L}{T_H} . \quad (20\text{--}2)$$

Therefore, the efficiency of a Carnot engine (maximum efficiency) is

$$e_{\text{ideal}} = 1 - \frac{T_L}{T_H} \qquad (20\text{--}3)$$

The efficiency, e_{ideal}, depends only on the temperatures T_H and T_L.

According to **Carnot's theorem**:

- *All reversible engines operating between the same two temperatures have the same efficiency.*

- *An irreversible engine operating between the same two temperatures will have less efficiency than that of a reversible engine.*

Since $T_L < T_H$, clearly, e_{ideal} is always less than unity (or 100 %). This is expressed in the Kelvin–Planck statement of the second law of thermodynamics as:

- *No device is possible whose sole effect is to transform a given amount of heat completely into work.*

*The Otto Cycle

The operation of an automobile internal combustion engine can be approximated by a reversible cycle, called the *Otto cycle*, as shown. The gasoline–air mixture enters the cylinder at point *a*, and is compressed adiabatically to the point *b*. The ignition takes place at *b* adding heat Q_H to the system. In the power stroke *cd* the gas expands adiabatically. In the exhaust stroke *da* heat Q_L is rejected to the surroundings. For paths *bc* and *da* volume stays constant.

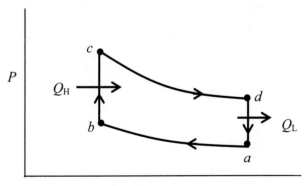

20–4 Refrigerators, Air Conditioners, and Heat Pumps

Refrigerators, air conditioners, and heat pumps are heat engines working in reverse.

A **refrigerator** uses work, W, to extract an amount of heat, Q_L, from the interior of a refrigerator (cold reservoir) and returns a larger amount of heat, Q_H, to the air in the kitchen (hot reservoir).

An **air conditioner** is also a refrigerator that uses electrical energy, W, to extract heat, Q_L, from a room (cold reservoir) and returns a larger amount of

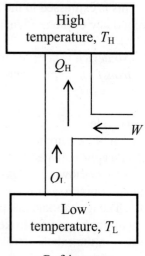

Refrigerator

heat, Q_H, to the warmer air outside (hot reservoir).

The effectiveness of refrigerators and air conditioners is expressed by their coefficient of performance, COP, defined as

$$\text{COP} = \frac{Q_L}{W} \qquad (20\text{–}4a)$$

Since, $Q_H = Q_L + W$,

$$\text{COP} = \frac{Q_L}{Q_H - Q_L} \qquad (20\text{–}4b)$$

For an ideal (Carnot) refrigerator,

$$\text{COP} = \frac{T_L}{T_H - T_L} \qquad (20\text{--}4c)$$

A **heat pump** uses work, W, to remove an amount of heat, Q_L, from outdoor air (cold reservoir) and returns a larger heat, Q_H, into the hot reservoir of air in the room. The coefficient of performance of a heat pump is

$$\text{COP} = \frac{Q_H}{W} \qquad (20\text{--}5)$$

20–5 Entropy

Like pressure, volume, temperature, and internal energy, **entropy**, S, is a thermodynamic quantity that can be used to describe the thermodynamic state of a system. The change in entropy is defined as

$$dS = \frac{dQ}{T} \qquad (20\text{--}6)$$

where dQ represents an infinitesimal reversible heat flow at temperature T.

For a Carnot cycle: $\dfrac{Q_L}{Q_H} = \dfrac{T_L}{T_H}$

Using the proper sign for Q (positive when heat is added and negative when heat is subtracted), this gives

$$\frac{Q_H}{T_H} + \frac{Q_L}{T_L} = 0 \qquad (20\text{--}7)$$

Any reversible process can be approximated as a series of Carnot's cycles (as shown below left, the approximation becomes better and better as the number of Carnot's cycles increases). Thus for all such Carnot cycles

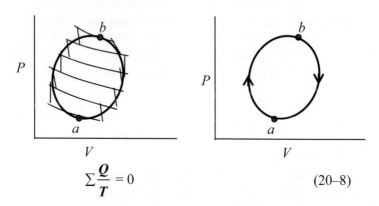

$$\sum \frac{Q}{T} = 0 \qquad (20\text{–}8)$$

But Q_L of one cycle is approximately equal to Q_H of the cycle below it and, hence, the heat flows on the inner paths of these Carnot cycles cancel out. Thus, for the whole cyclic process aba

$$\oint \frac{dQ}{T} = 0 \qquad (20\text{–}9)$$

$$\oint dS = 0 \qquad (20\text{–}10)$$

Thus, entropy change in a reversible cyclic process is zero.

$$\int_{\mathrm{I}a}^{b} \frac{dQ}{T} + \int_{\mathrm{II}b}^{a} \frac{dQ}{T} = 0$$

$$\int_{\text{I}a}^{b} \frac{dQ}{T} = \int_{\text{II}a}^{b} \frac{dQ}{T} \qquad (20\text{--}11)$$

Therefore, change in entropy does not depend on the path of the process. That is, entropy is a state variable.

$$\Delta S = S_a - S_b = \int_a^b dS = \int_a^b \frac{dQ}{T} \qquad (20\text{--}12)$$

The SI unit of S is J/K.

20–6 Entropy and the Second Law of Thermodynamics

Although definition of entropy applies to a reversible process, entropy changes ΔS of irreversible proceses are determined by figuring out some other reversible processes that take place between the same two states, since ΔS depends only on the initial and final states.

The total entropy is found to increase in any natural process, such as conduction of heat from one object to another because of temperature difference, or mixing of two liquids at different emperatures, etc. For example, when Q amount of heat flows from a hot body at temperature T_H to a cold body at temperature T_L, the change in entropy $\Delta S = -Q/T_H + Q/T_L > 0$ (since $T_H > T_L$).

The second law of thermodynamics is stated in terms of entropy as: *The total entropy of any system plus that of its environment increases as a result of any natural process.*

$$\Delta S = \Delta S_{\text{syst}} + \Delta S_{\text{env}} > 0 \qquad (20\text{--}13)$$

"Time's Arrow"

Entropy is called **time's arrow**, for it can tell us in which direction time is flowing. The principle of entropy increase is consistent with the Clausius and Kelvin–Planck statements of the second law.

20–7 Order to Disorder

Addition of heat to a system increases its entropy. From the molecular point of view, addition of heat increases molecular disorder. Entropy is thus a *measure of the disorder* of a system.

Also note that a jar containing separate layers of salt and pepper is more orderly than when the jar is shaken and the salt and pepper are mixed up. No amount of shaking brings the orderly layers back. In the conduction of heat, at the beginning there is more order since there are two classes of molecules: those with higher kinetic energy and those with lower kinetic energy. After the process, entropy and disorder increases because there is no orderly arrangement of two classes of molecules. In fact, a natural process always moves from a state of relative order to a state of disorder. The second law is stated as: *Natural processes tend to move toward a state of greater disorder.*

*Biological Evolution

Living systems use disordered raw materials in the environment to produce orderly structures; however, living organisms constantly waste heat to the atmosphere as a by-product of their metabolism. This heat increases the entropy of the surroundings by enough to produce a net increase in the entropy of the universe.

20–8 Unavailability of Energy; Heat Death

In the conduction of heat, initially the hot and cold objects can be used as heat reservoirs of a heat engine to obtain useful work. After the process the objects are at the same temperature and thus the objects cannot be used to obtain useful work. Thus, in conduction of heat, and similarly in any natural process, the energy becomes less available for mechanical work.

Therefore, as the disorder of the universe increases in natural processes, the energy available to do work decreases. This is known as "heat death" of the universe, which is a possible fate of the universe, in which every object will be at the same temperature and no work will be available.

*20–9 Statistical Interpretation of Entropy and the Second Law

The **microstate** of a system is specified by the position and velocity of every particle (or molecule) of the system. The **macroscopic** state of a system is specified by the macroscopic properties (such as temperature, pressure, number of moles, and so on) of the system. For example, if we repeatedly shake four coins in hand and drop them on a table, specifying the number of heads and the number of tails on a given throw is the macrostate. Specifying each coin as being a head or tail is the microstate. It is assumed that each microstate is equally probable.

The most probable state of a gas is one in which the molecules are disordered (moving in random directions) and speed distribution is Maxwellian. The orderly distribution of speeds in which all molecules have nearly the same speed is highly unlikely. Since entropy is related to the molecular disorder, when a system increases entropy its molecular disorder increases and, hence, the system passes from a lesss probable state to a more probable state. Similarly, when entropy decreases, the molecular disorder decreases and the system passes from a more probable state to a less probable state. Thus, entropy is a measure of probability of obtaining a state. Since entropy increases in any natural processs, only those processes occur in nature that are highly probable. Boltzmann showed that the entropy of a system can be written as:

$$S = k \ln W \qquad (20\text{–}14)$$

where k is the Boltzmann constant and W is the number of microstates corresponding tot the given macrostate.

The second law in terms of probability means that probability is extremely low for natural processes that decrease entropy.

*20–10 Thermodynamic Temperature Scale; Absolute Zero and the Third Law of Thermodynamics

The efficiency of a reversible engine depends only on the two temperatures but not on the working substance (such as helium, water, etc.). Kelvin used this fact to define absolute temperature scale so that the ratio of the two temperatures is the ratio of the heats exchanged with them by a revdrsible engine:

$$\frac{T_H}{T_L} = \frac{Q_H}{Q_L} \qquad (20\text{–}15)$$

Since T_{tp} (triple point of water) = 273.16 K, the definition of thermodynamic scale is:

$$T = (273.16 \text{ K}) \left(\frac{Q}{Q_{tp}} \right)$$

It is not possible to reach absolute zero by means of any finite number of processes. This is called the **third law of thermodynamics**.

*20–11 Thermal Pollution, Global Warming, and Energy Resources

The heat output from heat engines, from power plants to cars, is called **thermal pollution**. This heat is absorbed by the environment and increases the temperature of cooling water, altering the natural ecology of aquatic life or it raises the temperature of the atmosphere, affecting the weather. **Air pollution** is the chemicals released in the burning of the fossil fuels in cars, power plants, and industrial furnaces, giving rise to smog and other problems. The burning of fossils builds up CO_2 in the Earth's atmosphere. CO_2 absorbs infrared that the earth naturally emits, causing **global warming**.

Tips for Solving the Problems

1. The efficiency of an engine can be determined using the expression $e = W/Q_H$. $W = Q_H - Q_L$, so $e = 1 - Q_L/Q_H$.

2. The *coefficient of performance* (COP) of a refrigerator and an air conditioner is $COP = Q_L/W$. Note the difference in the definition of the *coefficient of performance* for a *heat pump*; $COP = Q_H/W$.

3. For a *reversible* (Carnot) engine or a refrigerator, an air conditioner, or a heat pump, $Q_L/Q_H = T_L/T_H$. Remember that this relationship is *not* true for an

irreversible engine, refrigerator, air conditioner, or heat pump. Also remember that all temperatures in this chapter are in kelvins.

4. You need to know the expressions in tips 1, 2, and 3 to solve problems related to heat engines, refrigerators, air conditioners, and heat pumps, such as problems in Sections between 20–2 and 20–4.

5. Note that the efficiency of a Carnot engine can be increased by decreasing T_L, or increasing T_H, or by both.

6. The change in entropy can be calculated from the expression $\Delta S = S_a - S_b = \int\limits_a^b dS = \int\limits_a^b \dfrac{dQ}{T}$. The sign of S is determined by the sign of Q. Remember that T may not be constant in the problems and cannot be taken out of the integral sign for those cases.

7. Note that ΔS of the universe is always positive for any natural process (irreversible process) such as heat conduction, mixing of substances at different temperatures etc. Usually in these cases entropy change is positive for one object while entropy change is negative for the other object(s) with the net entropy change being positive.

CHAPTER 21
Electric Charge and Electric Field

This chapter discusses electric charge, the building block of all atoms and molecules. The law of interaction between electric charges, concepts of electric field, and electric dipoles are discussed.

Important Concepts

    Electric charge
    Conservation of charge
    Conductor, semiconductor, and insulator
    Electroscope
    Coulomb's law
    Permittivity of free space
    Principle of superposition
    Electric field and field lines
    Electric dipole
    Dipole moment

21–1 Static Electricity; Electric Charge and Its Conservation

When an object (such as an amber rod or a glass rod) is rubbed with another (such as a piece of cloth), the object becomes electrically charged in the rubbing process and is said to possess a net **electric charge**.

  It is found that two charged plastic rulers or two charged glass rods repel one another whereas a charged glass rod attracts a charged plastic ruler. This shows that electric charges are of two types: *unlike charges attract, whereas like charges repel one another*.

  Charge is neither created nor destroyed in any process. It is usually transferred from one substance to the other.

Conservation of electric charge states that *the net amount of electric charge produced in any process is zero.*

21–2 Electric Charge in the Atom

Electric charge is an intrinsic property of the subatomic particles electrons and protons. Atoms have a small, dense, positively charged nucleus consisting of two types of particles—protons and neutrons—surrounded by a cloud of negatively charged particles, called electrons. A proton has a positive charge, an electron has an equal amount of negative charge, and neutrons have no electric charge. Atoms contain an equal number of electrons and protons and are electrically neutral.

The atom that loses an electron becomes a positive ion, and the atom that receives an electron becomes a negative ion. The charging of an object by rubbing can be explained by transfer of electrons from one object to another.

A water molecule is called a **polar** molecule because, although it is neutral, the charge is not distributed uniformly.

On humid or rainy days charge from an object (such as from a charged plastic ruler) can easily leak off on the water molecules in the air.

21–3 Insulators and Conductors

Substances such as wood that do not allow electric charges to move freely within them are called **insulators**; whereas substances such as any metal that allow electric charges to move freely within them are called **conductors**. Substances such as silicon or germanium whose properties are intermediate between those of insulators and conductors are called **semiconductors**.

In insulators the electrons are bound to the atoms and cannot move within the material. In the case of conductors, the outermost electrons, called *conduction electrons*, can be detached from the parent atoms. These electrons can move freely throughout the conductor.

21–4 Induced Charge; the Electroscope

A neutral metal rod acquires a charge when placed in contact with a charged metal object because of the direct transfer of electrons, as shown at the right.

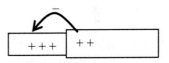

The neutral metal rod at the right acquires charge by contact

When a charged object (say, positively charged) is brought close to a neutral metal object, the electrons of the metal rod move within the metal toward the charged object, leaving the positive charges at the opposite ends. In this case, a charge is said to have *induced* at the two ends of the metal rod.

Separation of charge in the metal at right by induction

Another way to induce a net charge on a metal object is to connect it with a conducting wire to the ground, bring a charged object close, and then cut the wire from the metal.

An **electroscope** consists of a metal rod with a metal bulb at one end and a pair of hanging metal foil leaves at the other ends. When a charged object is brought close to the bulb, the electrons in the bulb are either attracted or repelled depending on the nature of the charge on the object. Since the electrons are conducted to or from the leaves, the leaves move apart because of the repulsive force between the same net charges on the leaves. An electroscope can be used to determine the sign of the charge if it is initially charged by conduction, say negatively. Now if a negatively charged object is brought close, more electrons will move down the leaves causing the leaves to separate further; whereas the separation between the leaves will decrease if a positively charged object is brought close.

21–5 Coulomb's Law

The magnitude of the electric force between two charges Q_1 and Q_2 is given by **Coulomb's law**:

$$F = k \frac{Q_1 Q_2}{r^2} \qquad (21\text{–}1)$$

where k is a proportionality constant. The SI unit of electric charge is the coulomb (C). In SI units, the constant $k = 8.988 \times 10^9 \ \text{N} \cdot \text{m}^2/\text{C}^2 \approx 9.0 \times 10^9 \ \text{N} \cdot \text{m}^2/\text{C}^2$ (up to two significant figures).

F_{12} = magnitude of force on 1 due to 2

F_{21} = magnitude of force on 2 due to 1

The magnitude of an electron's or proton's charge, e, is

$$e = 1.602 \times 10^{-19} \text{ C.}$$

A proton carries a charge of $+e$ and an electron carries a charge of $-e$. The charge of an electron or a proton is the smallest amount of free charge yet discovered. The charge of an object can be an integer multiple of this charge. Electric charge is **quantized**.

Coulomb's law can also be written as:

$$F = \frac{1}{4\pi \varepsilon_0} \frac{Q_1 Q_2}{r^2} \qquad (21\text{--}2)$$

where $\varepsilon_0 = 1/4\pi k$ ($= 8.85 \times 10^{-12}$ C^2/N $\cdot$ m^2), called the **permittivity of free space**.

Coulomb's law of electric force and Newton's law of gravitation look similar (both varies as $1/r^2$); however, the *force of gravity* is always *attractive*, whereas an *electric force* can be *attractive or repulsive*. The electric force between two subatomic particles is greater than the force of gravity between them by a factor of about 10^{39}.

*Vector Form of Coulomb's Law

$$\vec{F}_{12} = k \frac{Q_1 Q_2}{r_{21}^2} \hat{r}_{21},$$

where $\vec{F}_{12}$ is the vector force on charge Q_1 due to Q$_2$ and $\hat{r}_{21}$ is a unit vector directed from Q_2 toward Q_1.

21–6 The Electric Field

An *electric field* is defined as the force per unit charge. If a small positive test charge q experiences a force $\vec{F}$ at a given location, the electric field $\vec{E}$ at the location is

$$\vec{E} = \frac{\vec{F}}{q} \qquad (21-3)$$

The SI unit of electric field is newton/coulomb (N/C).

The electric field due to a point charge q at a distance r is

$$E = k\frac{q}{r^2} = \frac{1}{4\pi\varepsilon_0}\frac{q}{r^2}. \qquad (21-4)$$

The net electric field due to two or more charges is the vector sum of the electric fields due to the individual charges.

$$\vec{E} = \vec{E_1} + \vec{E_2} + \cdots. \qquad (21-5)$$

The force $\vec{F}$ on a charge q in an electric field $\vec{E}$ is

$$\vec{F} = q\vec{E}.$$

21–7 Electric Field Calculations for Continuous Distributions

In situations where charge is distributed uniformly, we first calculate the electric field dE due to an infinetismal charge dQ at a distance r from the equation:

$$dE = \frac{1}{4\pi\varepsilon_0}\frac{dq}{r^2}. \qquad (21-6a)$$

The total electric field $\vec{E}$ is then obtained from the integral

$$\vec{E} = \int d\vec{E} \qquad (21\text{–}6b)$$

21–8 Field Lines

Electric field is a vector and is sometimes called a *vector field*. Electric field lines are pictorial representations of an electric field.

- An electric field $\vec{E}$ is tangential at every point to the electric field lines.

- The number of field lines per unit area passing perpendicular to a surface is proportional to the magnitude of the electric field. Field lines are dense where $\vec{E}$ is large.

- The field lines are directed away from positive charges and toward negative charges.

- The number of lines entering or leaving a charge is proportional to the magnitude of the charge.

- No two electric field lines can intersect.

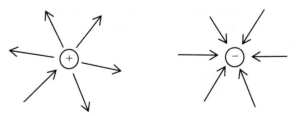

Electric field lines due to point charges

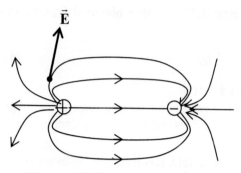

Electric field lines due to an electric dipole
(two equal and opposite charges separated by a
small distance)

Two metal plates with equal and opposite charges placed
parallel to each other and separated by a small distance
constitute a parallel plate capacitor.

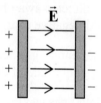

The electric field is uniform at any point between the
plates and is directed perpendicularly from the positively
charged plate to the negatively charged plate. Its value is
given by

$$E = \text{constant} = \frac{\sigma}{\varepsilon_0}.$$

The field is zero outside the plates.

Gravitational Field

Like an electric charge creates an electric field, the mass of an object (such as the earth) creates a **gravitational field**. The gravitational field is given by the force per unit mass.

21–9 Electric Fields and Conductors

Electric charges can move freely inside a conductor and experience no net electric force. That is, *electric field is zero inside a conductor. Any excess charge on a conductor distributes itself on the surface of the conductor.*

The electric field is always perpendicular to the surface outside of a conductor. If these were not the case, there would be a component of the field parallel to the surface, which would move the electrons until the parallel component vanished.

Nonconductors do not have free electrons inside, and an electric field can exist inside nonconductors.

21–10 Motion of a Charged Particle in an Electric Field

Knowing the force $\vec{F}$ on a charge q in an electric field $\vec{E}$ from the equation $\vec{F} = q\,\vec{E}$, the acceleration of the charge can be calculated from Newton's second law ($F = ma$). The distance and velocity of the charge can then be calculated from kinematic equations, such as $v^2 = v_0^2 + 2ax$, etc.

21–11 Electric Dipoles

Two equal and opposite charges, $+Q$ and $-Q$, separated by a small distance l, constitute an **electric dipole**. The **dipole moment** $\vec{p}$ is a vector of magnitude Ql that points from the negative to the positive charge.

Polar molecules have a net dipole moment, although the molecule as a whole is electrically neutral. The examples of polar molecules include CO (C has a small positive charge and O a small negative charge) and H_2O (O is negative and two H are positive).

Dipole in an External Field

When a dipole is placed in an uniform electric field, it experiences a torque (because of two equal and opposite forces on the charges), whose magnitude is given by

$$\tau = pE \sin \theta. \qquad (21\text{–}7a)$$

In vector notation:

$$\vec{\tau} = \vec{p} \times \vec{E}. \qquad (21\text{–}7b)$$

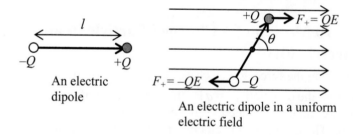

An electric dipole

An electric dipole in a uniform electric field

The effect of the torque is to try to turn the dipole so that $\vec{p}$ is parallel to $\vec{E}$.

Work done by the electric field decreases the potential energy, U, of the dipole. If $U = 0$ when $\vec{p}$ is perpendicular to $\vec{E}$,

$$U = -W = -pE \cos \theta = \vec{p} \cdot \vec{E}. \qquad (21\text{–}8)$$

Electric Field Produced by a Dipole

For a point P at a distance r on the perpendicular bisector of a dipole, the total electric field is:

$$\vec{E} = \vec{E}_+ + \vec{E}_-,$$

where $\vec{E}_+$ and $\vec{E}_-$ are the fields due to the + and − charges, respectively.

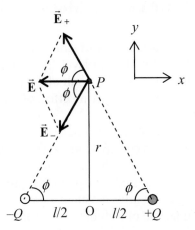

The magnitudes E_+ and E_- are equal. Their y-components cancel each other while their x-components are added up. Thus,

$$E = 2E_+ \cos \phi$$

$$= \frac{1}{4\pi\varepsilon_0} \frac{p}{\left(r^2 + l^2/4\right)^{\frac{3}{2}}} \qquad (21\text{–}9)$$

where $p = Ql$.

269

Far from dipole, $r \gg l$, this reduces to:

$$E = \frac{1}{4\pi\varepsilon_0} \frac{p}{r^3} \qquad (21\text{–}10)$$

Thus, E decreases more rapidly for a dipole than for a single point charge ($1/r^3$ versus $1/r^2$).

*21–12 Electric Field in Molecular Biology; DNA

Molecular biology is the study of structure and functioning of a living cell at the molecular level. The many processes that occur within a cell are the result of the random molecular motion and the ordering effect of the electrostatic force.

The genetic information that passes from one generation to others is contained in the chromosomes, which is made of genes. Each gene contains the information needed to produce a particular protein molecule. The genetic information in a gene is built into the principal molecule of a chromosome called the DNA (deoxyribonucleic acid). A DNA molecule has a long chain of many small molecules called nucleotide bases, which are of four types: adenine (A), cytosine (C), guanine (G), and thymine (T).

The DNA consists of two long DNA chains wrapped about one another in the form of a double helix. A in one strand is always opposite to a T on the other strand and similarly a G is always opposite to C. A and T attract each other and G and C attract each other through electrostatic forces to hold the double helix together. There are two weak bonds between A and T and three between C and G. The arrangement of A opposite T and G opposite C ensures that the genetic information passes accurately to the next generation. Thus, electrostatic forces hold the two chains together and also operate to select bases in proper order during replication so that genetic information passes accurately to the next generation.

*21–13 Photocopy Machines and Computer Printers Use Electrostatics

Both photocopy machines and laser printers use electrostatic attraction to print an image. In a *photocopier*, lenses and mirrors focus the image of an original sheet onto a drum. The toner is given a negative charge and brushed on the drum (which is selenium coated and positively charged in the dark areas where there is image) as it rotates. Thus, toner particles are only attracted to the positive areas and stick only there. The rotating drum finally presses against a piece of paper which has been stongly positively charged to transfer tonal particles to the paper forming the final image. In a color copier or printer, the process is repeated for each of black, blue, red, and yellow colors. A *laser printer* uses a computer output to program the intensity of a laser beam onto the selenium-coated drum. A movable mirror sweeps the laser beam in horizontal lines across the drum. The light parts of the selenium become conducting and lose electric charge whereas toner sticks only to the dark areas. In an *inkjet printer* nozzles spray tiny droplets of ink directly at the paper. The nozzles are swept across the paper and on each sweep the ink makes dots on the paper except for those points where no ink is desired.

Tips for Solving the Problems

1. Coulomb's law is similar to Newton's law of gravitation. *Be mindful of the direction of the electric force* as given by Coulomb's law. The force can be *attractive* or *repulsive*. Draw the sketch showing the direction of the force. Take right and upward as positive directions.

2. When using Coulomb's law for problems such as in Sections 21–5 make sure that the charges are in *Coulombs* and distance is in *meters*. Convert the units, if necessary.

3. The electric field due to a point charge can be calculated from the expression $E = kQ/r^2$. Watch the direction of the field. Use the method of components to determine the total electric field due to more than one charge, such as in the problems in Sections 21–6. For continuous charge distribution, first calculate dE for an elementary charge dQ and then integrate for the total electric field.

4. For more than two charges present in a problem, the net electric force (or net electric field) is the *vector sum* of the forces (or fields) between any pair. Use a convenient coordinate system, and sketch the forces (or fields). Use the method of components when adding two or more forces (fields). Recall that the components of the resulting force (or fields) vector are sum of the components of the individual force (or field) vectors.

5. If you need to draw a sketch of an electric field $\vec{\mathbf{E}}$, remember that electric field lines are directed away

from a positive charge and toward a negative charge. The force on a *positive charge* is in the *direction of the field*, whereas the force on a *negative charge* is *opposite* to the field.

6. Electric field due to a dipole can be obtained by calculating the electric fields due to the two charges ($+Q$ and $-Q$) in the dipole and then vector summing the individual fields. Resolving the fields into components may help. For distance $r \gg l$ (distance between the charges), the result can be further simplified by substituting r^2 for $(r^2 \pm l^2)$ or $(r^2 \pm l^2/4)$ in the expression.

CHAPTER 22
Gauss's Law

This chapter introduces the concept of electric flux and Gauss's law. It shows the use of Gauss's law to calculate electric field where there is symmetry.

Important Concepts

Electric flux

Gauss's law

Gaussian surface

22–1 Electric Flux

If a constant electric field $\vec{E}$ crosses an area A at an angle θ with the normal to the area, the **electric flux** through the area is

$$\Phi_E = EA \cos \theta = E A_\perp \qquad (22\text{–}1\text{a})$$

$$= \vec{E} \cdot \vec{A} \qquad (22\text{–}1\text{b})$$

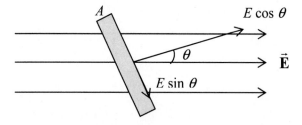

where $A_\perp = A \cos \theta$ = projection of A perpendicular to $\vec{E}$.

The flux through an area is proportional to the number of lines passing through the area. The SI unit of Φ_E is $N \cdot m^2/C$.

In a situation where electric field is not a constant, the total flux can be obtained from the integral

$$\Phi_E = \int \vec{E} \cdot d\vec{A} \tag{22–2}$$

For a *closed* surface (that is, a surface that completely encloses a volume, such as a sphere, the net flux through the surface is

$$\Phi_E = \oint \vec{E} \cdot d\vec{A} \tag{22–3}$$

where $\oint$ indicates that the integral is taken over a closed surface. Note that:

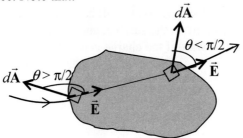

- The flux $\oint \vec{E} \cdot d\vec{A}$ entering an enclosed volume is *negative* ($\theta > \pi/2$),

- The flux $\oint \vec{E} \cdot d\vec{A}$ leaving an enclosed volume is *positive* ($\theta < \pi/2$),

- The net flux $\oint \vec{E} \cdot d\vec{A} = 0$, if each field line that enters the volume also leaves the volume. Since electric field lines start and stop only on electric charges, the net flux will be nonzero only if the surface encloses a net charge.

22–2 Gauss's Law

Gauss's law relates the electric flux passing through a surface to the net charges Q_{encl} enclosed by the surface. It can be expressed mathematically as

$$\oint \vec{\mathbf{E}} \cdot d\vec{\mathbf{A}} = \frac{Q_{encl}}{\varepsilon_0} \qquad (22\text{--}4)$$

To calculate the electric field at a distance r due to a single positive point charge $+Q$, we choose an imaginary "Gaussian surface" of radius centered on the charge. Because of spherical symmetry, $\vec{\mathbf{E}}$ will have the same magnitude at any point on the surface, and it is pointed radially outward (inward if the charge is negative).

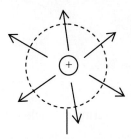

A Gaussian spherical
surface of radius r

Thus, from Gauss's law,

$$\frac{Q}{\varepsilon_0} = \oint \vec{\mathbf{E}} \cdot d\vec{\mathbf{A}} = \oint E \, dA = E \oint dA = E \, (4\pi r^2)$$

or $E = \dfrac{1}{4\pi\varepsilon_0}\dfrac{Q}{r^2},$

same as obtained from Coulomb's law

Gauss's law is useful for calculating the electric field for symmetrical charge distributions. Gauss's law follows from Coulomb's law and vice versa.

22–3 Applications of Gauss's Law

Note that Gauss's law is always true but is useful to determine electric field *only* when there is *symmetry*. To calculate electric field using Gauss's law, choose an imaginary closed surface called the "Gaussian surface" surrounding the charge or charge distribution.

When there is a *spherical symmetry*, make your Gaussian surface a *concentric sphere*; when there exists a *cylindrical symmetry*, make your Gaussian surface a *coaxial cylinder*; and when there exists a *plane symmetry*, use a Gaussian *pillbox* that straddles the surface.

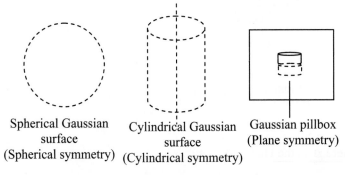

Spherical Gaussian surface
(Spherical symmetry)

Cylindrical Gaussian surface
(Cylindrical symmetry)

Gaussian pillbox
(Plane symmetry)

For example, for a solid sphere (of radius r_0) of charge, because of the spherical symmetry, electric field depends only on the distance and is directed radially outward.

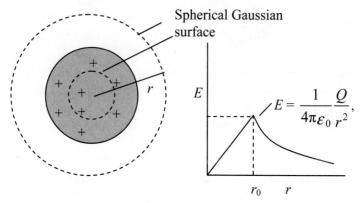

Spherical Gaussian surface

$$E = \frac{1}{4\pi\varepsilon_0}\frac{Q}{r^2},$$

We choose a spherical Gaussian surface of radius r concentric with the sphere. For a point outside the sphere, from Gauss's law:

$$\frac{Q}{\varepsilon_0} = \oint \vec{E} \cdot d\vec{A} = E\,(4\pi r^2)$$

or $\qquad E = \frac{1}{4\pi\varepsilon_0}\frac{Q}{r^2}, \qquad (r > r_0)$

For a point inside, $Q_{\text{enclosed}} = \left(\dfrac{\frac{4}{3}\pi r^3 \rho}{\frac{4}{3}\pi r_0^3 \rho} \right)Q = \left(\dfrac{r^3}{r_0^3} \right)Q,$

where ρ is the charge density.

This gives $\qquad E = \frac{1}{4\pi\varepsilon_0}\frac{Q}{r_0^3}r. \qquad (r < r_0)$

Using Gauss's law we find that the electric field due to:

- A long line charge of charge per unit length λ (cylindrical symmetry), $E = \dfrac{1}{2\pi\varepsilon_0} \dfrac{\lambda}{r}$,

- An infinite plane of charge of charge per unit area σ (plane symmetry), $E = \dfrac{\sigma}{2\varepsilon_0}$, and

- A good conductor near its surface of charge per unit area σ (plane symmetry), $E = \dfrac{\sigma}{\varepsilon_0}$.

*22–4 Experimental Basis of Gauss's and Coulomb's Law

Gauss's law predicts that any net charge on a conductor will lie on its surface. This has been verified experimentally. In the experiment, a charged metal ball is lowered into an insulated metal can carrying no net charge. The charge ball is then touched to the can and all of its charge quickly flows to the outer surface of the can. When the ball is then removed it carries no net charge as verified by an electroscope.

Tips for Solving the Problems

1. To draw a sketch of $\vec{E}$, remember that lines are directed away from a positive charge and toward a negative charge. The force on a *positive charge* is in the *direction of* $\vec{E}$, whereas the force on a *negative charge* is *opposite* to $\vec{E}$.

2. Use $\phi_E = EA \cos \theta$ to calculate electric flux. Remember that θ is the angle made by the electric field with the *normal* to the area. In a situation where electric field is not a constant, determine total flux from the integral $\Phi_E = \int \vec{E} \cdot d\vec{A}$.

3. Gauss's law $\oint \vec{E} \cdot d\vec{A} = Q_{encl}/\varepsilon_0$ is very useful for determining the electric field in situations having simple symmetry. Remember that if the net charge enclosed is zero, the electric field is also zero.

4. To determine an electric field using Gauss's law, such as in the problems in Section 22–3, determine the total electric flux passing through a suitably chosen Gaussian surface. Choose the symmetrical surface so that the electric field is normal to the surface. Calculate the total charge enclosed by the surface, and then apply Gauss's law. Depending on the symmetry, the Gaussian surface may be chosen as a spherical surface, a cylindrical surface, or a small pillbox.

CHAPTER 23
Electric Potential

This chapter introduces the concept of electric potential. It shows the relationship between electric potential and electric field and calculates electric potential in simple situations.

Important Concepts

> Electric potential
> Potential difference
> Volt
> Equipotential surface
> Electron volt
> Potential due to a point charge
> Potential due an electric dipole
> Cathode ray tube

23–1 Electric Potential Energy and Potential Difference

Electric Potential Energy
Like the force of gravity, the electric force is a conservative force. As a result, there is an electric potential energy, associated with the electric force $\vec{F}$.

The change in electric potential energy is given by $\Delta U = -W$, where W is the work done by the electric field, $\vec{E}$.

If a positive charge q moves through a distance d from point a to a point b in the electric field of magnitude E,

$$U_b - U_a = -W_{ba} = -qEd \qquad (23–1)$$

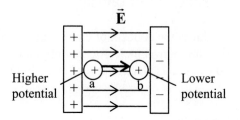

Electric Potential and Potential Diference
Thus, **electric potential** (V) at a point is the *electric potential energy of a unit charge*. Thus, at a point a

$$V_a = \frac{U_a}{q} \qquad (23\text{--}2a)$$

The potential difference (also called voltage) between the points b and a is

$$V_{ba} = V_b - V_a = -\frac{W_{ba}}{q} \qquad (23\text{--}2b)$$

where W_{ba} is the work done by the electric force to move the charge q from point b to point a. Clearly, the change in potential energy

$$U_b - U_a = qV_{ba} \qquad (23\text{--}3)$$

The SI unit of electric potential is the volt (V). 1 V = 1 joule/coulomb (J/C).

Practical sources, such as batteries and generators, are used to maintain a particular potential difference.

23–2 Relation Between Electric Potential and Electric Field

Since electric field is a conservative force, the difference in potential energy between any two points a and b is given by

$$U_b - U_a = -\int_a^b \vec{F} \cdot d\vec{l},$$

where $d\vec{l}$ is an infinitesimal increment of displacement.

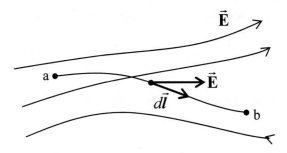

Since $V_{ba} = V_b - V_a = U_b - U_a /q$, we get

$$V_{ba} = V_b - V_a = -\int_a^b \vec{E} \cdot d\vec{l}. \qquad (23\text{–}4a)$$

This is the relation between electric field and potential difference.

From equation (23–4a), we get for an uniform electric field E,

$$V_{ba} = -Ed \qquad (23\text{–}4b)$$

where d is the distance, parallel to the field lines, between points a and b. The SI unit of E is V/m or N/C (V/m and N/C are equivalent).

23–3 Electric Potential Due to Point Charges

Since $E = \dfrac{1}{4\pi\varepsilon_0}\dfrac{Q}{r^2}$, we get

$$V_b - V_a = -\int_a^b \vec{\mathbf{E}} \cdot d\vec{l} = \frac{1}{4\pi\varepsilon_0}\left(\frac{Q}{r_b} - \frac{Q}{r_a}\right).$$

If we choose the electric potential to be zero at infinity, the electric potential due to a charge Q at a distance r is

$$V = \frac{1}{4\pi\varepsilon_0}\frac{Q}{r} \tag{23–5}$$

The following sketch shows how V changes with r.

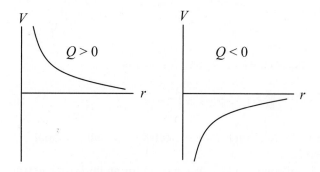

Electric potential is a scalar quantity. The total potential at any point due to two or more charges is equal to the algebraic sum (with proper sign) of the potentials due to the individual charges.

23–4 Potential Due to Any Charge Distribution

For n individual point charges, the potential at some point a,

$$V_a = \sum_{i=1}^{n} V_i = \frac{1}{4\pi\varepsilon_0} \sum_{i=1}^{n} \frac{Q_i}{r_{ia}} \qquad (23\text{–}6a)$$

where r_{ia} is the distance of the ith charge (Q_i) to the point a. For a continuous charge distribution,

$$V = \frac{1}{4\pi\varepsilon_0} \int \frac{dq}{r}, \qquad (23\text{–}6b)$$

where r is the distance of the point from an element of charge dq.

23–5 Equipotential Lines

An **equipotential surface** is a surface on which electric potential at any point is constant. Different equipotential surfaces have different electric potentials. Equipotential surfaces cannot intersect one another. The electric field is always perpendicular to the equipotential surface and points in the direction of decreasing electric potential. In two-dimensional drawings we show *equipotential lines*, which are intersections of equipotential surfaces with the plane of the drawing.

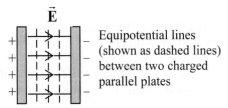

Equipotential lines (shown as dashed lines) between two charged parallel plates

Charges are free to move in an ideal conductor, and no work is done. Thus, every point on or within the conductor is at the same potential. Ideal conductors are equipotential surfaces. The electric field, therefore, is directed perpendicularly to the surface of an ideal conductor.

23–6 Electric Dipole Potential

Two equal and opposite point charges, each of magnitude Q, separated by a small distance l constitutes an **electric dipole**.

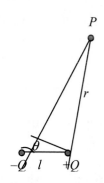

The electric potential at an arbitrary point P is the sum of the potentials due to each of the two charges. It can be shown that at a distance $r \gg l$, the potential due the dipole is given by

$$V = \frac{1}{4\pi\varepsilon_0} \frac{Ql\cos\theta}{r^2}$$

The product Ql is called the **dipole moment**, p, of the dipole. Thus,

$$V = \frac{1}{4\pi\varepsilon_0} \frac{p\cos\theta}{r^2} \tag{23–7}$$

Note that for a dipole V decreases as r^2 while for a single point charge V decreases as r.

The SI units of p is coulomb·meters (C · m). Another unit, *debye*, is also used, where 1 debye = 3.33 × 10^{-30} C · m.

Some molecules (such as water) have permanent dipole moments and are called **polar molecules**.

23–7 $\vec{E}$ Determined from V

Equation (23–4a) can be written in differential form as:

$$dV = -\vec{E} \cdot d\vec{l} = -E_l \, dl,$$

where E_l is the component of $\vec{E}$ in the direction of $d\vec{l}$. This gives,

$$E_l = -\frac{dV}{dl}. \qquad (23\text{–}8)$$

Thus, *the component of the electric field in any direction is equal to the negative of the rate of change of potential with distance in that direction.* In Cartesian coordinates:

$$E_x = -\frac{\partial V}{\partial x}, \qquad E_y = -\frac{\partial V}{\partial y}, \qquad E_z = -\frac{\partial V}{\partial z}. \qquad (23\text{–}9)$$

23–8 Electrostatic Potential Energy; the Elecron Volt

If a second point charge Q_2 is brought close to a point charge Q_1, the potential due to Q_1 at the position of the second charge is

$$V = \frac{1}{4\pi\varepsilon_0} \frac{Q_1}{r_{12}}$$

where r_{12} is the distance between the two charges. Thus, the potential energy of the two charges (relative to $V = 0$ at $r = \infty$) is given by

$$U = Q_2 V = \frac{1}{4\pi\varepsilon_0} \frac{Q_1 Q_2}{r_{12}}. \qquad (23\text{--}10)$$

For a system consisting of three charges, Q_1, Q_2, and Q_3,

$$U = \frac{1}{4\pi\varepsilon_0} \left(\frac{Q_1 Q_2}{r_{12}} + \frac{Q_1 Q_3}{r_{13}} + \frac{Q_2 Q_3}{r_{23}} \right).$$

Electron Volt Unit

In atomic and nuclear physics it is common to use electron volt (eV) as the unit of energy. An **electron volt** is defined as the energy acquired by an electron as it moves through a potential difference of 1 V.

$$1 \text{ eV} = 1.6 \times 10^{-19} \text{ J}.$$

*23–9 Cathode Ray Tube: TV and Computer Monitors, Oscilloscope

A **cathode ray tube** is used for displaying a voltage. Electrons are emitted inside the tube by **thermionic emission** when a cathode is heated to glowing and the electrons are attracted to a positively charged anode. The electrons pass out of this "electron gun" through a small hole in the anode. The electron beam strikes a fluorescent screen and makes a bright spot where it hits. Two horizontal and two vertical plates deflect the electron beam when a voltage is applied to them. By changing the voltage on the vertical deflection plates, the bright spot can be placed at any point on the screen.

In a picture tube or computer monitor the electron beam is swept horizontally by the horizontal deflecting plates (or magnetic coils). As the deflecting field changes from maximum to minimum, the beam moves from one edge of the screen to the opposite edge and then the voltage (or current) abruptly changes to move the beam back to the opposite side of the screen. At the same time vertical deflection plates deflect the beam downward and another horizontal sweep is made. In the U.S., 525 lines constitutes a complete sweep over the entire screen in 1/30 s. The picture is seen because the image is retained by the fluorescent screen and by our eyes for about 1/20 s. The picture consists of the varied brightness of the spots on the screen.

An **oscilloscope** is used for measuring and visually observing an electric signal which is displaced on the screen of a cathode ray tube. The electron beam is swept horizontally at a uniform rate by horizontal deflection plates whereas the signal to be displayed is applied to the vertical deflection plates. The visible trace on the screen is a plot of signal voltage (vertically) versus time (horizontally).

Tips for Solving the Problems

1. Both the electric potential and its unit of measurement are represented by the letter V. Also the letter C is used for the unit of charge (coulomb) and for capacitance. Make sure you understand these notations.

2. When calculating potential energy of a charge, be mindful of its sign. An electron (negative charge) at a positive potential has a negative potential energy. A positive charge falls to a lower potential, whereas a negative charge falls to a higher potential.

3. Note that the unit of electric field can be N/C or V/m. The units are equivalent.

4. The directions of electric field and equipotential surfaces are always perpendicular. Use this information when drawing equipotential surfaces or when finding the direction of the electric field.

5. Electric potential is a scalar quantity. As a result, when calculating the total electric potential due to several point charges algebraically add the individual potentials, using the *appropriate signs*.

6. Energies for electron, atoms, or molecules are often expressed in electron volts (eV). Remember the conversion 1 eV $= 1.6 \times 10^{-19}$ J.

7. To determine electric field from the potential, such as needed in some of the problems in Section 23–7, the following relations are useful: $E_x = -\partial V/\partial x$, $E_y = -\partial V/\partial y$, and $E_z = -\partial V/\partial z$.

CHAPTER 24
Capacitance, Dielectrics, Electric Energy Storage

This chapter introduces the concept of capacitance and dielectrics. It shows that capacitors store electric energy and discusses the series and parallel wiring of capacitors.

Important Concepts

Capacitors
Dielectrics and dielectric constant
Series and parallel wiring of capacitors
Electric energy density

24–1 Capacitors

A **capacitor** is a device that stores electric charges. It consists of two conductors separated by a finite distance. The amount of charge, Q, on each plate is proportional to the potential difference, V, between the plates. The constant of proportionality C is the **capacitance**:

$$Q = C V \qquad (24\text{–}1)$$

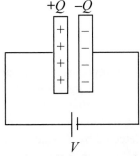

A parallel-plate capacitor connected to a battery

The SI unit of capacitance is coulomb/volt, which is called **farad** (F). 1F = 1 C/V.

A farad is a large unit. Usually, capacitors with a capacitance of microfarad (1 μF = 10^{-6} F) or picofarad (1 pF = 10^{-12} F) are used in electric circuits. Large capacitors such as 1 F or 2 F are used as power back ups, such as in computer memory and VCRs.

24–2 Determination of Capacitance

The capacitance of a capacitor can be determined experimentally from Equation (21–1) by measuring Q and V.

Value of C of a capacitor depends on the structure and dimension of the capacitor. For a parallel plate capacitor, if the area of each plate is A and the separation between the plates is d, it can be shown that its capacitance is

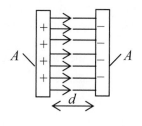

$$C = \varepsilon_0 \frac{A}{d} \qquad (24\text{–}2)$$

One type of computer keyboard uses a parallel plate capacitor, one plate of which is movable. Pressing the key reduces the capacitor spacing, thus increasing the capacitance. The change in capacitance becomes an electric signal that is detected by an electric circuit.

24–3 Capacitors in Series and Parallel

Capacitors in Parallel

When capacitors are connected across the *same potential difference*, they are said to be in **parallel**. The equivalent capacitance, C_{eq}, of capacitors C_1, C_2, C_3 connected in parallel is

$$C_{eq} = C_1 + C_2 + C_3. \qquad (24\text{–}3)$$

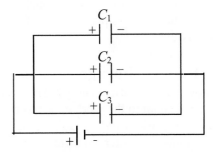

Capacitors in parallel

The equivalent capacitor C_{eq} will hold the same amount of charge as hold by this combination for the same applied voltage.

Capacitors in Series

When capacitors are connected one after another so that there is the *same charge* on each, they are said to be in **series**. The equivalent capacitance, C_{eq}, of capacitors C_1, C_2, C_3 connected in series is

$$\frac{1}{C_{eq}} = \frac{1}{C_1} + \frac{1}{C_2} + \frac{1}{C_3}. \qquad (24\text{–}4)$$

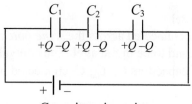

Capacitors in series

Other combinations of capacitors can be analyzed using charge conservation and simply in terms of series and parallel combinations.

24–4 Electric Energy Storage

A charged capacitor stores electrical energy. The energy stored in a capacitor is equal to the work done to charge the capacitor. Since the work to add a small amount of charge, dq, is $dW = V\,dq$ (V = potential difference), the electrical energy stored in a capacitor is given by

$$U = W = \int_0^Q V\,dq = \frac{1}{C} \int_0^Q q\,dq = \frac{1}{2}\frac{Q^2}{C}. \text{ (since } q = CV\text{)}$$

We can also write

$$U = \frac{1}{2}\frac{Q^2}{C} = \frac{1}{2}CV^2 = \frac{1}{2}QV \qquad (24\text{–}5)$$

For a parallel-plate capacitor, $Q = \varepsilon_0 E A$, and $V = E\,d$. Thus, from Equation (24–5), the **energy density** (energy per unit volume) is given by

$$u = \text{energy density} = \frac{1}{2}\,\varepsilon_0 E^2 \qquad (24\text{--}6)$$

If a dielectric is present, ε_0 is replaced by ε.
Equation (24–6) is general and holds for any electric field.

Health Effects
A charged capacitor can be a potential hazard even when the external power is off. A heart defibrillator is primarily a capacitor charged to a high voltage. It is allowed to discharge very rapidly through a pair of paddles that spread out the current throughout the chest. In a heart attack, this may cause complete heart stoppage sometimes followed by the resumption of normal beating of the heart.

24–5 Dielectrics

An electrically insulating material (such as plastic or paper) that usually is inserted between the plates of a capacitor is called a **dielectric**. Dielectrics allow higher voltages to be applied, and the plates to be placed closer without touching. It is found that if a dielectric fills the space between the two conductors, its capacitance increases by a factor K, called the **dielectric constant**. Thus,

$$C = KC_0, \qquad (24\text{--}7)$$

where C and C_0 are the capacitances in the presence and absence of the dielectric, respectively. The factor K is greater than one.

In the presence of a dielectric, the capacitance of a parallel plate capacitor is:

$$C = K\varepsilon_0 \frac{A}{d}. \qquad (24\text{--}8)$$

This is also written as

$$C = \varepsilon \frac{A}{d} \text{ where}$$
$$\varepsilon = K\varepsilon_0, \qquad\qquad (24\text{--}9)$$

called the **permittivity** of the material.

Note that when a battery of voltage V_0 is connected to a capacitor, it is found that the charge on the plates is increased by K when a dielectric is inserted, that is, $Q = KQ_0$.

If a charged capacitor is isolated so that the charge remains on the plates, when a dielectric is inserted, the potential difference drops to $V = V_0/K$. As a result the electric field ($E = V/d$) is also reduced by the factor K. Thus, in the presence of a dielectric.

$$E_D = \frac{E_0}{K} \qquad\qquad (24\text{--}10)$$

*24–6 Molecular Description of Dielectrics

The applied electric field polarizes the molecules of the dielectric. This results in the alignment of positive charges on the surface of the dielectric near the negative plate, and negative charges on the surface near the positive plate. As a result, fewer electric field lines exist within the dielectric.This reduction of electric field, and hence the voltage across the capacitor (since $V = Ed$) is expressed by the **dielectric constant**, K:

$$V = \frac{V_0}{K}$$

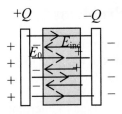

where V_0 and V are the voltages across the capacitor before and after the dielectric has been added, respectively. The value of K depends on the nature of the material. If C_0 (= Q/V_0) is the capacitor in the absence of a dielectric, the capacitance of a parallel-plate capacitor in the presence of dielectric is

$$C = \frac{Q}{V} = \frac{Q}{V_0/K} = KC_0.$$

The electric field, E_{ind}, due to the induced charges on the surfaces of the dielectric is

$$E_{ind} = E_0\left(1 - \frac{1}{K}\right).$$

This gives the surface density, σ_{ind}, of the induced charges and induced charge Q_{ind} (also called bound charge) as:

$$\sigma_{ind} = \sigma_0\left(1 - \frac{1}{K}\right) \qquad (24\text{--}11a)$$

$$Q_{ind} = Q_0\left(1 - \frac{1}{K}\right). \qquad (24\text{--}11b)$$

297

Tips for Solving the Problems

1. The capacitance, C, is always a positive quantity. Always use the magnitudes of Q and V to determine C from the expression $C = Q/V$.

2. Capacitance is usually expressed in microfarads or picofarads. Make sure to convert them to farads before solving any problem, such as those in Section 24–2 and 24–3.

3. Energy stored in a capacitor (U) will be in Joule when C is in farads, V is in volts, and Q is in coulombs. $U = \frac{1}{2} Q^2/C = \frac{1}{2} CV^2 = \frac{1}{2}QV$.

4. The capacitance of a parallel-plate capacitor with dielectric is $C = K\varepsilon_0 A/d$, where K is always greater than 1. The charge on the plates is $Q = CV$, the electric field between the plates is $= V/d$, and the capacitance without dielectric is $= C/K$.

5. When a dielectric is introduced in a capacitor the induced charge on the dielectric is $Q_{ind} = Q (1 - 1/K)$, which is always less than the free charge Q (since $K > 1$).

CHAPTER 25
Electric Currents and Resistance

This chapter introduces the basic concepts of electric circuits such as electric current, resistance, Ohm's law, and electric power.

Important Concepts

Electric cell
Electric current
Ampere
Resistance
Ohm's law
Ohm
Resistivity
Temperature coefficient of resistivity
Superconductivity
Electric energy and power
Watt and Kilowatt-hour
Direct current and alternating curent
Peak voltage and peak current
Rms value
Average power
Drift speed

25–1 The Electric Battery

Electric Cell and Battries

A battery produces electricity by converting chemical energy into electrical energy. A simple battery contains two plates or rods made of two dissimilar metals (one can be carbon) called **electrodes**. The electrodes are immersed in a solution called *electrolyte*. Such a device is called an **electric cell** and several cells connected together is called a **battery** (although today even a simple cell is also called a battery). With the suitable combination of electrodes and an electrolyte, a potential difference develops between the electrodes because of chemical reaction.

For example, in one kind of a simple cell, carbon and zinc rods are used as electrodes and sulfuric acid as electrolyte. Electric connections are made to the **terminals** (part of the electrodes outside the solution). Because of the chemical reaction the carbon rod becomes positively charged and the zinc rod negatively charged and a particular potential difference (voltage) is maintained between the two terminals.

When batteries are added in *series* (that is, the negative terminal of one battery is connected to the positive terminal of the next), the voltagers are added. For example, a 12 volt automobile battery consists of six 2-volt cells connected in series.

25–2 Electric Current

When a battery is connected to a conducting path (such as to a light bulb using metal wires) so that charges move in a closed path from one terminal of the battery, it is called an **electric circuit**.

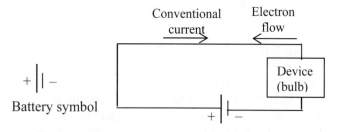

The flow of electric charges is called an **electric current**. In most situations, the flow of negative charges (electrons) through a metal constitutes electric current. Precisely, if a

charge of amount ΔQ passes a given point in a time
interval ΔT, the average electric current, $\overline{I}$, is defined as

$$\overline{I} = \frac{\Delta Q}{\Delta t} \qquad (25\text{–}1a)$$

The instantaneous current is

$$I = \frac{dQ}{dt}. \qquad (25\text{–}1b)$$

The SI unit of current is coulomb/second (C/s), or **ampere**
(A). 1 A = 1C/s.

When charges flow through a closed path, the closed
path is called a **complete circuit** whereas if there is a
break it is called an **open circuit.**

The direction of **conventional current** is the direction
that positive charges would follow through the circuit.
Conventional current flows through the circuit from the + to
– terminal of a battery, which is equivalent to a negative
(electron) current flowing from the – to + terminal of the
battery.

25–3 Ohm's Law: Resistance and Resistors

A potential difference is needed to produce an electric
current, and the current I in a metal conductor is found to
be linearly proportional to the potential differerence V
applied to it ends. Also, as electrons move through a
conductor they suffer collisions with the atoms creating a
resistance or opposition to the motion of electrons. The
higher the resistance R, the less is the current for a given
potential difference V. Thus, the current I is given by

$$R = \frac{V}{I} \qquad \text{(25–2a)}$$

That is, $V = IR$ (25–2b)

This is known as **Ohm's law**. Ohm's law is not a fundamental law, it provides the definition of resistance, $R = V/I$. For metals R is a constant (that is, $I \propto V$), but for devices such as diodes, transistors, etc. (known as *non-ohmic* devices), R is not a constant and depends on the applied voltage.

The unit of resistance is the **ohm** (Ω). $1\ \Omega = 1\text{V/A}$.

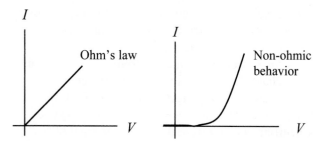

All electric devices such as heaters and light bulbs offer resistance. In electric or electronic circuits, **resistors** are used to control current. The resistance of a resistor is usually given as a color code.

 Symbol for resistor

Some Helpful Clarifications
- Batteries do not provide a constant current, they maintain a constant potential difference.

- The current through an electric device depends on its resistance, which is a property of the device.

- The current is *not* a vector, The direction of conventional current is from + to −.

- The charges that go in at one end come out at the other end and do not get "used up."

25–4 Resistivity

The resistance of a given material depends on its area, A, and length, L, in the following manner:

$$R = \rho \frac{L}{A} \qquad (25\text{--}3)$$

The quantity ρ is a characteristic of the material called **resistivity**. The greater the resistivity of a material, the greater is the resistance of a given piece of the material. The SI unit of resistivity is ohm·m (Ω·m). The reciprocal of resistivity is conductivity is:

$$\sigma = \frac{1}{\rho} \qquad (25\text{--}4)$$

Temperature Dependence of Resistivity

The resistivity of a conductor increases almost linearly with its temperature. That is,

$$\rho_t = \rho_0[1 + \alpha(T - T_0)] \qquad (25\text{--}5)$$

where ρ_0 and ρ_t are resistivity at some reference temperature T_0, and at T, respectively, and α is called the *temperature coefficient of resistivity*. α is positive for a metal but is negative for a semiconductor. Thus, resistance of a semiconductor can decrease with temperature.

25–5 Electric Power

Electric energy can easily be transformed into other useful forms of energy (such as heat and light). The rate at which electrical energy is transformed is given by the electric power. If I is the amount of current in a circuit as a result of an applied potential difference, V, then the electric power, P, is

$$P = IV \qquad (25\text{–}6)$$

The SI unit of electric power is the watt (W) (1 W = 1 J/s). Since $V = IR$, the power can also be written as

$$P = I^2R \qquad (25\text{–}7a)$$

$$= \frac{V^2}{R} \qquad (25\text{–}7b)$$

A practical unit of electric energy is the **kilowatt-hour** (kWh). 1 kilowatt-hour = 1000 W $\times$ 3600 s = 3.60 $\times$ 10^6 J. We pay electric bills based on the number of kilowatt-hours of energy used.

25–6 Power in Household Circuits

The electric wires that carry electricity produce heat at the rate I^2R. Fuses and electric circuit breakers are used in electric circuits to prevent overheating. These are basically switches that open the circuit when the current exceeds some particular value. A "short" in a circuit means the path of current is shortened, meaning that two wires have crossed.

Household appliances are connected in *parallel circuits* to the voltage main, which is 120 V in the U.S., so that each receives the same voltage. If a fuse blows or a circuit breaker opens, the total current in the circuit to be checked.

25–7 Alternating Current

Electric currents are of two types: direct current (dc) or alternating current (ac). If a current always flows in the same direction, it is called a **direct current**. If the current periodically changes its magnitude and direction, it is called an **alternating current**.

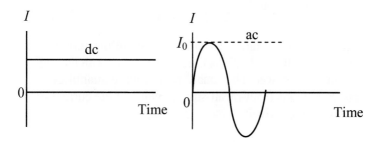

An alternating voltage can be expressed mathematically as

$$V = V_0 \sin 2\pi ft$$

where V_0 is the maximum value of the voltage during a cycle, called **peak voltage**.

The resulting current, I, flowing through a resistor (such as a light bulb), R, is

$$I = \frac{V}{R} = \frac{V_0}{R} \sin 2\pi ft = I_0 \sin 2\pi ft \qquad (25\text{–}8)$$

where $I_0 = \dfrac{V_0}{R}$.

The voltage and the current have the same time variation, and they are in **phase** with each other.

The power, P, dissipated in a resistor, R, is

$$P = I^2 R = I_0^2 R \sin^2 2\pi f t.$$

The average power is given by

$$\overline{P} = \frac{1}{2} I_0^2 R = \frac{1}{2} \frac{V_0^2}{R}.$$

The mean or average value of the square of the current $(I_0^2/2)$ or voltage $(V_0^2/2)$ is thus important for calculating the average power. The square-root of these quantities are called **rms** (root-mean-square) value of the current or voltage, respectively. Thus,

$$I_{rms} = \frac{1}{\sqrt{2}} I_0 = 0.707 \, I_0, \qquad (25\text{–}9a)$$

$$V_{rms} = \frac{1}{\sqrt{2}} V_0 = 0.707 \, V_0, \qquad (25\text{–}9b)$$

In terms of rms values, the average power is given by

$$\overline{P} = I_{rms} V_{rms} \qquad (25\text{–}10a)$$

$$= I_{rms}^2 R \qquad (25\text{–}10b)$$

$$= \frac{V_{rms}^2}{R}. \qquad (25\text{–}10c)$$

Clearly, a direct current whose I and V values equal the rms values of I and V for an alternating current will produce the same power. Rms values are sometimes called "effective values."

25–8 Microscopic View of Electric Current: Current Density and Drift Velocity

Current density j is defined as the electric current per unit cross-sectional area of the conductor at any point. Thus,

$$I = jA \qquad (25\text{–}11)$$

If the current density is not uniform,

$$I = \int \vec{\mathbf{j}} \cdot d\vec{A} \qquad (25\text{–}12)$$

In a conducting wire the free electrons move in random directions and collide with the atoms of the wire. When a potential difference is applied across the ends of the wire, the electrons initially begin to accelerate in the direction of the applied field and soon reach a steady average speed, called the **drift speed**, v_d, due to the collisions with the atoms. The current I is given by

$$I = \frac{\Delta Q}{\Delta t} = neAv_d \qquad (25\text{–}13)$$

where n, e, and A represent the number of electrons per unit volume, electronic charge, and area of cross section of the wire, respectively.

This gives, $j = nev_d \qquad (25\text{–}14)$

In vector form, $\qquad \vec{\mathbf{j}} = -ne\vec{\mathbf{v}}_d \qquad (25\text{–}15)$

The drift velocity v_d is less than 1 mm/s, which is very small. However, when a potential difference is applied, the associated electric field in the conductor moves at a speed comparable to the speed of light (3×10^8 m/s). This electric field influences the motion of electrons throughout the wire and a current is established almost instantaneously in the circuit.

*Electric Field inside a Wire

We assume that the electric field E is uniform throughout the length of the wire. Then, from Ohm's law and definition of resistivity ρ, it follows that

$$j = \frac{1}{\rho} E = \sigma E \qquad (25\text{--}16)$$

In vector form,

$$\vec{j} = \sigma \vec{E} = \frac{1}{\rho} \vec{E}$$

*25–9 Superconductivity

There is an important class of materials whose resistivity is found to suddenly drop to zero below a certain temperature near absolute zero, called the *transition temperature*, T_C.
Below this temperature these materials are said to be **superconducting** (such as mercury below 4.2 K as first observed by H. K. Onnes in 1911). In superconductors current persists undiminished so long as the temperature remains below T_C.

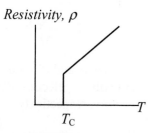

Superconductivity is a result of quantum effects. Much research has been done to find materials that superconduct at higher, more accessible temperatures to reduce the cost and inconvenience of refrigeration at the required very low temperature. Superconductivity at temperatures as high as 160 K has been reported in fragile compounds.

A major use of superconductors is to carry current in electromagnets: a common use for magnetic resonance imaging (MRI). These are conventional metal superconductors cooled by liquid nitrogen. These have advantages since non-superconducting electromagnets require much more energy, which is wasted as heat. Research is being done to use high T_C superconductors as wires for carrying currents for practical use.

*25–10 Electrical Conduction in the Nervous System

Messages are transmitted in the form of electric signals through the *neuron*, the basic element of the nervous system. Neurons are living cells that contain several small appendages, called *dendrites*, which are attached to the main cell body, and a long tail, called the *axon*. Signals are received by the dendrites and are transmitted along the axon. A signal, when it reaches the nerve endings, is transmitted to the next nerve or muscle at a connection called the *synapse*. The central nervous system constitutes the brain and the spinal cord. "Sensory neurons" carry messages from the eyes, ears, skin, and other organs to the central nervous system. "Motor neurons" carry signals from the central nervous system to muscles and direct them to contract. These two types of neurons constitute the "peripheral system."

Neurons contain K^+, Na^+, and Cl^- ions and there are large differences in the concentrations of these ions inside and outside a cell. As a result, the ions tend to diffuse across the cell membrane. Because of diffusion, both K^+

and Cl^- ions tend to charge the interior surface of the membrane negatively and the outside positively. In equilibrium, the tendency to diffuse is balanced by the electric potential difference across the membrane. When a neuron does not transmit a signal, this potential ($V_{inside} - V_{outside}$) is called the *resting potential*, which is about −60 mV to −90 mV.

When a stimulus (such as *thermal* from a hot source, *pressure* variation from eardrum, *electric* from the brain, etc.) occurs at a point, the membrane suddenly becomes more permeable to Na^+ ions than to K^+ and Cl^- ions. The Na^+ ion moves quickly into the cell and makes the inner surface of the cell positive. This creates a potential called the *action potential*, which is about +30 mV to +40 mV compared to the resting potential of about −70 mV. The adjacent negative charges are attracted toward this region, creating an action potential in their location. The process repeats and the action potential moves through the axon. The action potential lasts for about 1 ms and travels with a speed of about 30 m/s to 150 m/s.

Tips for Solving the Problems

1. The electric current, I, is a treated as positive quantity. When using the definition $\overline{I} = \Delta Q / \Delta t$ or $I = dQ/dt$ to determine current, use the magnitude of the charge, ΔQ.

2. Use of Ohm's law ($V/I = R$) and determination of power dissipated through a resistor is straightforward. Use Ohm's law and the expressions for power ($P = IV$) to solve problems in Sections 25–3 and 25–5.

3. The resistance of a wire increases with its length, L, but decreases with its cross-sectional area, A, since $R = \rho L/A$. The diameter of the wire is usually given in millimeters. Remember to convert all lengths to meters. Also, $\rho_t = \rho_0[1 + \alpha(T - T_0)]$. Use these tips to solve problems in Section 25–4.

4. Read the problem carefully to find out whether the given currents and voltages are their maximum values or their rms values in an ac circuit problem. If the maximum values are given, the rms values can be obtained by dividing by $\sqrt{2}$ (or multiplying by 0.707). If the rms values are given, maximum values can be obtained by multiplying by $\sqrt{2}$.

5. To calculate the average power, use the expression for dc power $\overline{P} = I V = I^2 R = V^2/R$ and replace I and V with their ac rms values. To find maximum power, use the maximum values of I and V. Tips 5 and 6 will be useful to solve problems in Section 25–7.

6. Equations $I = jA$, $j = nev_d$ will be useful to solve for the current density or drift velocity, etc. The equation $j = E/\rho = \sigma E$ will be useful to determine the electric field inside a wire.

CHAPTER 26
DC Circuits

This chapter introduces the basic concepts of direct-current electric circuits including series and parallel wirings of resistors and capacitors, Kirchhoff's rules, growth and decay of current in RC circuit.

Important Concepts
> Electromotive force
> Internal resistance
> Resistors in series
> Equivalent resistance
> Resistors in parallel
> Kirchhoff's rules
> *RC* circuit
> Time constant
> Ammeter
> Voltmeter
> Multimeter

26–1 EMF and Terminal Voltage

The potential difference between the terminals of a source (such as a battery) when no current flows through an external electric circuit is called its **electromotive force**, or *emf*, $\mathcal{E}$. The battery has some resistance, called its **internal resistance**, r. When a current I flows through the battery there is an internal drop in voltage equal to Ir, and hence the terminal voltage V_{ab} between the terminals a and b of the battery is

$$V_{ab} = \mathcal{E} - Ir \qquad (26-1)$$

The SI unit of electromotive force is the volt (V).

26–2 Resistors in Series and in Parallel

When resistors are connected one after another so that the *same current* flows through each of them, they are said to be in **series**.

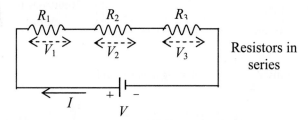

Resistors in series

For resistors R_1, R_2, and R_3 connected in series, the total voltage V is given by

$$V = V_1 + V_2 + V_3 = IR_1 + IR_2 + IR_3$$

$$= I(R_1 + R_2 + R_3) \qquad (26\text{–}2)$$

$$= IR_{eq}.$$

Thus, equivalent resistance, R_{eq}, is

$$R_{eq} = R_1 + R_2 + R_3 \qquad (26\text{–}3)$$

Clearly, the equivalent resistance is greater than the individual resistances.

When the resistors are connected across the *same potential difference*, they are said to be in **parallel**.
In this case, the current has three parallel paths through which to flow, and the total current is equal to the sum of the currents through the individual resistors:

$$I = I_1 + I_2 + I_3.$$

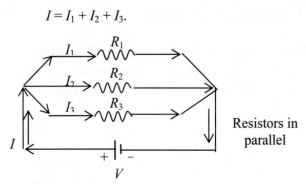

Resistors in parallel

The equivalent resistance, R_{eq}, is given by

$$\frac{1}{R_{eq}} = \frac{1}{R_1} + \frac{1}{R_2} + \frac{1}{R_3}. \qquad (26\text{--}4)$$

The equivalent resistance for resistors in parallel is less than the smallest resistance in the combination. Resistance is less because the current has additional paths to flow.

26–3 Kirchhoff's Rules

Kirchhoff rules are useful for analyzing complex electric circuits. The rules basically express conservation of charge (the junction rule) and conservation of energy (the loop rule) in a closed electric circuit.

Kirchhoff's first or junction rule

At any junction, the sum of the currents entering the junction must equal the sum of all currents leaving the junction.

Kirchhoff's second or loop rule

The algebraic sum (with proper sign) of all potential differences across each element around any closed path is zero.

Problem solving using Kirchhoff's rule:

- Label the current in each separate branch of the given circuit. Identify the unknown.

- Apply Kirchhoff's junction rule to one or more junctions.

- Apply Kirchhoff's loop rule for one or more loops. Follow each loop in one direction only. Note that the sign of the potential difference is *negative* if the chosen loop direction is the *same* as the chosen current direction through a resistor; otherwise the sign is positive. For a battery the potential difference is *positive* if the chosen loop direction is from the *negative terminal* toward the positive terminal; otherwise the sign is negative.

- Solve the resulting equations for unknowns.

26–4 Series and Parallel EMFs; Battery Charging

When two or more batteries are connected in series, the total voltage is the sum of the individual voltages. For example, in a flash light two 1.5-V batteries are connected in series to provide 3 V.

If the batteries are connected in reverse, the net voltage is the difference between two voltages. Such a reverse arrangement is needed to charge a battery. For example, if a 20 V battery is connected in reverse to charge a 12-V battery, because of the greater voltage of the 20-V battery, it forces charge back into the 12-V battery. An alternator keeps a car battery charged based on this principle.

Batteries are also connected in parallel. The arrangement is not used to increase voltage, but it provides more energy when large currents are needed.

26–5 Circuits Containing Resistor and Capacitor (*RC* Circuits)

Circuits containing both resistors and capacitors are called **RC circuits**. For a circuit containing a resistor, R, a capacitor, C, both connected in series with a battery of emf $\mathcal{E}$,

$$\mathcal{E} = RI + \frac{Q}{C} = R\frac{dQ}{dt} + \frac{Q}{C} \qquad (26\text{–}5)$$

The solution provides the charge on the capacitor as a function of time as:

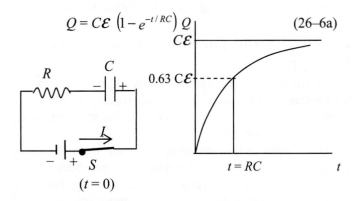

$$Q = C\mathcal{E}\left(1 - e^{-t/RC}\right) \qquad (26\text{–}6a)$$

The potential diference across the capacitor is:

$$V_C = \frac{Q}{C} = \mathcal{E}\left(1 - e^{-t/RC}\right), \qquad (26\text{–}6b)$$

and the current at any time t:

$$I = \frac{dQ}{dt} = \frac{\mathcal{E}}{R} e^{-t/RC}. \qquad (26\text{--}7)$$

The quantity RC has the dimension of time and is called the **time constant τ**,

$$\tau = RC. \qquad (26\text{--}8)$$

The variation of Q as a function of time is shown graphically.

The time constant is a measure of how quickly the capacitor becomes charged. When $t = \tau$, $Q = 0.63\ C\mathcal{E}$ and $V_C = 0.63\ \mathcal{E}$.

When a capacitor is already charged (to a voltage, say, V_0) and is allowed to discharge through a resistance R,

$$RI = \frac{Q}{C}, \quad \text{or,} \quad -R\frac{dQ}{dt} = \frac{Q}{C}$$

where $I = -dQ/dt$ is the discharging current.

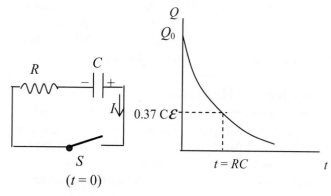

317

The charge on the capacitor decays *exponentially* according to the relation

$$Q = Q_0 e^{-t/RC}. \qquad (26\text{–}9)$$

Here Q_0 is the original charge on the capacitor. The current is

$$I = -\frac{dQ}{dt} = I_0 e^{-t/RC}. \qquad (26\text{–}10)$$

When $t = \tau$ ($= RC$), $Q = 0.37\,Q_0$, $V = 0.37 V_0$, and $I = 0.37\,I_0$.

* Applications of *RC* circuits

The charging and discharging of a capacitor is used to generate voltage pulses. An *RC* circuit coupled with a gas-filled switch can produce a repeating "sawtooth" voltage. An automobile turn signal indicator is an application of a sawtooth oscillator circuit. The windshield wiper of an automobile uses *RC* circuit where the time constant *RC* can be changed using a multipositioned switch.

An important medical use of *RC* circuits is the electronic heart *pacemaker*. These devices produce a regular voltage pulse that can start and control the frequency of the heartbeat. The electrodes are implanted in or close to the heart and the charge on the capacitor increases to a certain point and discharges; this repeats at a rate that depends on the values of *R* and *C*.

26–6 Electric Hazards

An electric current stimulates nerves and muscles and we feel a shock when the current is more than 1 mA. Currents above 10 mA cause severe contraction of muscles, and death from paralysis of the respiratory system can occur. When a current of more than 70 mA passes across the torso so that a portion of it passes through the heart at least for a second, heart muscles contract irregularly and blood is not pumped properly, a condition called "ventricular fibrillation." When it lasts longer, death results.

The resistance between two points on the opposite sides of a dry human body is about 10^4 to 10^6 Ω. If the skin is wet, the resistance is 10^3 Ω or less and, if a person in good contact with the ground touches a 120 V dc line, he suffers a current: 120 V/1000 Ω = 120 mA, which flows through his body to the ground. Obviously, this current could be lethal.

If the wire inside an electric appliance breaks its insulation and a person touches the case, the person could get a severe shock unless the case is grounded. If the case is grounded, most current flows through the low resistance ground wire instead of through the person. Grounding a metal case is done by a separate wire connected to a third round prong of a 3-prong plug and it is also done by connecting the case to the larger prong of a "polarized" 2-prong plug.

The human body can be modeled as a resistance and a capacitor in parallel, and dc current flows only through the resistance. But an ac current, since it passes through both the resistance and capacitor, for a given V_{rms} will be greater than for the same dc voltage. An ac current is, therefore, more dangerous than the same dc voltage.

*26–7 Ammeters and Voltmeters

Ammeters and **voltmeters** are used to determine current and voltage (potential difference), respectively, in an electric circuit. These can be (1) *analog* or (2) *digital* meters.

*Analog Ammeters and Voltmeters

The main component of an analog meter is a galvanometer, whose deflection of the needle is proportional to the current. An ammeter consists of a galvanometer in parallel with a resistor called a **shunt resistor**. A voltmeter consists of a galvanometer in series with a high resistor.

A single device that can function as an ammeter, a voltmeter, or an ohmmeter, is called a **multimeter**. An **ohmmeter** measures resistance. It contains a battery of known voltage in series with a resistor and an ammeter. The unknown resistance is connected to complete the circuit.

*How to Connect Meters

An ammeter is connected in series with the section. The resistance of an ammeter must be very low so that it does not alter the current to be measured.

A voltmeter is connected *in parallel* with the section. The resistance of a voltmeter must be very high; and otherwise, it will draw a significant amount of current and alter the potential difference it is intended to measure.

*Effects of Meter Resistance

The resistance of a meter can affect the result of a measurement. If the resistance of an ammeter is much less and if the resistance of a voltmeter is much higher than the resistance of the circuit the readings can only then be trusted.

*Digital Meters

The digital meters do not use galvanometers. The electronic circuitry and digital readout are more sensitive and have less effect on the circuit to be measured compared to analog meters. The internal resistance of a digital voltmeter is very high—about 10-100 MΩ. Digital meters are connected in the same way as analog meters.

Tips for Solving the Problems

1. A typical mistake made in calculating the equivalent resistance of two resistors in parallel is to compute $1/R_1$ and $1/R_2$ and then to add and present the sum as R_{eq} rather than as $1/R_{eq}$. Note that the sum of the reciprocals is not equal to the reciprocal of the sum. That is, $1/R_1 + 1/R_2$ is not equal to $1/(R_1 + R_2)$.

2. The equivalent resistance of a number of resistors in parallel is smaller than the smallest of the individual resistors. For a complex combination of resistors, reduce them to a single resistor by repeatedly applying the series and parallel laws for resistors. Keep these in mind when solving problems in Section 26–2.

3. Apply Kirchhoff's junction rule and loop rule to solve the problems in Section 26–3. Use the problem solving strategy mentioned in Section 26–3 when solving problems.

4. The rule of addition of capacitors *in series* is similar in form to the rule of addition of resistors *in parallel*. Also, the rule of addition of capacitors *in parallel* is similar in form to the rule of addition of resistor *in series*.

5. The time constant $\tau (= RC)$ has the dimension of time. If C is in farads and R is in ohms, τ will be in seconds.

6. The voltage and charge in a RC circuit vary as $V = \mathcal{E} \left(1 - e^{-t/RC}\right)$ and $Q = C\mathcal{E} \left(1 - e^{-t/RC}\right)$, respectively. The discharge of a charged capacitor is given by $Q = Q_0 e^{-t/RC}$. Use these expressions to solve the problems in Section 26–5.

CHAPTER 27
Magnetism

This chapter describes magnetism and the relationship between magnetism and electricity. In particular, it describes the force acting on a charge moving in a magnetic field, force and torque acting on a current-carrying coil in a magnetic field, force between two current carrying wires, and magnetic effects of an electric current.

Important Concepts
Magnetic field
Magnetic field lines
Magnetic declination
Angle of dip
Right hand rule
Force on a current carrying wire
Tesla and Gauss
Force on a charge moving in a magnetic field
Cyclotron frequency
Lorentz equation
Torque on a current loop
Magnetic dipole moment
Motor, loudspeaker, galvanometer
The Hall effect
Mass spectrometer

27–1 Magnets and Magnetic Field

A magnet attracts objects made of iron such as paper clip, nails, etc. Any magnet has two ends, called **poles**, where the magnetic effect is strongest. A freely-suspended magnet always points in the north–south direction. The end of the magnet pointing toward the north is called the **north pole**, and the end pointing toward the south is called the **south pole**. The same poles of two magnets repel each other, whereas opposite poles attract.

Magnets always have two poles. A magnetic monopole (a single north or south pole) was not observed. This behavior is different from that of electric charges, as two types of charges can exist separately.

Matrials such as iron, cobalt, and nickel that show strong magnetic effects are said to be **ferromagnetic**.

Just as an electric charge creates an electric field, $\vec{E}$, a magnet creates a **magnetic field**, $\vec{B}$. A magnetic field has magnitude and direction. The *direction* of the magnetic field at a point is given by the direction that the north pole of a magnetic needle would point when placed at that point. Like an electric field, a magnetic field can be represented by its **magnetic field lines,** so that (1) the direction of the magnetic field is tangent drawn to the magnetic field line at any point, and (2) the number of field lines per unit area is proportional to the magnitude of the magnetic field. Magnetic field lines can be visualized by placing a sheet of paper on top of a bar magnet and dropping iron filings onto the paper. The iron filings align with the field lines. The magnetic field lines leave the north pole of a magnet and enter at the south pole. The field lines continue within the body of the magnet. Unlike electric fields, magnetic fields always form closed loops.

Earth's Magnetic Field

Earth has its own magnetic field, which is similar to that of a giant bar magnet inside the Earth with its poles near the Earth, geographic poles but with its axis tilted from Earth's rotational axis that passes through the geographic poles. The north magnetic pole of the Earth is about 1300 km from the geographic north pole. The geographic north pole of the Earth is the south pole of the Earth's magnetic field.

The angular difference between the magnetic north (as indicated by a compass) and true geographical north is known as the **magnetic declination**. Also, the angle made by the Earth's magnetic field with the horizontal at any point is known as the **angle of dip**.

Uniform Magnetic Field

A magnetic field is called uniform if it does not change in magnitude or direction from one point to another. The field between two parallel pole pieces of a magnet is uniform in the central region of the gap, if the area of the pole faces is large compared to their separation.

27–2 Electric Currents Produce Magnetic Fields

A current-carrying conductor produces a magnetic field. This was first discovered by Oersted in 1820 who noticed that as soon as a current was established in a wire, a nearby compass needle was deflected from its usual orientation. This key observation unified the theories of electricity and magnetism.

A compass needle placed close to a current-carrying straight conductor experiences a force, causing the needle to align tangent to a circle around the wire. Thus, the magnetic field lines are in the form of circles for the conductor, as shown.

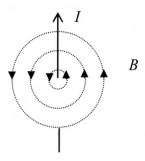

The direction of the magnetic field is given by the **right-hand rule**: *Point the thumb of your right hand in the direction of the conventional current and curl the fingers into the shape of a half-circle. The tip of the fingers will point in the direction of the magnetic field.*

27–3 Force on an Electric Current in a Magnetic Field; Definition of $\vec{B}$

Since a current carrying conductor exerts a force on a magnet, according to Newton's third law, a *magnet must exert a force on a current carrying conductor.* The direction of the force is found to be perpendicular to both the direction of the current and to the direction of the magnetic field. The direction of the force is given by the right hand rule:

Orient your right hand so that your outstretched fingers point along the direction of the current I, and when you bend your fingers they must point toward $\vec{B}$, then your thumb will point in the direction of the force, $\vec{F}$.

$\vec{F}$

Thumb points in the direction of the force

$\vec{B}$

I

Direction of curled fingers of right hand

The magnitude of the force on a wire of length l carrying a current I at an angle θ with the direction of a magnetic field, $\vec{\mathbf{B}}$, is given by

$$F = ILB \sin \theta \qquad (27\text{--}1)$$

- F is maximum when $\theta = 90°$; $F = 0$ when $\theta = 0°$ or $180°$.

$$F_{\max} = IlB \qquad (27\text{--}2)$$

- In the notation of vector cross product:

$$\vec{\mathbf{F}} = I\vec{l} \times \vec{\mathbf{B}} \qquad (27\text{--}3)$$

In a situation where $\vec{\mathbf{B}}$ is not uniform,

$$\vec{\mathbf{F}} = \int d\vec{\mathbf{F}} = \int I \, d\vec{l} \times \vec{\mathbf{B}} \qquad (27\text{--}4)$$

where $d\vec{\mathbf{F}}$ is the infinetesimal force acting on a differential length $d\vec{l}$ of the wire.

- The magnitude of the magnetic field, B, as defined by Equation (27–2) is, $B = F_{\max}/Il$, where $F_{\max}$ is the magnitude of the force on a straight length l of wire carrying current I when the wire is perpendicular to B.

- The SI unit of B is the **tesla** (T). $1 \text{ T} = \text{N}/(\text{A} \cdot \text{m})$. The tesla is a large unit. Another unit of magnetic fields is the **gauss** (G). $1 \text{ G} = 10^{-4} \text{ T}$.

27–4 Force on an Electric Charge Moving in a Magnetic Field

Since a conductor carrying electric current (which consists of moving electric charges) experiences a force in a magnetic field, it is expected that a moving charge experiences a force $\vec{F}$ in a magnetic field, $\vec{B}$.

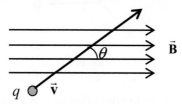

The force experienced by a charge q moving at a velocity $\vec{v}$ can be written as a *vector cross product*:

$$\vec{F} = q\ \vec{v} \times \vec{B} \qquad (27\text{–}5a)$$

- The magnitude of F is
  $$F = qvB \sin \theta. \qquad (27\text{–}5b)$$
  where θ is the angle between $\vec{v}$ and $\vec{B}$.

- If $\theta = 0°$ or $180°$, $F = 0$. Also, if $v = 0$, $F = 0$. F is maximum when $\theta = 90°$.

  $$F_{max} = qvB.$$

- The direction of $\vec{F}$ is perpendicular to both $\vec{v}$ and $\vec{B}$. The direction can be obtained from the *right-hand rule*: *Orient your right hand so that your outstretched fingers point along the direction of the current $\vec{v}$, and when you bend your fingers they must point toward $\vec{B}$, then*

*your thumb will point in the direction of the
force,* $\vec{F}$.

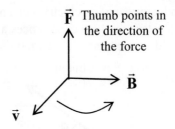

$\vec{F}$ Thumb points in
the direction of
the force

$\vec{B}$

$\vec{v}$

Direction of curled
fingers of right hand

- If q is negative, the direction of F is opposite to
  the direction given by the preceding rule.

The path of a charged particle moving in a plane
perpendicular to a uniform magnetic field is a circle. In
this case the radius of the circle can be obtained from the
relation $qvB = mv^2/r$. The frequency of rotation f, called the
cyclotron frequency, is given by

$$f = \frac{1}{T} = \frac{qB}{2\pi m} \qquad (27–6)$$

*Aurora Borealis

Charged particles from the sun enter the Earth atmosphere;
the velocity component of each particle perpendicular to
the Earth's magnetic field makes a circular orbit around
the field lines while the velocity component parallel to the
field carries the particles along the field lines toward the
poles. The high concentration of these particles ionizes the
air and as the electrons recombine, light is emitted, which
is called **aurora borealis** or northern lights in northern
latitudes.

Lorentz Equation

If a charge moves in the presence of both magnetic and electric fields, the force on the charge can be expressed by **Lorentz equation**:

$$\vec{F} = q\,(\vec{E} + \vec{v} \times \vec{B})\qquad (27\text{--}7)$$

- If a uniform electric field and a uniform electric field are arranged so that they are at right angles to each other and the net force on a moving charge in the arrangement is zero,

$$v = \frac{E}{B}\qquad (27\text{--}8)$$

27–5 Torque on a Current Loop; Magnetic Moment

A current-carrying loop when placed in a magnetic field experiences a torque. The torque, τ, exerted on N rectangular loops of area A carrying current I is

$$\tau = NIAB \sin\theta \qquad (27\text{--}9)$$

where B is the magnitude of the magnetic field, and θ is the angle between the magnetic field and the normal to the loop. Torque is maximum when $\theta = 90°$ and is zero when $\theta = 0°$ or $180°$. The SI unit of τ is $N \cdot m$.

The product NIA is called the **magnetic dipole moment** (vector) of the loop.

$$\vec{\mu} = NI\,\vec{A} \qquad (27\text{--}10)$$

In vector form,

$$\vec{\tau} = NI\,\vec{A} \times \vec{B} = \vec{\mu} \times \vec{B} \qquad (27\text{--}11)$$

Just as an electric dipole has a potential energy in an electric field, a magnetic dipole in a magnetic field has potential energy. The expression for potential energy U is similar in both cases and for a magnetic dipole

$$U = -\mu B \cos\theta = -\vec{\mu} \cdot \vec{B} \qquad (27-12)$$

*27–6 Applications: Motors, Loudspeakers, Galvanometers

*Electric Motors

An electric motor converts electric energy to mechanical energy. It works on the same principle as a galvanometer. But, for motors there is no spring so that the coil rotates continuously. The coil is larger and is mounted on a cylinder called the **rotor** or **armature**. The armature is mounted on a shaft or axle. To turn the motor continuously in one direction an alternation of current is necessary. In a dc motor this is done by using **commutators** and **brushes**. The brushes are stationary contacts that rub against the conducting commutators mounted on the shaft. At every half rotation the commutators change connections and the current in the coil reverses. An ac motor works without the use of commutators since the current itself alternates.

*Loudspeakers

In a loudspeaker a permanent magnet is mounted directly in line with a coil of wire, which is attached to the speaker cone. As ac current of an audio signal flows through the coil, the coil and the attached speaker cone experience a force due to the magnetic field of the magnet. The speaker cone moves back and forth with the frequency of the audio signal generating the original sound in air.

*Galvanometer

Galvanometers are the basic components of most meters including ammeters, voltmeters, and ohmmeters.

A galvanometer consists of a rectangular coil of wire (with an attached pointer) suspended in the magnetic field of a permanent magnet. As current flows through the coil it exerts a torque on the coil. This torque is opposed by a spring, which exerts a torque proportional to the angle ϕ through which the coil is turned. The deflection ϕ of the pointer is proportional to the current flowing through the coil.

27–7 Discovery and Properties of the Electron

When a high voltage was applied between two electrodes placed at the opposite ends inside a glass tube at very low pressure of air, rays were found to be emitted from the cathode and traveled through the opposite end of the tube. Since they originated from a cathode these were named **cathode rays**. Cathode rays consist of charged particles, called electrons, and they can be deflected by electric and magnetic fields. The direction of the deflection suggests that the charge of the electron is negative. An electron is a fundamental constituent of an atom.

As an electron moves in a direction perpendicular to a magnetic field, it follows a circular path because of the magnetic force that it experiences, so that

$$evB = \frac{mv^2}{r}$$

or $$\frac{e}{m} = \frac{v}{Br}.$$

The radius r and the magnetic field B can be measured. If the electron passes also through an electric field so that the force on the electron due to the electric and magnetic fields are equal and opposite, then it passes undeflected through the fields. In this case,

$$eE = evB \qquad \text{or} \qquad v = \frac{E}{B}$$

This gives,

$$\frac{e}{m} = \frac{E}{B^2 r}. \qquad (27\text{--}13)$$

This equation is used to determine the charge to mass ratio e/m of an electron. The accepted value is $e/m = 1.76 \times 10^{11}$ C/kg.

   Millikan determined the value of the electronic charge e by **oil-drop method**. In the method, tiny droplets of mineral oil carrying an electric charge (q) were allowed to fall under gravity between two plates. The electric field E between the plates was adjusted so that the downward force of gravity is balanced by the upward electric force on the droplet. This gives $qE = mg$ or $q = mg/E$. The mass m of the droplet was determined by measuring its terminal velocity in absence of the electric field. The result showed that the charge was an integral multiple of smallest charge, e, that was ascribed to an electron. The accepted values $e = 1.602 \times 10^{-19}$ C and $m = 9.11 \times 10^{-31}$ kg.

*27–8 The Hall Effect

When a current-carrying conductor is held in a magnetic field $\vec{B}$, the field exerts a sideways force (given by $\vec{F} = q\ \vec{v} \times \vec{B}$) on the moving charges and, as a result, the charges tend to move nearer to one face than the other. This builds up a potential difference between the two faces until the electric force $q\vec{E}$ becomes equal and opposite to the magnetic force. This effect is called the **Hall effect**, and the difference in potential is called the **Hall emf**.

For an electron moving to the right in a rectangular conductor in the presence of an inward magnetic field,

$$eE_H = ev_dB \quad \text{or} \quad E_H = v_dB$$

(E_H = electric field, v_d = drift velocity of electrons)

$$\text{The Hall emf } \mathcal{E}_H = E_H\,l = v_dB\,l, \qquad (27\text{–}14)$$

where l is the width of the conductor.

The direction of the Hall emf reveals that negative charges conduct electricity in metal conductors. Hall effect is used to determine the magnetic field B or the drift velocity when B is known.

*27–9 Mass Spectrometer

A mass spectrometer is used to determine masses of atoms accurately. In a spectrometer ions, after they are produced, enter a region where there are both electric and magnetic fields. The electric force is balanced by the magnetic force on the ion, if $qE = qvB$, or $v = E/B$. Only ions moving at speed v then enter a second region where there is only a magnetic field B' so that the ions follow a circular path, whose radius is given by $qvB' = mv^2/r$.

From the preceding equations, $m = \dfrac{qB'r}{v} = \dfrac{qBB'r}{E}$. The mass m can be determined using this relation.

Since the mass is proportional to the radius, spectrometers can be used to separate different elements, isotopes, and different molecules.

Tips for Solving the Problems

1. To solve problems involving magnetic force on a current-carrying wire such as those in Section 27–3 use the expression $F = ILB \sin \theta$ or $\vec{F} = \int I \, d\vec{l} \times \vec{B}$

    as appropriate. Here θ is the angle between the wire of length L and the magnetic field, B. B will be in Tesla, if F is in newtons, l in meters, and I in amperes.

2. The magnetic force on a charge is zero if the charge is at rest. A magnetic force occurs only when a charged particle moves in a direction different from the direction of the magnetic field lines.

3. Review the concepts of cross product of two vectors and how to expand a cross product in terms of the components of the two vectors. These will be very useful in this chapter since magnetic force and magnetic torque vectors are written as cross products.

4. The magnetic force on a charged particle is $\vec{F} = q \, \vec{v} \times \vec{B}$. The magnitude of the force is $F = qvB \sin \theta$. When both electric and magnetic fields are present $\vec{F} = q \, (\vec{E} + \vec{v} \times \vec{B})$. Use these equations to solve problems in Section 27–4.

5. Many equations in this chapter depend directly on the charge q on a particle. Make sure that the sign of the result corresponds to the sign on the charge.

6. The magnetic force right-hand rule provides the direction of magnetic force for a positively charged particle. The direction of the force will be opposite if the charge on the particle is negative.

7. The torque vector on a current carrying loop is written as a cross product $\vec{\tau} = NI\vec{A} \times \vec{B} = \vec{\mu} \times \vec{B}$ where $\vec{\mu}$ is called the dipole moment vector. The magnitude of the torque is $\tau = NIAB \sin \theta$. Recall that here θ is the angle between the magnetic field and the *normal* to the loop. (Note that the torque does not depend on the shape of the area.) The potential energy U of a dipole in a magnetic field is $U = -\mu B \cos \theta = -\vec{\mu} \cdot \vec{B}$. These expressions will be useful to solve problems in Section 27–5.

8. If a charge e is initially moving perpendicularly to a magnetic field B, it moves into a circular orbit of radius r so that its $evB = mv^2/r$. If it goes straight in a region where the magnetic and the electric fields are crossed, $eE = evB$.

9. The Hall emf $\mathcal{E}_H = v_d B \, l$, where v_d is the drift velocity, B is the magnetic field and l is the width of the conductor. Also note that $I = nev_d A$, where A is the cross-sectional area through which the current I flows and n is the density of free electrons. These expressions will be useful to solve problems in Section 27–8.

10. In a mass spectrometer a charge q first goes straight in crossed electric and magnetic fields, and its speed is $v = E/B$. The charge then moves in a circular orbit in a second magnetic field B', so that $qvB' = mv^2/r$. Problems about mass spectrometer can be solved from these two basic equations.

CHAPTER 28
Sources of Magnetic Field

This chapter describes magnetism and the relationship between magnetism and electricity. In particular, it describes the magnetic field due to a straight wire, force between two current-carrying wires, electromagnet, ferromagnetism, paramagnetism and diamagnetism.

Important Concepts

> Magnetic field due to a wire
> Force between two current-carrying wires
> Ampere
> Ampère's law
> Magnetic field of a solenoid
> Biot-Savart Law
> Ferromagnetism
> Domains
> Curie temperature
> Electromagnet
> Solenoid
> Magnetic permeability
> Hysteresis
> Paramagnetism and diamagnetism

28–1 Magnetic Field Due to a Straight Wire

Experiments show that the strength of a magnetic field due to a long, straight, current-carrying wire is directly proportional to the current, I, and inversely proportional to the distance, r. That is,

$$B \propto \frac{I}{r}.$$

The proportionality constant is $\mu_0/2\pi$.

Thus,

$$B = \frac{\mu_0 I}{2\pi r} \qquad (28\text{--}1)$$

The constant $\mu_0 = 4\pi \times 10^{-7}$ T · m/A is called the **permeability of free space**.

28–2 Force between Two Parallel Wires

A current-carrying wire experiences a force in a magnetic field. Also, a current-carrying wire generates a magnetic field. Thus, one current-carrying wire must exert a force on another. The magnitude of the force per unit length between two wires carrying currents I_1 and I_2 separated by a distance d is

$$\frac{F}{l} = \frac{\mu_0 I_1 I_2}{2\pi d} \qquad (28\text{--}2)$$

The force is *attractive* if the currents flow in the *same direction* (parallel currents), and *repulsive* if the currents flow in the *opposite directions* (antiparallel currents).

28–3 Definitions of the Ampere and Coulomb

Using Equation (28–2) one **ampere** is defined as the current flowing in each of two long parallel wires 1 m apart, which results in a force of exactly 2×10^{-7} N/m of length of each wire. 1 C = 1 A $\cdot$ s.

28–4 Ampère's Law

Ampère's law provides a way to calculate magnetic fields due to currents. According to this law,

$$\Sigma B_\parallel \, \Delta L = \mu_0 I \text{ enclosed}.$$

where $B_\parallel$ is the component of the magnetic field parallel to a segment of a closed path of length ΔL. In the limit $\Delta l = 0$, this relation becomes,

$$\oint \vec{B} \cdot d\vec{l} = \mu_0 I \text{ enclosed}, \qquad (28–3)$$

where $d\vec{l}$ is an infinetesimal length vector. Ampère's law is one of the fundamental laws of electricity and magnetism. The law is valid in general. $\vec{B}$ is the field at each point along the path chosen due to all sources (including the current I enclosed by the path and also due to other sources).

Fields Due to a Straight Wire

Applying Ampère's law for the magnetic field due to a long, straight current-carrying conductor, we get,

$$\oint \vec{B} \cdot d\vec{l} = B \oint dl = B \,(2\pi r) = \mu_0 I. \text{ Thus,}$$

$$B = \frac{\mu_0 I}{2\pi r}$$

28–5 Magnetic Field of a Solenoid and a Toroid

A long coil of wire is called a solenoid. When current flows, each loop produces a magnetic field. For a long solenoid with closely packed coils, the fields add up to produce a nearly uniform magnetic field parallel to the axis of the solenoid, whereas the field outside the solenoid is very small compared to the field inside (except near the ends).

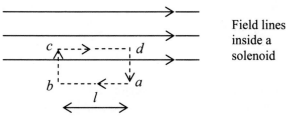

Field lines inside a solenoid

Considering a closed path *abcda*,

$$\oint \vec{\mathbf{B}} \cdot d\vec{l} = \int_a^b \vec{\mathbf{B}} \cdot d\vec{l} + \int_b^c \vec{\mathbf{B}} \cdot d\vec{l} + \int_c^d \vec{\mathbf{B}} \cdot d\vec{l} + \int_d^a \vec{\mathbf{B}} \cdot d\vec{l} \, .$$

$$= \int_c^d \vec{\mathbf{B}} \cdot d\vec{l} = Bl$$

since field is nearly zero for the first integral and perpendicular to the segments in the two other integrals. Here *B* is the field inside the solenoid of length *l*. Thus, and applying Ampère's law we get,

$$Bl = \mu_0 NI, \qquad (N \text{ is the number of loops})$$

$$B = \mu_0 NI/l. \qquad (28\text{–}4)$$

The field outside a solenoid is similar to a bar magnet and a current-carrying solenoid behaves like a magnet, each end acts like two poles of the magnet.

28–6 Biot-Savart Law

The Biot-Savart law expresses the magnetic field $d\vec{B}$ at a point P distant r from the small element $d\vec{l}$ of the wire along which the current I is flowing by means of the equation

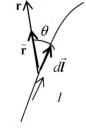

$$d\vec{B} = \frac{\mu_0 I}{4\pi} \frac{d\mathbf{l} \times \hat{\mathbf{r}}}{r^2}, \qquad (28\text{–}5)$$

where $\vec{r}$ is the displacement vector from the element $d\vec{l}$ of the wire to the point P, $\hat{\mathbf{r}}$ is the unit vector along $\vec{r}$, and μ_0 is a constant.

The magnitude of the field is

$$dB = \frac{\mu_0 I \; dl \; \sin\theta}{4\pi r^2}, \qquad (28\text{–}6)$$

where θ is the angle between $d\vec{l}$ and $\vec{r}$ and the direction of the field is given by the right-hand rule of the cross product.

The total magnetic field at the point P due to the wire $\vec{B} = \int d\vec{B}$.

28–7 Magnetic Materials—Ferromagnetism

Iron and other materials that can be used to make strong magnets are called **ferromagnetic**.

The origin of magnetic fields is circulating electric currents due to the motion of electrons in atoms. An electron orbiting the nucleus of an atom behaves like a tiny loop of current generating a tiny magnetic field. The spinning motion of an electron also gives rise to a magnetic field. For most substances the net magnetic field due to the orbital and spinning motions of different electrons is nearly zero, and the substance is nonmagnetic.

Ferromagnetic materials have magnetic fields due to the *spinning motion* of electrons that are nonzero. These materials have regions, called **domains**, of the order of 1 mm or less within the material in which electron spins are naturally parallel to each other. Each domain has a strong magnetic field, but the domains are oriented randomly, and the net magnetic effect is small. When the material is placed in an external magnetic field, the magnetic domains become aligned. They remain aligned for the most part, and the material becomes permanently magnetized. All ferromagnets lose their magnetic field if the temperature is high enough to orient the atoms in random directions. This temperatue is called the **Curie temperature** (1043 K for iron) and above this temperature a magnet cannot be made at all.

*28–8 Electromagnets and Solenoids— Applications

A **solenoid** is a long coil of wire wound into a succession of closely spaced loops. The magnetic field inside a solenoid is nearly uniform and directed along the axis of the solenoid. The field can be fairly strong since it is the sum of the fields due to each loops. A solenoid acts like a bar magnet with one end as north and the other as south pole. The magnetic field outside the solenoid is nearly zero. The magnetic field of a solenoid of length l having N loops and carrying current I is

$$B = \mu_0 (N/l) I \qquad (28\text{–}7)$$

The magnetic field of a solenoid can be increased by orders of magnitude by placing a piece of iron inside. Such an iron-core solenoid is called an **electromagnet**. Unlike the magnetic field of a permanent magnet, the magnetic field of a solenoid can be turned on or off. Solenoids are used as switches in many devices.

Electromagnets are used to generate high magnetic fields, such as in motors and generators. For some applications iron cores are not present and the current-carrying wires are made of superconducting materials and kept below the transition temperature to produce very high magnetic fields. Solenoids and electromagnets are used in doorbells, where, when the circuit is closed by pushing the button, the electromagnet exerts a force on a rod which strikes the bell.

*Magnetic Circuit Breakers
Modern circuit breakers contain a magnetic sensor so that, if the current exceeds a certain level, the magnetic field in it pulls an iron plate that breaks the circuit.

*28–9 Magnetic Fields in Magnetic Materials; Hysteresis

If a piece of iron or any other ferromagnetic material is placed inside a solenoid, the resulting magnetic field $\vec{\mathbf{B}}$ is the sum of that due to the current ($\vec{\mathbf{B}}_0$) and that due to the iron ($\vec{\mathbf{B}}_M$). This is because the domains in the iron become preferentially aligned by the external field. Thus,

$$\vec{\mathbf{B}} = \vec{\mathbf{B}}_0 + \vec{\mathbf{B}}_M \qquad (28\text{–}8)$$

In this case, the total field can be written as:

$$B = \mu n I, \qquad (28\text{–}9)$$

where μ is the characteristic of the material, called the **magnetic permeability** of the material. For ferromagnetic material, μ is much greater than μ_0. For other materials, μ is close to μ_0.

*Hystersis

In a ferromagnetic material, if the applied magnetic field (B_0) is increased, the total magnetic field (B) increases until all domains are aligned and the material reaches **saturation** (represented by b).

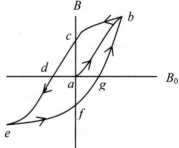

When the applied field is decreased to zero the residual magnetism is retained in the material (represented by the point *c*). As the applied magnetic field is reversed, the material approaches saturation in the opposite direction (point *e*). If the applied field is then reduced to zero and then increased in the original direction, the total field follows a curve *efgb*. The curve *bcdefgb* is called the **hystersis loop** and the fact that the curves do not retrace themselves on the same path is called **hystersis**. At points *c* and *f* the iron core is magnetized even though there is no current in the coils. For permanent magnets *ac* and *af* should be as large as possible and they are said to have high **retentivity**.

*28–10 Paramagnetism and Diamagnetism

Relative permeability K_m is defined as $K_m = \mu/\mu_0$ and **magnetic susceptibility** is defined as $\chi_m = K_m - 1$. Nonferromagnetic materials falls into two primary classes: *paramagnetic*, for which $K_m > 1$ and $\chi_m > 0$, and *diamagnetic* for which $K_m < 1$ and $\chi_m < 0$.

Paramagnetic materials have permanent magnetic dipole magnets. In the absence of external field, net magnetic effect is zero since the molecules are randomly oriented. In an external field, the magnetic dipoles experience torque and tend to align themselves parallel to the field. As a result, total magnetic field is slightly greater than B_0. The magnetic dipole moment per unit volume is called the magnetization vector (M) and for a paramagnetic substance, according to *Curie's law*,

$$M = C\,\frac{B}{T},$$

where B is external field (tending to align dipoles), T is Kelvin temperature (tending to randomize dipole directions), and C is a constant. Ferromangetic substances become paramagnetic above a characteristic temperature, called the *Curie temperature* (1043 K for iron).

Diamagnetic materials do not have permanent magnetic dipole magnets. In a magnetic field, the induced magnetic dipole moment acts in opposition to the applied field. Thus, the total field is slightly less than the external field. The external field increases the speed of the orbiting electrons in the atom moving in one direction and decreases speed for electrons moving in the other direction; the net result is a net dipole moment opposing the external field. Diamagnetism is present in all materials.

Tips for Solving the Problems

1. The magnetic field due to a long straight current-carrying wire is $B = \mu_0 \, I/2\pi r$. B will be in Tesla if I is in amperes and r is in meters.

2. The force between two current-carrying wires can be found using the relation $F = \dfrac{\mu_0 I_1 I_2}{2\pi \, d} \, l_2$. Remember that the force is attractive if the currents flow in the same direction and repulsive if the currents flow in opposite directions.

3. It may not be possible to remember all the equations in this chapter for the magnetic fields produced by currents, but try to remember Ampère's law $\oint \vec{\mathbf{B}} \cdot \vec{dl} =$

$\mu_0 I$ enclosed. The expressions for magnetic fields can be easily derived if you remember this law.

4. When applying Ampère's law, first determine the direction of the magnetic field from symmetry arguments and from the fact that field lines are closed. Choose an integration path that has the symmetry. Search for paths where B has constant magnitude along the entire path or along segments of the path.

5. The magnetic field B inside a solenoid is uniform and parallel to the axis of the solenoid. $B = \mu_0 NI/l$ where N is the number of loops, l is the length of the solenoid and I is the current. For an ideal solenoid the magnetic field outside the solenoid is zero. Tips 3–5 will be useful in solving problems in Sections 28–4 and 28–5.

6. To calculate magnetic field using the Biot-Savart law, first determine the magntitude of the elementary

   magnetic field $dB = \dfrac{\mu_0 I \; dl \; \sin\theta}{4\pi r^2}$ due to an element of

   current, where θ is the angle between $d\vec{l}$ and $\vec{r}$. Determine the total magnetic field by integrating over all the current elements.

7. When a ferromagnetic material is present inside a solenoid, the inside magnetic field, $B = \mu n I$, where μ is the permeability of the material, n is the number of turns per unit length, and I is the current.

CHAPTER 29
Electromagnetic Induction and Faraday's Law

This chapter describes magnetic flux and Faraday's magnetic induction, inductance, and ac circuits. The practical applications of magnetic induction such as electric generators and transformers are also described.

Important Concepts

> Induced current
> Electromagnetic induction
> Magnetic flux
> Weber
> Faraday's law of induction
> Lenz's law
> Electric generators
> Alternator
> Back emf
> Transformer
> Microphone
> Seismograph
> Ground Fault Current Interrupter

29–1 Induced EMF

Oersted discovered that an electric current generates a magnetic field. Faraday discovered the reverse effect.

Faraday found that when a magnet is quickly moved into a closed coil or is quickly removed, a current is induced in the coil in both the cases but in the opposite directions. If the magnet is held steady and the coil is moved toward or away, again current is induced in the coil. Such a current is called **induced current**. Thus **a**

changing magnetic field produces an induced emf. This phenomenon is called **electromagnetic induction**.

29–2 Faraday's Law of Induction; Lenz's Law

If the magnetic field $\vec{B}$ crosses an area A at an angle θ with the normal to the area, the **magnetic flux** through the area is given by

$$\Phi_B = BA \cos \theta = \vec{B} \cdot \vec{A} \qquad (29-1a)$$

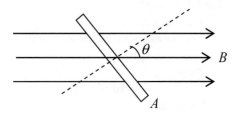

$\Phi = 0$ when $\theta = 90°$.
Φ is maximum ($= BA$), when $\theta = 0°$.

If $\vec{B}$ is not uniform,

$$\Phi_B = \int \vec{B} \cdot d\vec{A} \qquad (29-1b)$$

The SI unit of magnetic flux is the weber (Wb). 1 Wb = 1 T·m².

The basic feature of magnetic induction can be expressed using the concept of magnetic flux. Whenever magnetic flux through a closed coil changes, an induced emf is induced in the coil. The induced emf is proportional to the rate of change of magnetic flux with time, $d\Phi_B/dt$.

$$\mathcal{E} = -N \frac{d\Phi_B}{dt} \qquad (29-2)$$

where N is the number of loops in the coil. This is known as **Faraday's law of induction**.

Lenz's law gives a physical meaning to the negative sign in Faraday's law of induction. According to Lenz's law, *an induced emf always gives rise to a current whose magnetic field opposes the original change in flux.*

For example, when a magnet is brought close to a coil the magnetic flux increases. The direction of the current in the coil will be such that the magnetic field produced by the induced current in the coil will tend to oppose the increase in flux. Lenz's law is a consequence of the law of conservation of energy.

In most situations, the magnitude of the induced emf is calculated by using Faraday's law, and its direction is determined by using Lenz's law.

29–3 EMF Induced in a Moving Conductor

Induced emf generated due to the motion of a rod is called motional emf. If a conducting rod of length l that can slide horizontally without friction on a U-shaped wire is moved at right angles to a constant magnetic field $\vec{B}$ with a speed v, the area of the loop, and hence the flux passing through the loop changes with time. The change of flux, $\Delta\phi_B$, through the loop in time Δt is $Blv\,\Delta t$.

The induced emf (*motional emf*) in the conductor is

$$\mathcal{E} = Blv \qquad (29\text{–}3)$$

The emf equals the work done in moving the rod per unit charge.

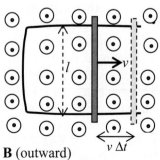

B (outward)

29–4 Electric Generators

An **electric generator** or **dynamo** transforms mechanical energy into electric energy. In an electric generator a closed coil containing many turns of wire wound on an armature is rotated in a constant magnetic field $\vec{\mathbf{B}}$. As the coil rotates, it changes the magnetic flux passing through the coil. As a result, an induced emf is generated in the coil based on Faraday's law of magnetic induction.

The induced emf produced by a generator is

$$\mathcal{E} = -N\frac{d\Phi_B}{dt} = -N\frac{d}{dt}\int\vec{\mathbf{B}}\cdot d\vec{\mathbf{A}} = -N\frac{d}{dt}\left(BA\cos\theta\right)$$

$$= NBA\omega\sin\omega t = \mathcal{E}_0\sin\omega t \quad (\mathcal{E}_0 = NBA\omega) \quad (29\text{--}4)$$

where N is the number of turns of the coil, B is the strength of the magnetic field, A is the area of the coil, and ω is the angular speed of rotation of the coil. Since ω is in radians per second, $\omega = 2\pi f$, where f is the frequency.

It is clear from the preceding expression that the induced emf, and hence the induced current, is sinusoidally alternating as shown at the right. This type of generator is called an alternating-current generator or ac generator.

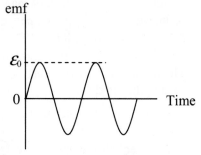

The frequency f is 60 Hz in the U.S. and Canada and 50 Hz in many other countries.

A **dc generator** is very much the same as an ac generator, except in this case the current is made unidirectional,

which is smoothed out by placing a capacitor in parallel
with the output.

Ac generators or **alternators** are used in automobiles
where current from a battery produces a magnetic field of an
electromagnet, called the *rotor,* which is rotated by a belt
from the engine. There is a set of stationary coils called a
stator that surround the rotating rotor. When the rotor
rotates, the field through the stator coils changes, thus
producing an alternating current. Using semiconductor
diodes, which allow current to flow in one direction only,
the ac output is changed to dc for charging the battery.

*29–5 Back EMF and Counter Torque; Eddy Currents

*Back emf
When the armature of a motor turns, the magnetic flux
through the coil changes. This generates an induced emf
that acts to oppose the motion and is called back emf or
counter emf. If there were no load, the motor's speed
would increase until the counter emf equals the input
voltage. In the presence of a load counter, emf is less than
the external voltage. The greater the load, the slower the
motor rotates and the lower is the counter emf.

*Counter Torque
In a generator the situation is opposite. When the generator
is connected to a device, a current flows in the coils in the
armature. Since the coil is placed in a magnetic field, it
exerts a torque that opposes the motion. This is called a
counter torque.

*Eddy Currents
When a conductor moves across a magnetic field or if the
magnetic flux through a conductor changes, an induced

current is developed that opposes the cause for which it is developed. This current is called an eddy current. Eddy currents can be used as a smooth braking device. In this case an electromagnet can be turned on that applied magnetic field to the wheels or to the moving steel rails. Eddy currents are also used to dampen the oscillation of a system. Eddy current developed in the armature of a motor or generator generates heat, which is wasted. To reduce eddy current in these situations the armatures are laminated consisting of thin sheets of iron that are well insulated from one another. This increases total resistance thus decreasing current and heat loss.

29–6 Transformers; Transmission of Power

A **transformer** is an electrical device that can increase or decrease an ac voltage using the principle of Faraday's magnetic induction.

A transformer consists of an iron core on which two coils are wound: a **primary** coil with N_P number of turns and a **secondary** coil with N_S number of turns. The primary coil is connected to the ac voltage that needs to be increased or decreased. As the alternating current in the primary coil changes, the flux through the primary and secondary also changes. The following **transformer equation** gives the relationship between voltage, V, and number of turns, N, of the primary and secondary circuits:

$$\frac{V_S}{V_P} = \frac{N_S}{N_P} \qquad (29\text{–}5)$$

- If $N_S > N_P$, $V_S > V_P$, and the transformer is called a **step-up transformer**.

- If $N_S < N_P$, $V_S < V_P$, and the transformer is called a **step-down transformer**.

Since the input power is essentially equal to the output power: $V_P I_P = V_S I_S$, or

$$\frac{I_S}{I_P} = \frac{N_P}{N_S} \qquad (29\text{–}6)$$

Transformers are used in transmission of electricity over long distances. The greater the voltage to be transmitted the less is the current and less power is transmitted in the transmission lines. Therefore, the voltage is stepped-up using a transformer before it is transmitted. Upon arrival for distribution, the voltage is stepped down in stages at electric substations prior to distribution.

Changing voltage with dc is difficult and expensive.

29–7 A Changing Magnetic Flux Produces an Electric Field

A changing magnetic flux produces an electric field. The results hold for any region in space.

Faraday's Law—General Form

The emf induced in a circuit is equal to the work done per unit charge by the field around a closed path:

$$\mathcal{E} = \oint \vec{E} \cdot d\vec{l} \qquad (29\text{–}7)$$

Thus, the general form of Faraday's law is:

$$\oint \vec{E} \cdot d\vec{l} = -\frac{d\Phi_B}{dt} \qquad (29\text{–}8)$$

*Forces Due to Changing $\vec{B}$ are Nonconservative

Electric field produced by electric charges at rest start and stop on electric charges. In this case, for a closed path $\oint \vec{E} \cdot d\vec{l} = 0$. Thus, electrostatic force is a conservative force and potential energy could be defined. But, when the electric field is produced by a changing magnetic field, the integral around a closed path is not zero, since $\oint \vec{E} \cdot d\vec{l} = -\dfrac{d\Phi_B}{dt}$. Forces due to changing magnetic fields are thus nonconservative and no potential energy can be defined.

Electric field produced by static charge is conservative but electric field produced by changing magnetic field is **nonconservative**.

*29–8 Applications of Induction: Sound Systems, Computer Memory, Seismograph, GFCI

*Microphone

In a microphone, a small coil connected to a membrane is suspended close to a permanent magnet. As the membrane and the coil move back and forth inside the magnetic field according to an incoming sound, an emf is generated in the coil whose frequency is the same as the incoming sound.

*Read/Write on Tape and Disks

Recording and playback on tape (or disks) is done by magnetic heads. Recording tapes contain a thin layer of magnetic oxide on a plastic tape. In recording (or writing), the electric input signal to the head, which acts as an electromagnet, magnetizes the passing tape (or the disk). In play back or reading, the changing magnetic field of the passing tape (or the disk) induces a changing magnetic field to the head. This changing magnetic field induces in the coil an emf that is the output signal.

*Credit Card Swipe
When a credit card is swiped, the information about the account that is contained in the magnetic stripe is read and is evaluated after connecting by telephone for approval.

*Seismograph
A seismograph (used to measure earthquake intensity) contains a magnet and a coil of wire, one of which is fixed to the case placed in contact with the Earth. Because of the relative motion of the magnet and the coil (when the earth shakes) an emf is generated in the coil.

*Ground Fault Current Interrupter (GFCI)
A ground fault current interrupter (GFCI), which works based on the principle of electromagnetic induction, is used to protect humanas.

The two thick conductors of a power line leading to an electric device pass through a small iron core coil of wire. Under normal conditions, the current flowing in the hot wire is balanced by the current moving oppositely in the neutral wire. However, under faulty situations, the return current can be less and a net ac current through the GFCI coil will induce an emf in the coil. The emf after amplification is sent to its own solenoidal circuit breaker that opens the circuit.

Tips for Solving the Problems

1. Use the expression $\phi_B = \vec{B} \cdot \vec{A} = BA \cos \theta$ to calculate magnetic flux passing through an area. Note that θ is the angle made by the magnetic field B with the *normal* to the surface of area A. Remember that the magnetic flux through a coil can be changed by changing the magnetic field, by changing the area, or by changing both. If $\vec{B}$ is not uniform, $\Phi_B = \int \vec{B} \cdot d\vec{A}$.

2. Use Faraday's law, $\mathcal{E} = -N \dfrac{d\Phi_B}{dt}$, to determine the induced emf, and use Ohm's law to determine the induced current from the induced emf from problems such as those in Sections 29–2 and 29–3.

3. The direction of the induced emf follows from Lenz's law. Lenz's law is simple but often confusing. Practice by applying it to different problems.

4. Use $\mathcal{E} = Blv$ for the motional emf (as long as B, l, and v are mutually perpendicular) and then determine the induced current from Ohm's law. Assuming a uniform electric field E in the rod, $E = \mathcal{E}/l = Bv$. For a rotating coil, such as in a generator (problems in Section 29–4), induced emf can be obtained from $\mathcal{E} = NBA\omega \sin \omega t$.

5. For a transformer, the voltage ratio (secondary to primary) is the same as the turns ratio (secondary to primary) of the coils. The current ratio is the reciprocal of the voltage ratio. This information can be used to solve problems in Section 29–6.

CHAPTER 30
Inductance, Electromagnetic Oscillations, and AC Circuits

This chapter describes inductance, circuits containing inductance, electromagnetic oscillations and ac circuits.

Important Concepts

Mutual inductance
Henry
Self-inductance
Energy density
Time constant
Inductor
Inductive reactance
Impedance
Capacitive reactance
Phasor diagram
Power factor
Resonance in ac circuit
Resonant frequency
Electromagnetic oscillation
Impedance matching

30–1 Mutual Inductance

A change of current in one coil (coil 1) can induce a current in another coil (coil 2) because of a change of magnetic flux in the second coil. This phenomenon is known as mutual induction. If ϕ_{21} is the flux passing through each of N_2 turns in coil 2, the **mutual inductance** between the coils is

$$M = \frac{N_2 \phi_{21}}{I_1} \qquad (30\text{–}1)$$

where I_1 is the current in coil 1.

The induced emf in coil 2 due to a change in current I_1 in coil 1 is

$$\mathcal{E}_2 = -M\,\frac{dI_1}{dt} \qquad (30\text{–}2)$$

Similarly, the induced emf in coil 1 due to a change in current I_2 in coil 2 is

$$\mathcal{E}_1 = -M\,\frac{dI_2}{dt} \qquad (30\text{–}3)$$

The constant M is called the **mutual inductance**. The SI unit of mutual inductance is the henry (H).

$$1\ \text{H} = 1\ \text{V}\cdot\text{s/A} = 1\ \Omega\cdot\text{s}.$$

A transformer is a practical application of mutual inductance. Because of mutual induction a changing current in a circuit can induce an emf to another part of the circuit or to a different circuit creating a problem. In a shielded cable the inner conductor is surrounded by a cylindrical grounded conductor to reduce this problem.

30–2 Self-Inductance

A change of current in a coil induces a current in the same coil because of its change in magnetic flux. This is known as self-induction. Because of self-induction, a coil resists changes in its current. If ϕ_B is the amount of flux passing through each of N turns in the coil that carries a current I, the self-inductance of the coil is

$$L = \frac{N\phi_B}{I} \qquad (30\text{--}4)$$

The self-induced emf in a coil due to a change in current I is

$$\mathcal{E} = -L\,\frac{dI}{dt} \qquad (30\text{--}5)$$

The constant L is called the **self-inductance** (or simply **inductance**) of the coil. It is measured in henry.

A coil with finite inductance is called an **inductor** or a **choke coil**. Just as a mass resists changes in its velocity, an inductor resists changes in its current. For a given $\mathcal{E}$, if L is large, the change in current and, therefore, the current itself if it is ac, will be small. The greater the value of L, the less the ac current. This quality of the inductor is called the *reactance* or *impedance*.

30–3 Energy Stored in a Magnetic Field

Like a capacitor, an inductor stores energy. The energy stored in an inductor, L, carrying current, I, is

$$U = \frac{1}{2}LI^2 \qquad (30\text{--}6)$$

The energy in an inductor is stored in its magnetic field.
 Inductance of a solenoid of length l and area A with N turns can be shown to be $L = \mu_0 N^2 A / l$. Since inside a solenoid $B = \mu_0 NI/l$, this gives

$$U = \frac{1}{2\mu_0} B^2 Al$$

Since the volume $V = Al$, the magnetic energy density is

$$u = \frac{\text{magnetic energy}}{\text{volume}} = \frac{1}{2}\frac{B^2}{\mu_0} \qquad (30\text{--}7)$$

30–4 *LR* Circuits

A *LR* circuit consists of a resistor, R, an inductor, L, and a source of emf, V_0. Since an inductor resists changes in its current, after the switch is closed, the current takes a finite time to approach its equilibrium value, $I_{max} = V/R$. Applying Kirchhoff's loop rule to the circuit:

$$V_0 - L\frac{dI}{dt} - IR = 0 \qquad (30\text{--}8)$$

The solution shows that current increases with time according to the following equation:

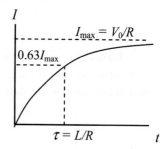

$$I = \frac{V_0}{R}\left(1 - e^{-t/\tau}\right) \qquad (30\text{--}9)$$

Here τ is the characteristic time over which this change in current occurs and is given by

$$\tau = L/R \qquad (30\text{--}10)$$

τ is called the **time constant**.

When $t = \tau$, $I = 0.63 I_{max}$.

If the battery is suddenly removed from this circuit, the decay of current is given by

$$I = I_{max}\, e^{-t/\tau}. \qquad (30\text{--}11)$$

When $t = \tau$, $I = 0.37 I_{max}$.

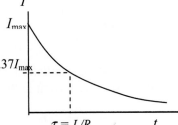

30–5 *LC* Circuits and Electromagnetic Oscillations

An ac circuit consisting of an inductor and a capacitor with no ac generator can display an oscillating electric current. From Kirchhoff's loop rule,

$$-L\,\frac{dI}{dt} + \frac{Q}{C} = 0.$$

Now, $I = -\dfrac{dQ}{dt}$; the sign is negative because charge Q on the positive plate of the capacitor decreases. Thus,

$$\frac{d^2Q}{dt^2} + \frac{Q}{LC} = 0 \qquad (30\text{--}12)$$

The solution is $Q = Q_0 \cos{(\omega t + \phi)}$ \qquad (30\text{--}13)

where Q_0 is the initial charge on one of the plates and ϕ is a constant that depends on the initial conditions. The charge flows back and forth from one plate of the capacitor to the other through the inductor. From the preceding equations,

$$\omega_0 = \sqrt{\frac{1}{LC}} \qquad (30\text{–}14)$$

This is called LC oscillation or an **electromagnetic oscillation**. The current in the inductor is

$$I = -\frac{dQ}{dt} = I_0 \sin(\omega t + \phi) \qquad (30\text{–}15)$$

The energy stored in the electric field of the capacitor:

$$U_E = \frac{1}{2}\frac{Q^2}{C} = \frac{Q_0^2}{2C}\cos^2(\omega t + \phi)$$

The energy stored in the magnetic field of the inductor:

$$U_B = \frac{1}{2}LI^2 = \frac{Q_0^2}{2C}\sin^2(\omega t + \phi)$$

The energy oscillates back and forth between being stored in the electric field of the capacitor and the magnetic field of the oscillator. The total energy

$$U = U_E + U_B = \frac{Q_0^2}{2C} \qquad (30\text{–}16)$$

30–6 *LC* Oscillations with Resistance (*LRC* Circuit)

If the capacitor in series *LRC* circuit is given a charge Q_0 and the source is then removed, from Kirchhoff's loop rule around the circuit, we get

$$-L\frac{dI}{dt} - IR + \frac{Q}{C} = 0, \text{ or,}$$

$$L\frac{d^2Q}{dt^2} + R\frac{dQ}{dt} + \frac{Q}{C} = 0. \qquad (30\text{–}17)$$

This represents the differential equation of a damped harmonic motion. The solution is

$$Q = Q_0\, e^{-(R/2L)t} \cos(\omega' t + \phi) \qquad (30\text{–}18)$$

where ϕ is a phase constant and angular frequency

$$\omega' = \sqrt{1/LC - R^2/4L^2} \qquad (30\text{–}19)$$

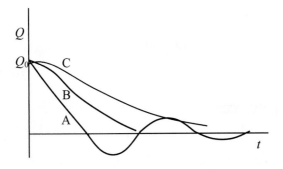

The preceding curves show the charge Q on the capacitor as a function of time. Curve A is for underdamped oscillation ($R^2 < 4L/C$); curve B is for critically damped (discharges in shortest time) situation, and curve C is for over damped situation.

30–7 AC Circuits with AC Source

Resistor

An alternating voltage can be expressed as $V = V_0 \cos \omega t$ where V_0 is peak voltage. When an ac source is connected to a resistor, the current

$$I = I_0 \cos \omega t, \qquad (30\text{–}20)$$

which is in *phase* with the voltage. The average power dissipated is given by $\overline{P} = I_{rms}{}^2 R = V_{rms}{}^2/R$.

Inductor

For an ac circuit containing an inductor L, from Kirchhoff's rule

$$V - L \frac{dI}{dt} = 0 \qquad (30\text{–}21)$$

This gives

$$V = V_0 \sin (\omega t + 90°) \qquad (30\text{–}22a)$$

where $\qquad V_0 = I_0 \omega L \qquad (30\text{–}22b)$

is the peak voltage.

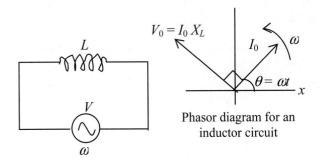

$V_0 = I_0 X_L$

I_0

ω

$\theta = \omega t$ x

Phasor diagram for an
inductor circuit

In an inductor, the *current lags the voltage by 90°*.
Since the phase difference, ϕ, is 90°, the power consumed
is zero.

The peak voltage can be written as

$$V_0 = I_0 X_L, \qquad (30\text{--}23a)$$

where $$X_L = \omega L. \qquad (30\text{--}23b)$$

X_L is called the **inductive reactance**, or impedance of the
inductor. X_L increases with frequency, because as the
frequency increases, the current changes more rapidly with
time, thus generating greater voltage across the inductor.
The SI unit of inductive reactance is the ohm, the same
unit for resistance.

Capacitor
For an ac circuit containing an inductor C, from
Kirchhoff's rule

$$V - \frac{Q}{C} = 0$$

This gives

$$V = V_0 \sin (\omega t - 90°) \qquad (30\text{–}24a)$$

where
$$V_0 = I_0 \left(\frac{1}{\omega C} \right) \qquad (30\text{–}24b)$$

is the peak voltage.

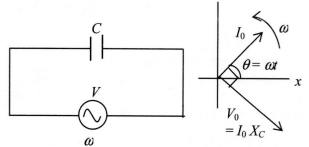

Phasor diagram for a
capacitor circuit

In a capacitor, the *current leads the voltage by 90°*.
Because of this the power consumed is zero. The peak
voltage can be written as

$$V_0 = I_0 X_C., \qquad (30\text{–}25a)$$

where
$$X_C = \frac{1}{\omega C}. \qquad (30\text{–}25b)$$

X_C is called the **capacitive reactance**, or impedance of the
capacitor. X_L decreases with frequency because at very
high frequency, voltage changes direction so rapidly that
there is not enough time to charge the capacitor. As a
result, a capacitor offers essentially no resistance to the

flow of charges onto its plates. The SI unit of capacitive reactance is the ohm.

Capacitors are used to prevent dc current and as filters in different circuits.

30–8 *LRC* Series AC Circuit

A circuit consisting of a resistor, R, an inductor, L, and a capacitor, C, in series with an ac generator is known as an *LRC* circuit. At any instant the voltage V supplied by the source is given by

$$V = V_R + V_L + V_C \qquad (30\text{–}26)$$

The behavior of an *RLC* circuit can be analyzed using a **phasor diagram**. The voltage of the resistor V_R is in phase with the current, whereas the voltage of the inductor V_L and capacitor V_C point in opposite directions (90° ahead of the current in the inductor, and 90° behind the current for the capacitor).

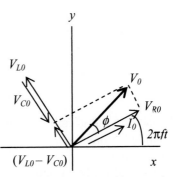

Phasor diagram for a series *LRC* circuit showing the sum vector V_0

The total rms voltage is

$$V_{\text{rms}} = I_{\text{rms}} Z \quad \text{or} \quad V_0 = I_0 Z \qquad (30\text{–}27)$$

The total impedance is

$$Z = \sqrt{R^2 + \left(X_L - X_C \right)^2} \qquad (30\text{–}28a)$$

$$= \sqrt{R^2 + \left(\omega L - 1/\omega C\right)^2} \qquad (30\text{--}28\text{b})$$

The impedance, Z, arises from the resistor, R, the inductive reactance, X_L, and the capacitive reactance, X_C.

The angle between the voltage and current is

$$\tan \phi = \frac{X_L - X_C}{R} \qquad (30\text{--}29\text{a})$$

If $X_L > X_C$, $\phi > 0$, the voltage leads the current.
If $X_L < X_C$, $\phi < 0$, the voltage lags the current.
If $X_L = X_C$, $\phi = 0$, the voltage and the current are in phase.

$$\cos \phi = \frac{R}{Z} \qquad (30\text{--}29\text{b})$$

The average power, $\overline{P} = I_{rms}^2 R$
$$= I_{rms} V_{rms} \cos \phi \qquad (30\text{--}30)$$

The factor $\cos \phi$ is called the **power factor** of the circuit. For an inductor and capacitor $\phi = +90°$ or $-90°$ and, power is zero.

The results for RC and RL circuits can be obtained as special cases of the results for an RLC circuit.

30–9 Resonance in AC Circuits

AC electric circuits can have natural frequencies of oscillation just like physical systems such as a pendulum or a mass attached to a spring. The rms current in an LRC circuit is

$$I_{\text{rms}} = \frac{V_{rms}}{Z} = \frac{V_{rms}}{\sqrt{R^2 + \left(\omega L - 1/\omega C\right)^2}} \quad (30\text{--}31)$$

The current will be maximum when $\omega L - 1/\omega C = 0$. Solving, we get the resonant frequency ω_0 as:

$$\omega_0 = \sqrt{\frac{1}{LC}} \qquad (30\text{--}32)$$

At this frequency $X_C = X_L$, impedance is purely resistive, current is maximum, and cos $\phi = 1$. A plot of I_{rms} versus f is shown for small R and large R.

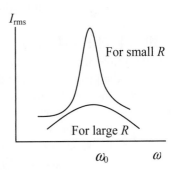

Electric resonance is used in radio and TV sets for tuning in a station. Although many frequencies reach the circuit, a significant current flows for those at or very close to the resonant frequency.

*30–10 Impedance Matching

When one electric circuit is connected to a second circuit it is important that maximum power be transferred from the one to the other. If the internal resistance of circuit 1 (called output impedance) is R_1 and the equivalent resistance (called input impedance) of circuit 2 is R_2, then it can be shown that the power P delivered to circuit 2 is

$$P = \frac{V^2}{R_1} \frac{\left(\dfrac{R_2}{R_1}\right)}{\left(1 + \dfrac{R_2}{R_1}\right)^2} \qquad (30\text{--}33)$$

P is maximum when $R_2 = R_1$. Thus, maximum power is transferred when output impedance of one device equals the input impedance of the second. This is known as **impedance matching**. The result also holds for ac circuits containing capacitors and inductors.

*30–11 Three-Phase AC

Transmission lines typically consist of four wires, one of which is the ground and the other three transmit three-phase ac power that is the superposition of three ac voltages 120 ° out of phase with each other:

$$V = V_0 \sin \omega t$$
$$V = V_0 \sin (\omega t + 120°)$$
$$V = V_0 \sin (\omega t + 240°)$$

It can be shown that the power delivered is a constant (that makes electric equipment run smoothly) and is equal to $3V_0^2/2R$.

Tips for Solving the Problems

1. Self-inductance and mutual inductance depend on the geometrical factors and not on currents. The self-inductance of a long solenoid of length l and cross-sectional area A containing N turns is $L = \mu_0 N^2 A/l$. The mutual inductance for a solenoid (of length l, cross-sectional area A, containing N_1 closely packed turns) and a coil (containing N_2 turns, wrapped around the solenoid so that all the flux from the solenoid passes through the coil) is $M = \mu_0 N_1 N_2 A/l$.

2. Energy stored in an inductor is $E = \frac{1}{2}LI^2$ (which is similar in form to the energy stored in a compressed spring).

3. Use Equation $I = (V_0/R)\left(1 - e^{-t/\tau}\right)$ or $I = I_0 e^{-t/\tau}$ (depending on the situation) to determine the change of current with time in RL circuits. The time constant $\tau = L/R$ is a characteristic time, which will be in seconds if R is in ohms, and L is in henries.

4. Read the problem carefully to find out whether the given currents and voltages are their maximum values or their rms values in an ac circuit problem. If the maximum values are given, the rms values can be obtained by diving by $\sqrt{2}$ (or multiplying by 0.707). If the rms values are given, maximum values can be obtained by multiplying by $\sqrt{2}$.

5. To calculate the average power disspated in a resistor, use the expression for power that holds for dc: $\overline{P} = I^2 R = V^2/R$. Replace I and V with their ac rms values.

6. It may be useful to draw phasors in solving problems. Draw a capacitor phasor at 90° *clockwise* with respect to the current phasor. Draw an inductor phasor at 90° *counterclockwise* with respect to the current phasor.

7. Usually C is given in microfarads (μF). Make sure to convert the value to farads. L may be given in millihenry (mH). Make sure to convert the value to henry.

8. To calculate power or the power factor using the expression $P_{av} = I_{rms}V_{rms} \cos \phi$, note that here ϕ is the angle between the voltage and the current phasors, not the angle between the voltage phasor and the x or y axis. Also, $\cos \phi = R/Z$.

9. For an LRC circuit, $V_{rms} = I_{rms} Z$. Here $Z = \sqrt{R^2 + (X_L - X_C)^2}$. The impedance arises from the resistor, R, the inductive reactance, X_L, and the capacitive reactance, X_C. The angle between the voltage and current is given by $\tan \phi = \dfrac{X_L - X_C}{R}$. Here, ϕ can be greater than, equal to, or less than zero, depending on the values of X_L and X_C. The power factor is $\cos \phi = R/Z$. For an RL circuit, assume $X_C = 0$, and for an RC circuit assume $X_L = 0$.

10. Remember that at resonance in an RLC circuit, the current and voltage are out-of-phase, and $X_L = X_C$. As a result, $Z = R$. The resonant frequency is $\omega_0 = (1/\sqrt{LC})$. It depends on L and C only. Tips 8 to 10 will be useful for solving problems in Sections 30–8 and 30–9.

CHAPTER 31
Maxwell's Equations and Electromagnetic Waves

This chapter discusses electromagnetic waves that carry energy and momentum, exert pressure, and can travel in a vacuum. Electromagnetic spectrum as well as transmission and reception of radio and television signals are described.

Important Concepts

>Maxwell's equations
>Displacement current
>Electromagnetic wave
>Wave equation
>The electromagnetic spectrum
>Radio and television waves
>Microwave
>Infrared rays
>Ultraviolet rays
>X-rays
>Energy in em waves
>Poynting vector
>Radiation pressure
>Radio
>Television
>Amplitude modulation
>Frequency modulation
>Demodulation

31–1 Changing Electric Fields Produce Magnetic Fields; Ampère's Law and Displacement Current

Ampère's Law

Ampère's law provides a way to calculate magnetic fields due to currents. According to the law, the magnetic field produced due to current is given by

$$\oint \vec{\mathbf{B}} \cdot d\vec{l} = \mu_0 I_{\text{encl}}.$$

Maxwell modified Ampère's law by introducing displacement current. Maxwell argued that since a changing magnetic field produces an electric field, a changing electric field or changing electric flux will produce a magnetic field. Ampère's law as generalized by Maxwell is

$$\oint \vec{\mathbf{B}} \cdot d\vec{l} = \mu_0 I_{\text{encl}} + \mu_0 \varepsilon_0 \frac{d\Phi_E}{dt} \qquad (31\text{–}1)$$

where $\Phi_E = EA$ (E is electric field and A is the area), called the electric flux.

Displacement Current

Maxwell interpreted the second term of the preceding equation as being equivalent to an electric current, called **displacement current**, I_D. The ordinary current is called **conduction current** and thus,

$$\oint \vec{\mathbf{B}} \cdot d\vec{l} = \mu_0 (I + I_D)_{\text{encl}} \qquad (31\text{–}2)$$

where

$$I_D = \varepsilon_0 \frac{d\Phi_E}{dt}. \qquad (31-3)$$

31–2 Gauss's Law for Magnetism

According to the Gauss law for electricity,
the electric flux passing through a surface is related to the
net charges Q_{encl} enclosed by the surface by

$$\oint \vec{E} \cdot d\vec{A} = \frac{Q_{encl}}{\varepsilon_0}$$

Since there is no evidence of magnetic monopoles (single
isolated pole), the similar relation, called the **Gauss's law
for magnetism**, is

$$\oint \vec{B} \cdot d\vec{A} = 0. \qquad (31-4)$$

This law means that as many magnetic field lines enter an
enclosed volume as leave it.

31–3 Maxwell's Equations

Maxwell's equations are basic equations of
electromagnetism. These are as fundamental as Newton's
law of motion or law of gravitation. The equations are:

$$\oint \vec{E} \cdot d\vec{A} = \frac{Q_{encl}}{\varepsilon_0} \qquad (31-5a)$$

$$\oint \vec{B} \cdot d\vec{A} = 0 \qquad (31-5b)$$

$$\oint \vec{E} \cdot d\vec{l} = -\frac{d\Phi_B}{dt} \qquad (31\text{–}5c)$$

$$\oint \vec{B} \cdot d\vec{l} = \mu_0 I + \mu_0 \varepsilon_0 \frac{d\Phi_E}{dt}. \qquad (31\text{–}5d)$$

The first two equations are Gauss's laws for electricity and magnetism, respectively. The third is Faraday's law and the fourth is Ampère's law as modified by Maxwell.

31–4 Production of Electromagnetic Waves

Electromagnetic waves were predicted and their properties were theoretically studied by Maxwell in 1864, before they were experimentally observed. *Electromagnetic waves* are traveling waves of oscillating electric and magnetic fields. Electromagnetic waves can be produced using an ac electric circuit and an antenna. An accelerated charge radiates energy in the form of electromagnetic waves.

The electric field $\vec{E}$ and magnetic field $\vec{B}$ in electromagnetic waves are perpendicular to each other, and they are also perpendicular to the direction of propagation of the waves. Clearly, electromagnetic waves are *transverse*. The electric and magnetic fields alternate in directions (that is, for example, into the page at some points and out of the page at others). They vary from a maximum in one direction, to zero, to a maximum in the other direction, and are in phase (that is, they each are zero at the same points and reach their maxima at the same point in space). The direction of propagation of electromagnetic waves can be determined from the following right-hand rule: *Point the fingers of your right hand in the direction of* $\vec{E}$ *, curl your fingers toward* $\vec{B}$*, and your thumb will point in the direction of propagation.*

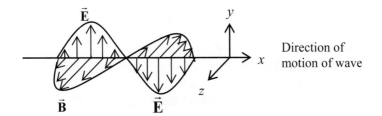

Very far from the antenna the field lines are flat over a large area, and the waves are called the **plane waves**.

31–5 Electromagnetic Waves, and Their Speed, from Maxwell's Equations

In a region of free space far from the source where there are no charges or conduction currents (that is $Q = 0$, $I = 0$), Maxwell's equations become

$$\oint \vec{E} \cdot d\vec{A} = 0 \qquad (31\text{–}6a)$$

$$\oint \vec{B} \cdot d\vec{A} = 0 \qquad (31\text{–}6b)$$

$$\oint \vec{E} \cdot d\vec{l} = -\frac{d\Phi_B}{dt} \qquad (31\text{–}6c)$$

$$\oint \vec{B} \cdot d\vec{l} = \mu_0\varepsilon_0 \frac{d\Phi_E}{dt}. \qquad (31\text{–}6d)$$

We assume that the electromagnetic wave is traveling in the x-direction, $\vec{E}$ is parallel to the y-axis, and $\vec{B}$ is parallel to the z-axis. If the wave is sinusoidal, then:

$$E = E_y = E_0 \sin (kx - \omega t) \qquad (31\text{–}7)$$

$$B = B_z = B_0 \sin (kx - \omega t)$$

where $k = 2\pi/\lambda$, $\omega = 2\pi f$, and $f\lambda = \omega/k = v$. $\qquad (31\text{–}8)$
Here λ, f, and v are the wavelength, frequency, and the
speed of the wave, respectively.

Using a small rectangle in the plane of the electric
field $\vec{E}$ of the electromagnetic wave and third Maxwell's
equation (which is Faraday's law), we find that E and B,
which are functions of position x and time t, are related by:

$$\frac{\partial E}{\partial x} = -\frac{\partial B}{\partial t} \qquad (31\text{–}9)$$

Also, using a small rectangle in the plane of the
magnetic field $\vec{B}$ of the electromagnetic wave and fourth
Maxwell's equation (which is modified Ampère's law), we
find that:

$$\frac{\partial B}{\partial x} = -\mu_0\varepsilon_0 \frac{\partial E}{\partial t} \qquad (31\text{–}10)$$

Using the preceding four equations we find that the speed
of an electromagnetic wave is

$$v = \frac{E}{B} \qquad (31\text{–}11)$$

and that the speed can be written as

$$v = \frac{1}{\sqrt{\varepsilon_0 \mu_0}} \qquad (31\text{–}12)$$

The last equation shows that speed v is a constant, independent of the wavelength or frequency. Since $\varepsilon_0 = 8.85 \times 10^{-12}$ $C^2/N \cdot m^2$ and $\mu_0 = 4\pi \times 10^{-7}$ $N \cdot s^2/C^2$, this gives $v = 3.00 \times 10^8$ m/s, which is the same as the speed of light, c.

From equations (31–9) and (31–10) we find the following *wave equations*:

$$\frac{\partial^2 E}{\partial t^2} = \frac{1}{\mu_0 \varepsilon_0} \frac{\partial^2 E}{\partial x^2} \qquad (31\text{–}13a)$$

and

$$\frac{\partial^2 B}{\partial t^2} = \frac{1}{\mu_0 \varepsilon_0} \frac{\partial^2 B}{\partial x^2}. \qquad (31\text{–}13b)$$

31–6 Light as an Electromagnetic Wave and the Electromagnetic Spectrum

Maxwell argued that light is an electromagnetic wave since electromagnetic waves move at the speed of light. The idea was accepted after electromagnetic waves were first generated and detected experimentally by Hertz in 1887. These waves were shown to move at the speed of light and had shown all characteristics of light such as reflection, refraction, and interference. The speed, c, of an electromagnetic wave is related to its frequency, f, and its wavelength, λ, by

$$c = f\lambda \qquad (31\text{–}14)$$

The ordered series containing electromagnetic waves of increasing frequency or wavelength is called the **electromagnetic spectrum**.

Although the boundary between different regions is not sharp, several bands of frequencies in the spectrum were given the following names:

Name	Approx Frequency range (Hz)
Radio and TV waves	$10^6 - 10^9$
Microwaves	$10^9 - 10^{12}$
Infrared	$10^{12} - 10^{14}$
Visible light	$4 \times 10^{14} - 7.5 \times 10^{14}$
Ultraviolet	$10^{15} - 10^{17}$
X-rays	$10^{17} - 10^{19}$
Gamma rays	Greater than 10^{19}

Radio waves and microwaves can be produced in the laboratory using electronic equipment. Radio waves that we receive with radios and televisions are produced by alternating currents in antennas. Molecules and accelerated electrons in space generate radio waves. X-rays are produced when fast moving electrons are rapidly decelerated upon striking a metal target. Infrared radiation is responsible for the heating effect. Visible light is emitted by an ordinary incandescent lamp due to electrons undergoing acceleration within the hot filament. Human eyes detect wavelength between 4 and 7×10^{-7} m (visible light) but skin detects longer wavelengths.

Electromagnetic waves can travel through vacuum and through a medium. In a medium of permittivity ε and permeability μ, its speed $v = 1/\sqrt{\varepsilon\mu}$.

31–7 Measuring the Speed of Light
The first serious attempt to determine the speed of light was made by Galileo showing that speed of light must be extremely high. Roemer's experiment showed that the

speed of light is finite. Michelson used a rotating mirror apparatus to determine speed of light with high precision. In his apparatus, light from a source was directed to the rotating mirror and the light after being reflected by a stationary mirror entered a telescope. From the speed of the rotating mirror and the geometry of the apparatus he determined the speed of light. The accepted value of the speed of light in a vacuum is

$$c = 2.99792458 \times 10^8 \text{ m/s}$$

31–8 Energy in EM Waves; the Poynting Vector

Electromagnetic waves carry energy from one region to another. This energy is associated with the moving electric and magnetic fields. The total energy per unit volume for an electromagnetic wave is

$$u = \frac{1}{2} \varepsilon_0 E^2 + \frac{1}{2\mu_0} B^2 \qquad (31\text{–}15)$$

$$= \varepsilon_0 E^2 \qquad (31\text{–}16a)$$

$$= \frac{1}{\mu_0} B^2 \qquad (31\text{–}16b)$$

This can also be written as

$$u = \sqrt{\varepsilon_0 / \mu_0} \, EB \qquad (31\text{–}16c)$$

The amount of energy crossing unit area in unit time is given by is a vector $\vec{S}$, called the **Poynting vector** . It can be shown that

$$S = c\varepsilon_0 E^2 = \frac{c}{\mu_0} B^2 = \frac{EB}{\mu_0} \qquad (31\text{--}17)$$

The direction of $\vec{S}$ is the direction in which energy is transported, which is the direction of propagation of the wave (that is, perpendicular to $\vec{E}$ and $\vec{B}$). Thus, the Poynting vector can be written as

$$\vec{S} = \frac{1}{\mu_0}\left(\vec{E} \times \vec{B}\right) \qquad (31\text{--}18)$$

If E and B vary sinusoidally, the average value

$$\overline{S} = \frac{1}{2} c\varepsilon_0 E^2 = \frac{1}{2\mu_0} cB^2 = \frac{E_0 B_0}{2\mu_0} \qquad (31\text{--}19a)$$

$$= \frac{E_{rms} B_{rms}}{\mu_0} \qquad (31\text{--}19b)$$

where E_0 and B_0 are the maximum value of the electric field and magnetic field, respectively.

31–9 Radiation Pressure

The momentum, p, transferred to an object by an EM radiation is

$$\Delta p = \frac{\Delta U}{c}$$

where ΔU is the total energy absorbed by the object. If the radiation is totally reflected, the momentum transfer is

$$\Delta p = \frac{2\Delta U}{c}.$$

The average rate at which energy is delivered is

$$\frac{dU}{dt} = \overline{S}A$$

The pressure exerted by electromagnetic waves is called **radiation pressure**. When an electromagnetic wave is shined on an object, the radiation pressure (which is F/A and $F = dp/dt$) is

$$P = \frac{\overline{S}}{c}$$

or, if the wave is fully reflected

$$P = 2\frac{\overline{S}}{c}$$

Radiation pressure can be significant for very small particles. For example, it is possible to move atoms and molecules around by steering them using a laser beam in "optical tweezers."

31–10 Radio and Television: Wireless Communication

Electromagnetic waves are used for transmitting information over long distances.

A radio station first converts an audio signal into an electric signal of the same frequencies using a microphone or a tape recorder head. The electric signal has frequencies 20 to 20000 Hz and is called the audiofrequency (AF) signal. The signal is then amplified and mixed with a radio-frequency (RF) signal. Each station has a particular radio frequency, called its **carrier frequency**. AM radio stations have carrier frequencies from about 530 KHz to 16000 kHz; FM radio stations have frequencies 88 MHZ and 108 MHZ; TV stations between 54 and 88 MHZ for channels 2 through 6 and between 174 and 216 MHz for channels 7 through 13; and UHF stations have frequencies between 470 and 890 MHz.

The mixing of the audio and carrier frequencies is done by (i) **amplitude modulation** (AM) in which case the amplitude of the carrier frequency is modulated or altered according to the amplitude of the audio signal, or (ii) **frequency modulation** (FM) in which the frequency of the carrier wave is altered by the audio signal. In a TV transmitter both the audio and video signals are mixed with the carrier frequencies in frequency modulation.

The EM waves sent by all stations are received by the antenna of the receiver. The detected signal contains frequencies from many different stations. At the receiver a particular station is "tuned-in" by adjusting L or C of a tuning circuit so that the resonant frequency of the circuit is the same as the RF carrier frequency. The signal containing both audio and carrier frequencies, after amplification, goes to a detector where the carrier frequency is separated from the audio frequency by a process called "demodulation." In a TV receiver both the audio and video signals are separated. The audio signal goes to a loudspeaker and the video signal to the picture tube.

Other EM Wave Communications

There are bands of radiation spectrum used for ships, airplanes, police, military, satellites and space, and radar. Cell phones are complete transmitters and receivers and in the U.S. function on two bands: 800 MHz and 1.9 GHz. Cable TV channels are carried as EM waves along a coaxial cable.

Tips for Solving the Problems

1. In an RC circuit during charging, charge changes with time as $Q = C\mathcal{E}\left(1 - e^{-t/RC}\right)$. Also, electric field between two closely spaced conductors is $E = \sigma/\varepsilon_0 = (Q/A)/\varepsilon_0$. Thus, $dE/dt = (dQ/dt)/A\varepsilon_0$.

2. Electric field between the plates of a capacitor depends on voltage, $E = V/d$. So, $dE/dt = (1/d) dV/dt$.

3. The displacement current $I_D = \varepsilon_0 \, d\Phi_E/dt = \varepsilon_0 A \, dE/dt$.

4. For an electromagnetic wave, the electric and magnetic fields are related by $E = cB$, where c is the speed of light.

5. The speed of an electromagnetic wave is $c = f\lambda = \omega/k$. $c = 3 \times 10^8$ m/s in a vacuum.

6. The energy crossing unit area in unit time, S, of an electromagnetic wave is given by $S = c\varepsilon_0 E^2 = c B^2/\mu_0 = EB/\mu_0$. If E and B vary sinusoidally, the average

   value $\overline{S} = \dfrac{1}{2} c\varepsilon_0 E_0^2 = \dfrac{1}{2\mu_0} cB_0^2 = \dfrac{E_0 B_0}{2\mu_0}$.

7. The total energy density for an electromagnetic wave is $u = \frac{1}{2}\varepsilon_0 E^2 + \frac{1}{2\mu_0}B^2 = \varepsilon_0 E^2 = \frac{1}{\mu_0}B^2 = \sqrt{\varepsilon_0/\mu_0}\, EB.$
Problems in Section 31–8 can be solved using the expressions in 6 and 7.

8. The unit of E is N/C, of B is tesla (T), of S is W/m^2, and of pressure is N/m^2.

9. Radiation pressure is given by $P = \overline{S}/c$ (or $= 2\overline{S}/c$, if the radiation is fully reflected) where $\overline{I}$ is the average intensity of the incident radiation.

CHAPTER 32
Light: Reflection and Refraction

This chapter discusses reflection of light by plane and spherical mirrors, as well as refraction of light. The laws of reflection and refraction are discussed.

Important Concepts

Law of reflection
Plane mirror
Convex mirror
Concave mirror
Principal axis
Focal point and focal length
Paraxial rays
Ray tracing
Mirror equation
Real image and virtual image
Magnification
Refraction of light
Snell's law
Total internal reflection
Critical angle
Fiber optics

32–1 The Ray Model of Light

Light travels in straight lines. The straight line paths followed by light are called **rays**. Although light is an electromagnetic wave, the ray model is successful in describing reflection, refraction, and image formation by mirrors and lenses. Studies of these is the objective of **geometric optics**.

32–2 Reflection; Image Formation by a Plane Mirror

When rays of light strike a surface and return to the same medium, it is called reflection of light. The direction of light is changed by reflection. According to the law of reflection:

Angle of incidence, θ_i, = Angle of reflection, θ_r.

Both angles are measured between the ray and the normal to the surface at the point of incidence. Also, *the incident and reflected rays lie in the same plane with the normal to the surface.*

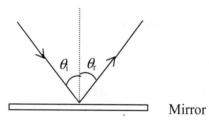

Mirror

If the surface is rough, the reflected rays move in different directions, producing **diffuse reflection**.

A mirror forms an **image** of an object by reflection of light, as shown below. The law of reflection is always followed in the formation of the image. Images formed by plane mirrors have the following characteristics:

(i)   The image is upright.
(ii) Object distance, d_0 (distance of the object from the mirror = Image distance, d_i (distance of the image behind the mirror).
(iii) The image is the same size as the object.

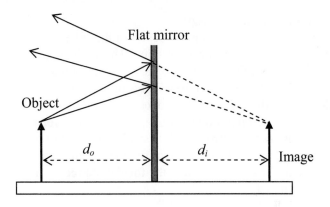

To see the image of a person's full body, the size of the mirror needs to be only half the height of the person.

An image is said to be **real** if the rays actually meet to form the image. If the rays do not actually meet but they appear to meet when the rays were extended, the image is called **virtual**. Virtual image will not appear in paper or film placed at the location of the image. The image formed by a plane mirror is virtual. Curved mirrors and lenses can form real images.

32–3 Formation of Images by Spherical Mirrors

The common curved mirrors are *spherical*, that is, sections of a sphere. If the outside (that is, the convex surface) is polished, it is a **convex** mirror. If the inside (that is, the concave surface) is polished, it is a **concave** mirror.

For an object very far away (such as the sun), the incoming rays are parallel.

Focal Point and Focal Length

The center of curvature (*C*) is the center of the sphere of radius *r* of which the mirror is a section. The straight line joining the center of curvature and the midpoint of the mirror is called the **principal axis**. Rays of light incident parallel to the principal axis of a concave mirror after reflection meet at a point on its principal axis, called **focal point** (as long as the mirror width is small compared to its radius of curvature). The distance of the focal point from the surface of the mirror is called the **focal distance,** *f*. It can be shown that,

$$f = \frac{r}{2} \qquad\qquad (32\text{--}1)$$

Concave mirror   Convex mirror

Image Formation—Ray Diagrams

Location, size, and orientation of images formed by spherical mirrors can be obtained qualitatively by ray tracing. The following components are typically used in a ray tracing:

- Ray 1 is drawn parallel to the principal axis so that after reflection it passes through the focal point (for a concave mirror) or seems to originate at the focal point (for a convex mirror).

- Ray 2 is drawn passing through the focal point (for a concave mirror) or directed toward the focal point (for a convex mirror) so that after reflection it becomes parallel to the principal axis.

- Ray 3 is drawn passing through the center of curvature (for a concave mirror) or directed toward the center of curvature (for a convex mirror) so that after reflection it follows the same path.

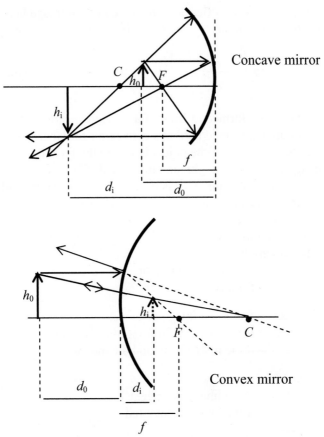

Concave mirror

Convex mirror

Mirror Equation and Magnification

The mathematical relation between the object distance, d_o, and the image distance, d_i, of a mirror is known as the **mirror equation**. The mirror equation for a spherical mirror (concave and convex) of focal length f is

$$\frac{1}{d_o} + \frac{1}{d_i} = \frac{1}{f} \qquad (32\text{-}2)$$

The **lateral magnification**, m, is the ratio of image height, h_i, to object height, h_o. The magnification is also given by

$$m = -\frac{d_i}{d_o} \qquad (32\text{-}3)$$

An image formed in front of a concave mirror is real, but an image formed behind a convex mirror is virtual. For a concave mirror, if the object is within the focal point, the image is virtual, upright, and magnified. This is how a shaving or makeup mirror is used as you place your head within the focal point.

Sign Conventions

Focal length:
> \+ for a concave mirror
> − for a convex mirror

Object distance and image distance:
> \+ if the image is in front of the mirror
> − if the image is behind the mirror

Magnification:
> \+ if the image is upright
> − if the image is inverted

32–4 Index of Refraction

The ratio of the speed of light in a vacuum c to the speed of light v in a given material is called the *index of refraction n* of the material. Thus,

$$n = \frac{c}{v} \qquad (32\text{–}4)$$

$n > 1$ (except for vacuum for which $n = 1$) and it depends on the wavelength of light used.

32–5 Refraction: Snell's Law

Refraction is the change of direction of light as it moves from one medium to another. Light refracts because of its change in speed in the second medium. The refracted ray bends *toward* the normal when it enters a medium where the speed of light is *less* (such as entering from air to water). The refracted ray bends *away* from the normal if the speed of the refracted ray in the second medium is *greater* than in the first (such as when entering from water to air).

Snell's law

Snell's law of refraction provides a relationship between the index of refraction and angle of incidence of the first medium to the index of refraction and angle of refraction of the second medium. According to this law,

$$n_1 \sin \theta_1 = n_2 \sin \theta_2 \qquad (32\text{–}5)$$

where θ_1 is the angle of incidence and θ_2 is the angle of refraction both measured with the normal to the surface at the point of incidence; n_1 and n_2 are the respective indices of refraction in the medium.

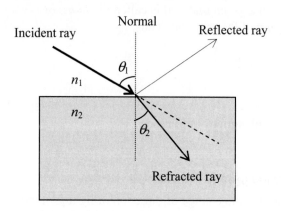

Because of refraction a straight object appears to be bent when placed in water. The depth of an object placed in water appears to be less than its actual depth.

When light passes through a flat glass, it first refracts towards the normal and then away from the normal. The net result is that the ray is displaced laterally without changing its direction.

32–6 Visible Spectrum and Dispersion

The **intensity** of light is the energy it carries in unit time and it is proportional to the square of the amplitude of the wave. The **color** of a light depends on the wavelength of light. **Visible spectrum** ranges between 400 nm to 750 nm. Wavelengths shorter than 400 nm are called **ultraviolet** (UV) and wavelengths greater than 750 nm are called **infrared** (IR).

White light is a mixture of all visible wavelengths. When it passes through a prism, different component colors refract by different amounts (because index of refraction is different for different colors) and as a result it is spread into its component colors. This spreading of light is called **dispersion**.

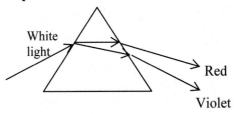

Rainbows are caused by the dispersion of sunlight in rain droplets. Diamonds show their brilliance from a combination of dispersion and total internal reflection.

32–7 Total Internal Reflection; Fiber Optics

Since a light ray bends away from the normal when it moves from a material to another material where index of refraction is less, at a certain angle of incidence the angle of refraction will be 90°. This angle of incidence is called the **critical angle**, θ_c.

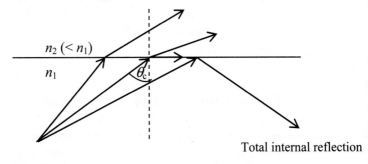

Total internal reflection

From Snell's law

$$\sin \theta_c = \frac{n_2}{n_1} \sin 90° = \frac{n_2}{n_1} \qquad (32\text{--}6)$$

When the angle of incidence is greater than the critical angle, *all the light is reflected.* This is called **total internal reflection**.

Total internal reflection by prisms is used in binoculars, periscopes, and other optical instruments. The advantage is that very nearly 100 % of the light is reflected.

Fiber Optics

Light is transmitted through optical fibers using total internal reflection. An *optical fiber* is composed of a core with a high index of refraction and a cladding with a low index of refraction. As light travels through the fiber it undergoes many internal reflections until it emerges from the opposite end of the fiber. Important applications of fiber optic cables are in telecommunications and medicine. In medical instruments such as bronchoscopes (for viewing lungs), colonoscopes (colon), and endoscopes (stomach) light is sent using a set of fibers to illuminate the appropriate internal part of the human organ and the reflected light is returned using another set of fibers to view the image of the part directly or on a TV monitor or film.

*32–8 Refraction at a Spherical Surface

Refraction of light occurs at the spherical boundary of a transparent material, such as one face of a lens or the cornea of the eye. In general, if an object is located in a medium of index of refraction n_1, rays from each point of the object will enter a medium whose index of refraction is n_2. All rays leaving a point O on the object will be focused at a single point I if we consider only paraxial rays (rays that make small angles with the axis). In this case, the following equation is found to hold:

$$\frac{n_1}{d_o} + \frac{n_2}{d_i} = \frac{n_2 - n_1}{R} \qquad (32\text{--}7)$$

where d_o and d_i are object and image distances, respectively, and R is the radius of curvature of the spherical baoundary.

Tips for Solving the Problems

1. An image formed by a plane mirror can easily be drawn using the law of reflection (angle of incidence = angle of reflection). Keep in mind that the image is virtual, has the same size as the object, and is formed on the other side of the mirror at a distance that is the same as the distance of the object from the mirror.

2. When using the expression $\frac{1}{d_o} + \frac{1}{d_i} = \frac{1}{f}$, a common mistake that students make is to determine the value of the left-hand side and use that as the result for f. Remember that the left-hand side gives the value for $1/f$, so you need to invert the result to find f. Also remember that you cannot add d_o and d_i and take the reciprocal of the sum to get the value of the left-hand side. Note $f = r/2$.

3. Keep in mind the sign conventions for mirrors. The focal length for a concave mirror is positive, whereas

the focal length for a convex mirror is negative. Object and image distances are positive in front of the mirror and negative if behind the mirror.

4. The magnification is positive if the image is upright, and negative if the image is inverted. The magnitude of m is equal to h_i/h_o or d_i/d_o.

5. The speed of light in a medium $v = c/n$ where n is the index of refraction of the medium.

6. A common mistake made when using the condition $\sin \theta_c = n_2/n_1$ to determine the critical angle or an unknown index of refraction is to use the wrong values for n_2 and n_1. Remember that the left-hand side is less than 1 (since the magnitude of the sine of an angle is always less than 1 except when the angle is 90° in which case the value is 1). Thus, n_2 must be less than n_1 in this expression.

7. Use the relation $\dfrac{n_1}{d_o} + \dfrac{n_2}{d_i} = \dfrac{n_2 - n_1}{R}$ for problems involving refraction at a single spherical boundary.

CHAPTER 33
Lenses and Optical Instruments

This chapter discusses image formation by lenses. In particular, lens equation, human eye, telescope, microscope, and aberration in images are described.

Important Concepts

Thin lens equation
Lens combination
Lensmaker's equation
The human eye
Near point
Far point
Nearsightedness
Farsightedness
The camera
Digital camera
Magnifying glass
Angular magnification
Telescope
Astronomical telescope
Reflecting telescope
Terrestrial telescope
Compound microscope
Aberration
Spherical aberration
Coma
Curvature of the field
Distortion
Chromatic aberration
Achromatic doublet

33–1 Thin Lenses; Ray Tracing

A lens is a transparent piece of a material that can form an image by refracting light. The two faces can be convex, concave, or plane. Lenses are used in eyeglasses, cameras, magnifying glasses, telescopes, binoculars, microscopes, and medical instruments.

The **axis** of a lens is a straight line that passes through the center of the lens and is perpendicular to its two surfaces. If rays parallel to the axis are incident on a thin lens they will meet (or diverge but appear to meet when extended backward) at a point on the axis, called the **focal point**, F. The distance of F from the lens is called the **focal length**.

Convex lenses are called **converging lenses**, since they can converge incident parallel rays of light to the focal point of the lens. Concave lenses are called **diverging lenses**, since they diverge incident parallel rays of light. When the refracted rays are extended back they appear to originate at the focal point of the concave lens.

The **power**, P, of a lens is given by

$$P = \frac{1}{f} \qquad (33\text{--}1)$$

The unit of power is diopter (D), $1 \text{ D} = 1 \text{ m}^{-1}$.

The following three components are used in a ray tracing:
- Ray 1 is drawn so that an incident ray parallel to the principal axis (1) that after refraction by the lens passes through the focal point (for a converging lens) or seems to originate at the focal point (for a diverging lens)

- Ray 2 is drawn so that an incident ray (2) passing through the focal point (for a converging lens) or directed toward the focal point (for a diverging lens) that after refraction by the lens becomes parallel to the principal axis

- Ray 3 is drawn so that an incident ray passes through the center (3) of the lens without being refracted.

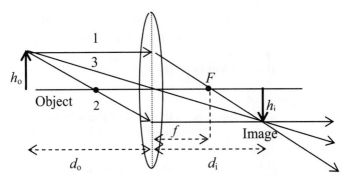

The ray tracing can be used to determine the image formed by a diverging lens. In this case the image is virtual.

33–2 The Thin Lens Equation; Magnification

The mathematical relation between the object distance, d_o, and the image distance, d_i, of a lens is called the **lens equation**. The equation for a lens of focal length f is

$$\frac{1}{d_o} + \frac{1}{d_i} = \frac{1}{f}$$ (33–2)

The lateral magnification is given by

$$m = \frac{h_i}{h_o} = -\frac{d_i}{d_o} \qquad (33\text{--}3)$$

Sign Conventions

Focal length:
> \+ for a converging lens
> − for a diverging lens

Object distance:
> \+ if the object is on the left of the lens (that is on the side of the lens from which light is coming)
> − if the object is on the right of the lens

Image distance:
> \+ if the image is on the right of the lens (that is on the opposite side of the lens from where light is coming)
> − if the image is on the left of the lens

Magnification:
> \+ if the image is upright
> − if the image is inverted.

Problem Solving for Lenses

1. Draw a ray diagram as precisely as possible, showing at least two, and preferably three rays.
2. Follow the sign conventions.
3. Use Equations (33–2) and (33–3) to solve for unknowns.
4. Check that the analytic results agree with those from your ray diagram.

33–3 Combinations of Lenses

Note the following points when solving problems involving an image formed by a lens combination:

- The image formed by the first lens alone becomes the *object* for the second lens.

- The total magnification is the *product* of the individual magnifications of each lens.

*33–4 The Lensmaker's Equation

The **lensmaker's equation** relates the focal length of a lens to the radii of curvature of its two surfaces and the index of refraction. The equation is

$$\frac{1}{f} = (n - 1)\left(\frac{1}{R_1} + \frac{1}{R_2}\right) \qquad (33\text{–}4)$$

where R_1 and R_2 are radii of curvature of the two surfaces of the lens. R_1 and R_2 are considered positive if both surfaces are convex, but for a concave surface the corresponding radius must be considered *negative*. Equation (23–10) is symmetrical in R_1 and R_2; that is, the focal length f is the same if the lens is turned around so that the light is incident on the other surface.

33–5 Cameras: Film and Digital

The basic elements of a **camera** are a lens, a light-tight box, a shutter, and photographic film (in a traditional camera) or electronic sensor (in a digital camera). When the shutter is opened, light from objects in the field of view passes through the lens to form an image on the film or sensor.

Digital Cameras, Electronic Sensors (CCD, CMOS)

In digital cameras two types of sensors are used: CCD and CMOS. A CCD (charged-coupled device) sensor consists of millions of tiny picture elments, called "pixels." Light reaching any pixel liberates electrons depending on the light intensity. Conducting electrodes carry each pixel's charge to a central processor that allows reformation of the image later on a computer screen or printer. CMOS (complementary metal oxide sensor) incorporates some electronics within each pixel, allowing parallel readout. Color is achieved by red, green, and blue filters over alternating pixels, similar to a color CRT or LCD screen.

*Digital Artifacts

Digital cameras can produce artificial effects in the image resulting from the imaging process. One example is using the "mosaic" of pixels. In a more advanced technology a semitransparent silicon semiconductor layer system is used wherein different wavelengths of light penetrate silicon to different depths. Each pixel is a sandwich of partly transparent layers, one for each color. The top layer can absorb blue light, the second layer absorbs green and the bottom layer detects the red. All three colors are detected by each pixel resulting in better color resolution and fewer artifacts.

Camera Adjustments

The three main adjustments on good-quality cameras are: shutter speed, f-stop, and focusing.

Exposure Time or Shutter Speed: This refers to how quickly the digital sensor can take an accurate reading or how long a film camera shutter is open and the film is exposed. Shutter speed varies from a second or more to 1/100 s or less.

f-**stop**: The amount of light reaching the film must be carefully controlled to avoid underexposure or overexposure. To control the exposure, the size of the opening of a "stop" or iris diaphragm, placed behind the lens, is varied for different shutter speeds. The **f-stop** of a lens relates the diameter, D, of the opening to the focal length, f, of the lens: f-stop $= f/D$. The faster the shutter speed or darker the day, the greater must be the opening for proper exposure, thus, giving the smallest *f*-stop number.

Focusing: A camera is focused for sharpest image by moving the lens closer to or farther away from the film. The image is formed at the focal distance for objects at infinity and for closer objects the image distance is more than the focal length.

When a lens is focused on a nearby object, the distant object will be out of focus and produces an image consisting of circles, called **circles of confusion**. For a given distance setting, the circle of confusion will be small enough over a range of distances that the image is reasonably sharp. This is known as the **depth of the field**. The depth of the field depends on the lens opening—the smaller the opening, the greater the depth of field.

***Picture sharpness**: The sharpness of pictures depends not only on accurate focusing and short exposure times but also on the graininess of the film or the number of pixels for a digital camera.

***Telephoto and wide-angled:** Depending on the focal length and film size, camera lenses are called normal, telephoto, and wide angle. A **normal lens** covers the film with a field of view corresponding approximately to that of normal vision. The focal length of a normal lens is about

50 mm for 35-mm film. A **telephoto lens** magnifies images and they have longer focal lengths than a normal lens. A wider field of view is included and objects appear smaller for a **wide-angle lens**, whose focal length is shorter than a normal lens. For a zoom lens its focal length can be changed so that you zoom up to, or away from, the subject as its focal length is changed. For a digital camera **optical zoom** means the ability of the camera to change its focal length maintaining resolution; hower, the **digital zoom** just enlarges the dots (pixels) with loss of sharpness.

In a **single-view reflex** (SLR) camera the image is viewed through the lens with the help of a mirror and prism. The mirror hangs at 45° behind the lens and flips up out of the way just before the shutter opens.

33–6 The Human Eye; Corrective Lenses

The optical function of the human eye is similar to that of a camera. Light enters the human eye through the **cornea** and *lens*. The eye forms a real inverted image behind the eye on the **retina** by refraction. The eye can control the amount of light reaching the retina by changing an opening, called the **pupil**, at the center of the iris. The retina is light-sensitive and consists of millions of structures called *rods* and *cones*. When stimulated by light, these structures send electric impulses through the optic nerves to the brain.

Ciliary muscles change the shape, and hence focal length, of the eye lens to focus the image on the retina. This process is called **accommodation**. When the eye is viewing a distant object, the ciliary muscles are relaxed, but when the eye is viewing a nearby object, the ciliary muscles are tensed to give a greater curvature to the lens.

The **near point** is the closest distance of an object from the eye that still produces a clear sharp image on the retina. For an average young person the near point distance is about 25 cm. The **far point** is the greatest distance at which a fully relaxed eye can focus. The normal far point distance is infinity.

Nearsightedness or *myopia* is a defect of the human eye in which clear vision is restricted to nearby objects. The far point is not at infinity but is instead at a finite distance from the eye. This defect can be corrected by using a suitable divergence lens in front of the eye.

Farsightedness or *hyperopia* is a defect of the human eye which cannot focus nearby objects. The near point is much farther from the eye than the typical value of 25 cm. This defect can be corrected by using a suitable convergence lens in front of the eye. *Presbyopia* refers to the lessening ability of the eye as one ages. Presbyopia is also compensated using converging lenses.

Astigmatism is caused by out-of-round cornea or lens so that point objects are focused as short lines, thus blurring the image. An astigmatic eye focuses light at a shorter distance in one pane (say, vertical plane) than it does for lights in its orthogonal plane (say, horizontal plane). Astigmatism is corrected by using a compensating cylindrical lens.

33–7 Magnifying Glass

A **magnifying glass** is a simple converging lens that can make objects appear to be many times larger than their actual sizes. A magnifying glass works by moving the near point closer to the eye. Since the distance is reduced, the angular size is increased. The factor by which an image is enlarged is called the **angular magnification**, or **magnifying power**, *M*.

Angular magnification is defined as

$$M = \frac{\theta'}{\theta} \qquad (33\text{--}5)$$

where θ' is the angular size of the final image produced by an optical instrument, and θ is the reference angular size of the object seen without the optical instrument.

The angular magnification produced by a magnifying glass of focal length f for a person with near point distance N:

$$M = \frac{N}{f} \qquad \text{image at infinity} \qquad (33\text{--}6a)$$

$$M = 1 + \frac{N}{f} \qquad \text{image at near point} \qquad (33\text{--}6b)$$

33–8 Telescopes

A **telescope** is an optical instrument that is used for seeing and magnifying distant objects such as stars and galaxies. The common type of telescope, called **refracting** or sometimes called **Keplerian**, contains two converging lenses located at opposite sides of a long tube. The lens closest to the object is called the **objective lens**, which focuses the light from distant objects at its focal length, f_0. The **eyepiece** magnifies the image formed by the objective and produces an inverted, virtual image. If the viewing eye is relaxed, the eye piece is adjusted so that the final image is at infinity meaning that the first real image formed by the objective is at its focal length f_e. The length of a telescope is the sum of the focal lengths of the objective and eyepiece: $f_0 + f_e$.

The total angular magnification of a telescope is

$$M = \frac{\theta'}{\theta} = -\frac{f_o}{f_e}. \qquad (33\text{--}7)$$

The objective of a telescope should be large, since the objects to be viewed are usually very dim. A **reflecting telescope** uses concave mirrors instead of a converging lens for the objective. Mirrors can be much thinner and lighter, and these have many advantages (such as they do not have chromatic aberration). Either a lens or a mirror can be used as the eyepiece. The largest telescopes are all reflectors.

A **terrestrial telescope** provides an upright final image. In the **Galilean** type, a diverging lens is used as an eyepiece which intercepts the converging rays from the objective before they reach a focus and acts to form a virtual upright image. In the other type, often called a **spyglass,** a third lens (field lens) is used to form an upright image. In a **prism binocular**, the objective and eyepiece are converging lenses. Two prisms are used to reflect rays by total internal reflection and to produce an upright image.

*33–9 Compound Microscope

The **compound microscope** uses two converging lenses in combination—an objective and an eyepiece. The design is different from a telescope because a microscope is used to magnify objects that are very close. The objective has a relatively short focal length. The object to be viewed is placed just outside the focal length of the objective. The real, inverted, and enlarged image formed by the objective serves as the object of the eyepiece. The eyepiece is a magnifier and gives additional magnification.

The lateral magnification by the objective

$$M_0 = \frac{l - f_e}{d_o} \qquad (33\text{–}8)$$

where d_o is the object distance and l is the distance between the lenses.

The angular magnification by the eye piece is

$$M_e = \frac{N}{f_e} \qquad (33\text{–}9)$$

where N is the near point distance.

The total magnification is given by

$$M = M_e m_o \approx \frac{Nl}{f_e f_o} \qquad (33\text{–}10)$$

*33–10 Aberrations of Lenses and Mirrors

An **aberration** is a defect in the formation of images by a lens that deviates from its ideal behavior.

For a point object on the axis of a lens, the rays after passing through the outer regions of the lens are brought to focus at a different point from those that pass through the center of the lens. This is called **spherical aberration** and consequently the image of the point object will not be a point but a tiny circular patch of light. At some point on the axis of the lens the circle will have the smallest diameter, which is called the **circle of least confusion**. This aberration is related to the shape of the lens, so a lens must be polished to a precise shape to prevent spherical aberration. It is usually reduced using lenses in combination and using only the central part of the lenses.

For objects off the lens axis, other aberrations occur. These include **coma** (image of a point is comet shaped rather than a circle) and **off-axis astigmatism**. Also, the image points for objects off the axis but at the same distance from the lens fall on a curved surface instead of a flat plane. This aberration is called the **curvature of the field**. **Distortion** is another aberration that arises because of the variation of magnification at different distances from the lens axis.

Chromatic aberration arises when different colors are focused at different points on the principal axis of the lens. This happens because the index of refraction of the material from which the lens is made is different for different colors. As a result, the lens forms a fringe of colors around the image.

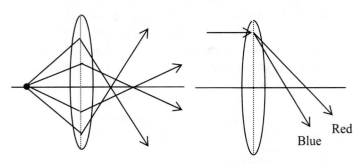

Spherical aberration   Chromatic aberration

Chromatic aberration can be prevented or reduced by combining two lenses to form a compound lens. A suitable combination of converging and diverging lenses can make the effects of the divergent refractions cancel, and bring different colors to the same focus. Such a lens combination is said to be **achromatic doublet**.

411

Spherical aberration is minimized for the human eye because (i) the cornea is less curved at the edges than at the center, and (ii) the lens is less dense at the edges than at the center. Chromatic aberration is partially reduced because the lens absorbs the shorter wavelengths appreciably and the retina is less sensitive to the blue and violet wavelengths.

Spherical mirrors suffer spherical aberration but do not show chromatic aberration.

Tips for Solving the Problems

1. When using the expression $\dfrac{1}{d_o} + \dfrac{1}{d_i} = \dfrac{1}{f}$, a common mistake that students make is to determine the value of the left-hand side and use that as the result for f. Remember that the left-hand side gives the value for $1/f$, so you need to invert the result to find f. Also remember that you cannot add d_o and d_i and take the reciprocal of the sum to get the value of the left-hand side.

2. The magnification is positive if the image is upright, and negative if the image is inverted.

3. Remember that the focal length of a convex lens is considered to be positive, and the focal length of a concave lens is considered to be negative. The sign conventions for object and image distances as well as for magnification for image formation by lenses are the same as those for images formed by mirrors.

4. It is a good habit to draw the ray diagram when solving a problem. If you draw the sketch to scale, your result from the sketch should agree with the numerical solution of the problem.

5. Keep in mind that the total magnification of an object by a lens combination (such as compound microscopes and telescopes) is the product of the magnifications by the individual lenses.

6. Also keep in mind that for a lens combination, the image formed by the first lens acts as the object for the second lens in the combination.

7. The focal length of a lens is usually given in centimeters. Remember to convert the focal length to meters. The refractive power will be in diopters only if the focal length is expressed in meters.

8. For an astronomical telescope the total angular magnification $M = -f_o/f_e$ and the length of the telescope $= f_o + f_e$. This information will be useful to solve problems in Section 33–8.

CHAPTER 34
The Wave Nature of Light; Interference

This chapter discusses the wave properties of light, including interference, diffraction, and polarization. These properties are responsible for a variety of optical phenomena.

Important Concepts
>Huygens' principle
>Diffraction
>Constructive interference
>Destructive interference
>Double slit experiment
>Coherence
>Interference by thin films
>Newton's rings
>Michelson interferometer

34–1 Waves Versus Particles; Huygens' Principle and Diffraction

Light is an electromagnetic wave. A surface drawn at a particular time through all points of a wave having the same phase of motion is called a wave front. The future position of a wavefront can be obtained using **Huygens' principle** which states that: *Every point on a wave front can be considered as a source of tiny wavelets that spread out in the forward direction at the speed of the wave. The new wave front is the envelope of all the wavelets obtained by drawing a tangent to all of them.*

Bending of light as it passes through a small hole or around an obstacle is called **diffraction**. Diffraction is a property of all waves. Huygens' principle is useful in explaining diffraction. Diffraction of sound waves is obvious, as one can hear a person talking even when the person is out of sight. Diffraction of light is very small compared with that of sound, since its wavelength is very short (about 10^{-7} m for light compared with about a meter for sound).

34–2 Huygens' Principle and the Law of Refraction

Huygens' principle can be used to explain the law of refraction. For example, when light travels from air into water, it bends toward the normal. The effect can be shown by constructing wave front for the incoming light if we assume that the speed of light in the second medium (water) is less ($v_2 < v_1$). Snell's law of refraction follows directly given that the speed of light v in any medium is related to the speed in a vacuum c and the index of refraction by the relation $v = c/n$. By constructing the wave front, we can show from the geometry that

$$\frac{\sin \theta_1}{\sin \theta_2} = \frac{v_1}{v_2}$$

where θ_1 and θ_2 are angles of incidence and angle of refraction, respectively. Since $v_1 = c/n_1$ and $v_2 = c/n_2$,

$$n_1 \sin \theta_1 = n_2 \sin \theta_2.$$

If λ_1 and λ_2 are the wavelengths in the two media, respectively,

$$\frac{\lambda_2}{\lambda_1} = \frac{v_2 t}{v_1 t} = \frac{v_2}{v_1} = \frac{n_1}{n_2}.$$

If medium 1 is a vacuum, $n_1 = 1$, $v_1 = c$, then the wavelength in another medium of index of refraction n (= n_2), is

$$\lambda_n = \frac{\lambda}{n} \tag{34-1}$$

On a hot day a mirage of water is formed on a road. Hot air is less dense than cooler air and the index of refraction is slightly lower in the hot air. We can explain how images are formed by constructing the wave fronts, which shows that the rays are bent and seem to come from below as if reflected off the road.

34–3 Interference—Young's Double-Slit Experiment

The interference of light can be seen in Young's experiment. In the experiment a monochromatic beam of light, after passing through a small slit, illuminates two other slits on a screen. The light passing through the two slits forms bright and dark interference "fringes" on a distant screen, showing constructive interference and destructive interference of light at different points.

Constructive interference (bright fringe) occurs when the two waves travel the same distance (such as at the center of the screen) so that they are in phase and when the path of one wave differs by one wavelength or any whole number of wavelengths. **Destructive interference** (dark fringe) occurs when the paths of two rays differ by one half wavelength or any odd number of half-wavelength, so that the waves are exactly out of phase. Constructive and destructive interferences of two traveling waves are shown.

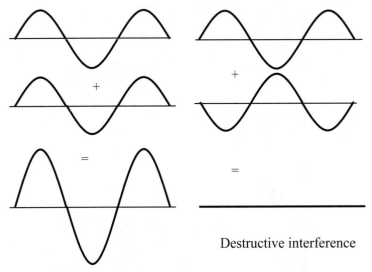

Constructive interference

Destructive interference

In the double slit experiment, constructive interference occurs at different angles θ given by

$$d \sin \theta = m\lambda, \qquad m = 0, 1, 2,\dots \qquad (34\text{–}2a)$$

where λ is the wavelength of the light, and d is the separation between the two slits. Different values of m correspond to different bright fringes.

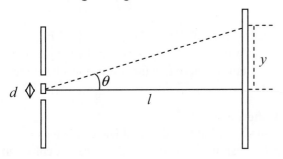

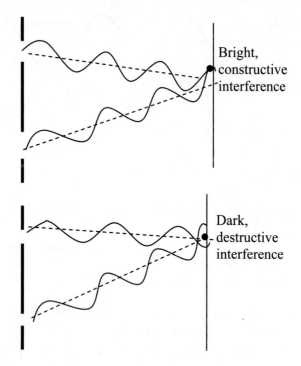

Bright, constructive interference

Dark, destructive interference

Destructive interference occurs at angles:

$$d \sin \theta = \left(m + \frac{1}{2} \right) \lambda, \, m = 0, 1, 2,\ldots \quad (34\text{--}2b)$$

The location of a given bright or dark fringe in terms of its linear distance, x, from the central bright fringe is given by $x = L \tan \theta$ where L is the distance of the screen from the slits.

Coherence
Sources of light where light waves have a constant phase relationship with one another are called **coherent sources**. Light sources that are not coherent are called **incoherent**

sources. Ordinary light source, such as incandescent lamps are incoherent sources. A laser and the two slits in Young's experiment are coherent sources.

*34–4 Intensity in the Double-Slit Interference Pattern

The intensity of light at a point is proportional to the square of electric field E. In a double-slit experiment, the electric field at a point P (at an angle θ) is

$$E_\theta = E_1 + E_2$$

where E_1 and E_2 (which are essentially parallel on a screen far away compared to the slit separation) are the electric fields from the two slits, and, which given by

$$E_1 = E_{10} \sin \omega t$$

$$E_2 = E_{10} \sin (\omega t + \delta) \qquad (34\text{–}3)$$

The phase difference δ can be shown to be related to θ by

$$\delta = \frac{2\pi}{\lambda} d \sin \theta \qquad (34\text{–}4)$$

E_θ can be determined by adding E_1 and E_2 using a phasor diagram (note that E_1 and E_2 are sine functions differing in phase by δ). The result shows that

$$E_\theta = 2E_0 \sin\left(\omega t + \frac{\delta}{2}\right)\cos\frac{\delta}{2} \qquad (34\text{–}5)$$

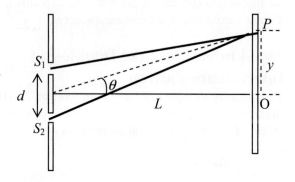

Let I_0 be the intensity at point O, the center of the screen, where $\theta = \delta = 0$. Then, $I_0 \propto (E_{10} + E_{20})^2 = (2E_{10})^2$. This gives

$$\frac{I_\theta}{I_0} = \frac{E_{\theta0}^2}{(2E_0)^2} = \cos^2 \frac{\delta}{2}$$

Thus, the intensity of light I_θ at any point is given by

$$I_\theta = I_0 \cos^2 \frac{\delta}{2}$$

$$= I_0 \cos^2 \frac{\pi d \sin \theta}{\lambda} \qquad (34\text{–}6)$$

This shows that I_θ is maximum for $\delta = 0, 2\pi, 4\pi,\dots$ etc., giving, from Equation (34–4), the condition of maxima as:

$$d \sin \theta = m\lambda, \qquad m = 0, 1, 2,\dots$$

Similarly, I_θ is minimum for $\delta = \pi, 3\pi, 5\pi, \ldots$ etc., giving the condition of minima as:

$$d \sin \theta = \left(m + \frac{1}{2} \right) \lambda, \qquad m = 0, 1, 2, \ldots$$

In situations where L (the distance between the slit and the screen) $\gg d$, if $L \gg y$, $\sin \theta = y/L$, giving

$$I_\theta = I_0 \cos \left[\frac{\pi d \sin \theta}{\lambda} \right]^2 \qquad (34\text{–}7)$$

The intensity pattern expressed by the preceding equation is shown below.

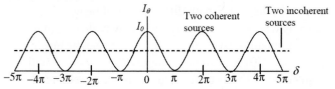

In reality, the resulting pattern will show that the center maximum is strongest and each succeeding maxium on each side is less strong because of the diffraction effect.

34–5 Interference in Thin Films

Interference of light is observed in many everyday phenomena such as the bright colors reflected from soap bubbles and from thin films on water. The colors are the result of constructive interference between light reflected from the two surfaces of the thin film.

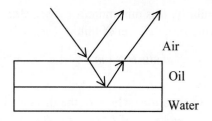

If the path difference between two interfering rays is one or a whole number of λ (where λ is the wavelength in the film), the two waves (shown above) will interfere constructively and the light will be bright. But if the path difference equals $\lambda/2$, $3\lambda/2$, ···, the two waves interfere destructively.

For white light, the path difference will be equal to λ or $m\lambda$ (m is whole integer) only for one wavelength at a given viewing angle and this color will be seen as bright. For a slightly different viewing angle different color will undergo constructive interference. Thus for an extended white source, a series of bright colors will be seen next to one another. The thickness of the film may be different and will have an effect in the interference pattern.

If a curved piece of glass with a spherical cross section is placed on a flat glass, the interference pattern formed by reflection consists of concentric circular rings of alternate darkness and brightness. These are called **Newton's rings**. The center of the rings, where the glass plates meet, is a dark spot. This is because one of the waves undergoes a change of phase of $180°$ upon reflection. In general, *a beam of light reflected by a material whose index of refraction is greater than that of the material in which it is traveling changes phases by* $180°$. Fringes are more closely spaced as their distance from the center increases, since the curved surface of the upper glass moves away from the lower glass plate at a progressively faster rate.

When white light is incident on the thin wedge a colorful series of fringes are seen because of constructive interference in the reflected light at locations along the wedge for different wavelengths.

*Color in a Thin Soap Film

A soap bubble is a thin spherical shell or film. There is air on both sides of the film and the variation of the thickness of soap bubble gives rise to bright colors from the soap bubble.

In a thin film of soapy water that has stood vertically we see a series of separated colors like a rainbow because of the constructive (and destructive) interference of different colors at different thickness.

Lens coating

A glass surface reflects about 4 % of incident light. A thin coating on the surface can reduce reflection. The thickness of the film is chosen so that light reflected from the front and rear surfaces interfere destructively (at least for one wavelength). The coating material has an index of refraction (for that wavelength) which is the geometric mean of air and glass. A single coating cannot eliminate reflection for all colors. Lenses containing two or three separate coatings more effectively reduce reflection for wider wavelengths.

*34–6 Michelson Interferometer

In a **Michelson interferometer** monochromatic light from source is split into two by a **beam splitter**, which is oriented at 45° with respect to the incident light.

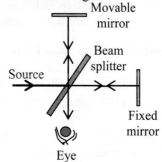

The light reflected by the beam splitter hits a movable mirror and is reflected. The transmitted light is reflected by another fixed mirror. These two reflected beams interfere to produce an interference pattern.

If the two path lengths are identical, the two coherent beams constructively interfere (producing a brightness), but if the movable mirror is moved a distance of $\lambda/4$ one beam travels an extra distance of $\lambda/2$ (because it travels back and forth over the distance $\lambda/4$) and they interfere destructively (producing a darkness). Wavelength of the source as well as other length measurements can be precisely made with a Michelson interferometer by counting the number of fringes shifted. .

*34–7 Luminous Intensity

Intensity of any electromagnetic wave (including light) is expressed in W/m^2 and its total power output is expressed in watts. For visible light, since the eye is not equally sensitive for all visible wavelengths, brightness is expressed by **luminous flux**, F_l, whose unit is *lumen*. One lumen is defined as the brightness of $1/60$ cm^2 of platinum surface at its melting temperature (1770°C).

Luminous intensity is defined as the luminous flux per unit solid angle (sr). It is expressed in **candela** (cd); 1

cd = 1 lm/sr. The **illuminance** is defined as the luminous flux per unit area of surface.

Tips for Solving the Problems

1. For *constructive interference*, the path difference between the two waves is $m\lambda$, where $m = 0, 1, 2,...$; and for *destructive interference*, the path difference is $(m + \frac{1}{2})\lambda$, where $m = 0, 1, 2,\cdots$. This information will be useful for solving many problems in this chapter.

2. You may not need to remember all the equations for the double-slit experiment; the conditions for bright and dark fringes can be easily derived from tip 1.

3. The intensity at a point in a double-slit experiment is given by $I_\theta = I_0 \cos\left[\dfrac{\pi d \sin\theta}{\lambda}\right]^2$. $\sin\theta \approx \tan\theta = y/l$.

4. To solve problems on interference in reflected beams remember that the phase of a light wave changes by half its wavelength (phase difference of 180°) when it is reflected from a region with a higher index of refraction. Phase does not change when light is reflected from a region with a lower index of refraction.

5. In a Michelson interferometer keep in mind that when a mirror moves by a distance d, the beam travels an extra distance of $2d$ (because it travels back and forth over the distance d).

CHAPTER 35
Diffraction and Polarization

This chapter discusses the wave properties of light, including diffraction and polarization. These properties are responsible for a variety of optical phenomena.

Important Concepts

        Diffraction
        Single-slit diffraction
        Diffraction grating
        Raleigh criterion
        Resolving power
        Line spectrum and continuous spectrum
        Polarization
        Polaroid
        Polarizer
        Analyzer
        Polarizing angle or Brewster's angle
        X-ray diffraction
        Bragg's equation
        Liquid crystal display
        Scattering

35–1 Diffraction by a Single Slit or Disk

Because of diffraction of light, the shadow of an object (such as a solid disk) does not have sharp boundaries. Instead, the edges of the shadow show diffraction fringes. At the center of the shadow the light interferes constructively and produces a bright spot.

    A beam of monochromatic light is diffracted as it passes through a single narrow slit. According to Huygen's principle, each point on the slit behaves as a source of a new light wave that radiates energy. The interference between these new light waves gives rise to the diffraction pattern.

The first *minimum* or dark fringe occurs at an angle θ in single-slit diffraction for

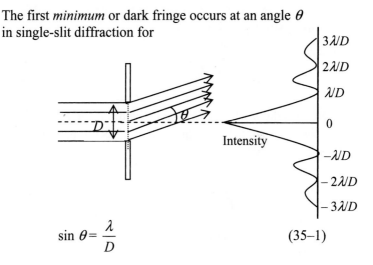

$$\sin \theta = \frac{\lambda}{D} \qquad (35\text{–}1)$$

In this expression, D is the width of the slit, and λ is the wavelength of the light. In general, minima occurs at angles given by

$$D \sin \theta = m\lambda, \qquad m = \pm 1, \pm 2, \pm 3... \qquad (35\text{–}2)$$

Bright fringes appear approximately halfway between successive dark fringes. The central fringe is also bright and is about twice as wide as the other bright fringes. The diffraction pattern is symmetrical about the central bright fringe.

*35–2 Intensity in Single-Slit Diffraction Pattern

It can be shown from geometric analysis that the phase difference between the rays from the top and bottom edges of a single slit is given by

$$\beta = \frac{2\pi}{\lambda} D \sin \theta \qquad (35\text{--}3)$$

where D is the total width of the slit and θ is the angle of diffraction. Also, the wave amplitude E_θ as a function of θ is given by

$$E_\theta = E_0 \frac{\sin \beta/2}{\beta/2}. \qquad (35\text{--}4)$$

Since intensity is proportional to the wave amplitude

$$I_\theta = I_0 \left[\frac{\sin \beta/2}{\beta/2} \right]^2. \qquad (35\text{--}5)$$

$$= I_0 \cdot \left[\frac{\sin \left(\dfrac{\pi D \sin \theta}{\lambda} \right)}{\left(\dfrac{\pi D \sin \theta}{\lambda} \right)} \right]^2 \qquad (35\text{--}6)$$

$$\text{where} \quad I_0 \propto E_0^2 \qquad (35\text{--}7)$$

Clearly when I_θ is a minimum

$$\frac{\pi D \sin\theta}{\lambda} = 0 \qquad\qquad (35\text{–}8)$$

This gives

$$D \sin\theta = m\lambda, \qquad m = 1, 2, 3\ldots$$

*35–3 Diffraction in the Double-Slit Experiment

It can be shown that in the double-slit experiment the intensity is given by

$$I_\theta = I_0 \left(\frac{\sin \beta/2}{\beta/2}\right)^2 \left(\cos\frac{\delta}{2}\right)^2. \qquad (35\text{–}9)$$

The first term in the parentheses is called the "diffraction factor" and the second one the "interference factor." The product of these two factors gives the actual intensity wherein the diffraction factor acts as a sort of envelope that limits the interference peaks.

Interference vs. diffraction

Interference and diffraction both arise from the superposition of coherent waves of different phase. The term diffraction is used when considering the superposition of many infinitesimal sources (such as when we subdivide a source into many parts), whereas the term interference is used when considering the superposition of a finite (usually small) number of coherent sources.

35–4 Limits of Resolution; Circular Aperture

The ability to visually distinguish closely spaced objects is called the **resolution** of a system. The closer the two images are and still can be seen as distinct, the higher the resolution. The one factor that limits the resolution is the aberrations. The other factor is the diffraction, which places a natural limit on the resolution of a visual system because of the wave nature of light.

A **circular aperture** produces a diffraction pattern consisting of concentric alternate dark and bright fringes. The central maximum has an angular half width given by

$\theta = \dfrac{1.22\,\lambda}{D}$ where D is the diameter of the aperture.

According to **Raleigh's criterion** of resolution, *two images are just resolvable when the center of the diffraction pattern of one is directly over the first minimum in the diffraction pattern of the other*.

Since the first minimum occurs at an angle $\theta = 1.22$ λ/D from the central maximum, two objects will appear as separate only if their angular separation is greater than θ, given by

$$\theta = \frac{1.22\,\lambda}{D} \qquad (35\text{–}10)$$

35–5 Resolution of Telescopes and Microscopes; the λ Limit

A telescope or microscope cannot magnify an object to any desired value because of diffraction. In a microscope, the minimum separation s between two objects that can just be resolved is given by its **resolving power** (RP):

$$RP = s = f\,\theta = \frac{1.22\,\lambda f}{D} \qquad (35\text{--}11)$$

Diffraction sets an ultimate limit on the detail that can be seen of an object. The focal length f of the lens can be at best about $D/2$. In this case,

$$RP \approx \frac{\lambda}{2} \qquad (35\text{--}12)$$

Thus, we can say that it is not possible to resolve the details of objects smaller than the wavelength of the light.

*35–6 Resolution of the Human Eye and Useful Magnification

For $\lambda = 550$ nm (where eye sensitivity is greatest), the diffraction limit is $\theta = 1.22\ \lambda/D$, which is about 8×10^{-5} rad to 6×10^{-4} rad. The eye is about 2 cm long. This corresponds to a resolving power s of about 2 μm (at best) to 10 μm (at the worst). Spherical and chromatic aberration limit the resolution to about 10 μm. The result is that the eye can resolve objects whose angular separation is about 5×10^{-4} rad at the best, corresponding to a separation of 1 cm at a distance of 20 m.

At the near point (25 cm) the eye can just resolve objects that are about 0.1 mm apart.

35–7 Diffraction Gratings

A large number of parallel, closely spaced slits through which a beam of light can pass is called a **diffraction grating**. Gratings containing 10 000 lines per cm are common. The diffraction pattern consists of sharp and widely spaced bright fringes.

Bright fringes (constructive interference) are located at different angles θ relative to the central fringe. These bright fringes are called *principal maxima*. These are created by constructive interference between light waves from different slits in the grating.

The locations of the principal maxima are given by

$$\sin \theta = \frac{m\lambda}{d} \qquad m = 0, 1, 2,\ldots \qquad (35\text{–}13)$$

where d is the separation between two consecutive slits.

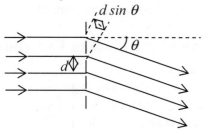

If the incident consists of more than one wavelength, then for all orders other than $m = 0$ each wavelength will produce a maximum at a different angle. If the incident light is white, the central maximum will be a sharp white peak but there will be a distinct spectrum of colors spread out over a certain angular width, producing a **spectrum** of light.

35–8 The Spectrometer and Spectroscopy

In a **spectrometer** or **spectroscope**, light from a source passes through a narrow slit and parallel rays of light formed by an inside lens fall on a diffraction grating to be separated into its component colors. A movable telescope brings the rays to a focus and the angle θ that corresponds to a diffraction peak (usually first order) is measured. The wavelength of the line can then be accurately determined from the relation

$$\lambda = \frac{d}{m} \sin \theta \qquad (35\text{–}14)$$

A reflection grating or prism is also used in some spectrometers.

Spectroscopy is used to identify atoms and molecules. A gas, when excited, gives a characteristic **line spectrum** consisting of discrete lines or colors. A solid, when heated, produces a **continuous spectrum** including a wide range of wavelengths. Atoms and molecules can absorb light at the same wavelengths at which they emit light. As a result, when a continuous spectrum of light passes through a gas, it absorbs the colors of the spectrum that it can emit, showing *dark lines* in the resulting spectrum. The solar spectrum contains a number of dark lines, called **absorption lines**, due to the absorption by atoms and molecules in the cooler atmosphere of the sun and by the atoms and molecules in the Earth's atmosphere. The presence of elements in the atmosphere of planets, interstellar space, the Sun and other stars are determined by spectroscopy.

*35–9 Peak Widths and Resolving Power of a Diffraction Grating

If $\Delta \theta_0$ is the angular position of the minimum next to the peak at $\theta = 0$, it can be shown that for a diffraction grating

$$\text{Sin } \Delta \theta_0 = \frac{\lambda}{Nd} \qquad (35\text{–}15)$$

433

Since N is usually large, $\Delta\theta_0$ is very small, so $\sin\Delta\theta_0 \approx \Delta\theta_0$. Thus,

$$\Delta\theta_0 = \frac{\lambda}{Nd} \qquad (35\text{--}16)$$

The **resolving power** R of a grating is defined as

$$R = \frac{\lambda}{\Delta\lambda} \qquad (35\text{--}17)$$

It can be shown that

$$R = \frac{\lambda}{\Delta\lambda} = Nm \qquad (35\text{--}18)$$

If R is known, the minimum separation $\Delta\lambda$ between two wavelengths near l is

$$\Delta\lambda = \frac{\lambda}{R} \qquad (35\text{--}19)$$

35–10 X-Rays and X-Ray Diffraction

X-rays are emitted when electrons emitted by a heated filament in a vacuum tube are accelerated by high voltage (30 kV to 150 kV) and strike the surface of the anode, called the target. X-rays are electromagnetic radiation and they cannot be deflected by electric and magnetic fields. The wavelengths of X-rays are about the same order as the spacing of atoms (about 0.1 nm). Since the atoms in a crystal are arranged in a regular array, crystals serve as grating for X-ray diffraction. Laue first observed X-ray diffraction pattern from zinc sulfide crystal that consisted of diffraction spots on the plate. X-ray diffraction confirms that X-rays have a wave nature of wavelength about 10^{-2} nm to 10 nm.

X-ray diffraction (or **crystallography**) has been proven to be effective for examining the microscopic world of atoms and molecules. In a simple crystal such as NaCl the atoms are arranged in an orderly cubic fashion with the atoms spaced d apart. When a beam of X-rays is incident on the crystal at an angle ϕ to the surface, the two rays reflected from the two subsequent planes of atoms will constructively interfere if the path difference between the rays is a whole number of wavelengths (λ). Thus, for constructive interference

$$m\lambda = 2d \sin \phi \quad m = 1, 2, \dots \qquad (35\text{--}20)$$

This is known as the **Bragg equation**. Knowing λ and ϕ, d can be determined. This is the basis of X-ray crystallography.

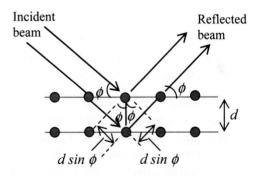

If the substance is not a single crystal but a mixture of many tiny crystals such as in a powder, then instead of a series of spots, a series of circles is obtained. Each circle corresponds to a diffraction of a particular order m.

X-ray diffraction is very useful in determining biologically important molecules such as the double helix structure of DNA, the structure of protein molecules and others.

35–11 Polarization

The **polarization** of light refers to the direction of vibration of its electric field. Ordinary light is

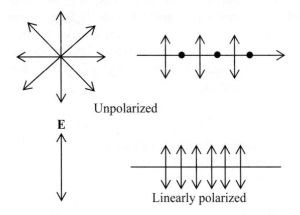

Unpolarized

E

Linearly polarized

unpolarized, which means that it has vibrations in many planes at once. The light is said to be **plane-polarized** when the vibrations are confined to a plane.

Polaroids (Polarization by Absorption)

Plane-polarized light can be obtained from unpolarized light using crystals such as tourmaline or a **polaroid sheet** that transmit the components of the electric field along a preferred direction and absorb components perpendicular to this direction. The preferred direction is called the transmission axis of the polarizer sheet or the crystal.

If a light with intensity I_0 passes through a polarizer with its transmission axis at an angle θ relative to its polarization, the transmitted intensity, I, is given by

$$I = I_0 \cos^2 \theta \qquad\qquad (35\text{–}21)$$

This relation is known as the law of Malus. The transmitted light is polarized in the direction of the polarizer.

When an unpolarized light passes through a polarizer, the transmitted intensity is reduced by half ($I = I_0/2$).

Once polarized light has been produced using a polarizing material, it is possible to use a second polarizing material to change the intensity and direction of polarization of the polarized light. The second polarizing material is called an **analyzer**. Polarizers with their transmission axes at right angles are called *crossed* polarizers ($\theta = 90°$). The transmitted intensity through crossed polarizers is zero.

Polarization by Reflection
When light strikes a nonmetallic surface at any angle other than perpendicular, the reflected light is polarized preferentially in the plane parallel to the surface. Polarized glasses are made with their axes vertical so that it eliminates the more strongly reflected horizontal component and thus reduces glare. The amount of polarization in the reflected light increases with angle varying from no polarization at normal incidence to 100 percent polarization at an angle, called the **polarization angle**, θ_p, which is given by

$$\tan \theta_p = \frac{n_2}{n_1} \qquad\qquad (35\text{–}22a)$$

where n_1 is the index of refraction of the incident medium and n_2 is that of the medium beyond the reflecting boundary. If the incident medium is air $n_1 = 1$, and

$$\tan \theta_{\mathrm{p}} = n \qquad\qquad (35\text{–}22\mathrm{b})$$

The polarization angle is also called **Brewster's angle** and the equation is called *Brewster's law*.

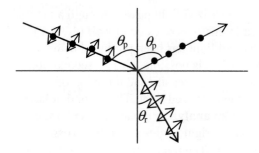

It can be shown that at Brewster's angle the reflected and transmitted rays make $90°$ to each other. That is, $\theta_{\mathrm{p}} + \theta_{\mathrm{r}} = 90°$.

*35–12 Liquid Crystal Displays (LCD)

Liquid crystals are organic materials, which at room temperature arenot full solids, or liquids. They show some orderliness of a solid, but in one dimension. Liquid crystal molecules tend to align parallel to each other but have random position (left-right, up-down).

In an LCD, each *pixel* contains a liquid crystal placed between two glass plates whose inner surfaces are brushed to form nanometer-wide parallel scratches. The two plates have scratches at $90°$ with each other and a twisted pattern of rod-like molecules is formed since the weak forces between them tend to align them with their nearest neighbors. The outer surface of the glass plate contains thin film polarizer oriented at $90°$ with each other. An incident unpolarized light becomes plane-polarized parallel to the molecules and the plane of polarization of light rotates with the molecules gradually by 90° and the light emerges through the second polarizer easily making the LCD pixel bright. If the voltage is applied to the transparent electrodes, the molecules are disoriented by electric field. Plane of polarization is not changed and light does not pass through, making the pixel dark. LCD displays are used in calculators, watches, computer screens, and others.

*35–13 Scattering of Light by the Atmosphere

Light scattered by atoms or molecules in the atmosphere is polarized when viewed at right angles to the Sun. When viewed at an intermediate angle, the light has an intermediate amount of polarization.

Scattering of light by molecules is most effective for light of short wavelength because scattering decreases as $1/\lambda^4$. Red and orange light have longer λ and are thus scattered much less than blue and violet, and that is why the sky appears blue. The sunset appears red when the Sun is viewed through the atmosphere because most of the blue light has been scattered off in different directions.

The dependence is valid for scattering objects smaller than λ (such as oxygen and nitrogen molecules whose diameters are about 0. 2 nm). Cloud contains water droplets or crystal much larger than λ, and they scatter nearly uniformly all wavelengths, making them white (or gray, if shadowed).

Tips for Solving the Problems

1. For *constructive interference*, the path difference between the two waves is $m\lambda$, where $m = 0, 1, 2,...$; and for *destructive interference*, the path difference is $(m + \frac{1}{2})\lambda$, where $m = 0, 1, 2, \cdots$.

2. To solve problems on interference in reflected beams remember that the phase of a light wave changes by half its wavelength (phase difference of 180°) when it is reflected from a region with a higher index of refraction. Phase does not change when light is reflected from a region with a lower index of refraction.

3. For problems of diffraction by a single slit use the equation $D \sin \theta = m\lambda$, where $m = \pm 1, \pm 2, \pm 3, \cdots$. Remember that this equation provides the conditions for the dark fringes and *not* for the bright fringes.

4. For a single slit the intensity at an angle θ is $I_\theta = I_0 \left[\dfrac{\sin \beta/2}{\beta/2} \right]^2$. Note that the position of the maxima can be determined by differentiating this expression.

5. The density of lines on a grating is usually provided as the number of lines per centimeter. Make sure you convert the density of lines to *number of lines per meter*, N. Also, note that the distance d between lines is $d = 1/N$ meters. The equation $\sin \theta = m\lambda/d$, $m = 0, 1, 2,...$ gives the locations of the *principal maxima*.

6. In the equation $\theta = 1.22\ \lambda/D$, λ and D must have the same unit and θ will then be in radians. A smaller value of θ means a better resolution.

7. Keep in mind that the Bragg's equation for X-ray diffraction, $m\lambda = 2d \sin \phi$ $m = 1, 2, \ldots$, provides the conditions for constructive interference.

8. When using the relation to calculate transmitted intensity $I = I_0 \cos^2 \theta$ remember θ is the angle between the polarizer axis and the plane of polarizer of the incoming light.

CHAPTER 36
The Special Theory of Relativity

This chapter describes the special theory of relativity. The theory gives the true interpretation of space and time as well as of mass and energy. The theory is needed to understand the dynamics of objects moving at speeds comparable to the speed of light.

Important Concepts

Inertial reference frame
Relativity principle
Postulates of the special relativity
Relativity of simultaneity
Time dilation
Twin paradox
Length contraction
Four-dimensional space-time
Relativistic momentum
Rest mass
Relativistic energy
Rest energy
Relativistic kinetic energy
Relativistic addition of velocities
Relativistic Doppler shift
Correspondence principle

36–1 Galilean–Newtonian Relativity

Einstein's theory of relativity deals with how events in one inertial reference frame are related to the events observed from other **inertial reference frames**. An inertial reference frame is one in which Newton's first law of motion is obeyed. All inertial reference frames move with constant velocity with respect to the others.

Galileo and Newton were aware of the **relativity principle**: *The laws of physics are same in all inertial frames of reference.* For example, if a coin is dropped by a person in a moving car, in the reference frame of the car, the coin falls straight whereas in the reference frame fixed on the earth the coin follows a parabola. This is consistent with the relativity principle since in the reference frame of the car the coin does not have any initial velocity and the law of physics predicts that the coin will fall straight down. But, in the reference frame of the earth the car has an initial velocity (equal to the velocity of the car) and the law of physics predicts that the path of the coin will be a parabola.

Galilean–Newtonian relativity assumes that both space and time are **absolute**; that is, their measurements do not change from one reference frame to another. The position and velocity are different in different frames, but, the acceleration is the same in any inertial frame. Since the force (F), mass (m), and acceleration (a) do not change, the Newton second law $F = ma$ does not change, in agreement with the relativity principle. Since the laws of mechanics are the same in any inertial frame, *all inertial frames are equivalent.*

In the last half of the nineteenth century Maxwell presented his electromagnetic theory and showed that light is an electromagnetic wave that travels with a speed of $c = 3 \times 10^8$ m/s in vacuum. It was assumed that this speed must be referenced to a particular frame called ether that permeates all space. It appeared that Maxwell's electromagnetic equations did not satisfy the relativity principle because they were not the same in all reference frames because of the relativity of velocity. They were simplest in the reference frame where $c = 3 \times 10^8$ m/s, that is, in a reference frame that could be considered absolutely at rest in ether.

*36–2 The Michelson–Morley Experiment

The objective of the Michelson–Morley experiment was to determine the speed of ether, the medium through which light was then assumed to travel with respect to the earth. The experiment used a precision Michelson interferometer. In the interferometer one light beam was assumed to travel parallel to the "ether wind" and the other beam traveled crosswise to the ether wind. These would cause a time difference of arrival of the two beams and the two beams returned out of phase. Michelson and Morley realized that they could detect the difference in phase if they rotated the apparatus by 90°, because the interference pattern between the two beams then would change. The theory predicted a movement of 0.4 fringe, which the experiment could have easily detected. But no fringe shift was observed in the experiment. The null result of the experiment is explained by the postulate of special relativity.

36–3 Postulates of the Special Theory of Relativity

Einstein's special theory is based on two postulates:

First postulate (the relativity principle): *The laws of physics have the same form in all inertial reference frames.*

Second postulate (constancy of the speed of light): *The speed of light is the same in all inertial frames of reference and is independent of the motion of the source or the observer.*

The constancy of the speed of light agrees with the result of Michelson and Morley's experiment. Recent accurate experiments confirm the constancy of the speed of light.

36–4 Simultaneity

Two events are said to be simultaneous if they occur at the same time. Einstein found from the constancy of the speed of light that the concept of simultaneity is relative. That is, if two events are simultaneous in one inertial frame it may not be simultaneous as measured in a different inertial frame. The concept of the relativity of simultaneity violates our everyday experience. The effect is noticeable only when the relative speed of the two reference frames is very large (comparable to the speed of light).

36–5 Time Dilation and the Twin Paradox

The fact that two events simultaneous to one observer may not be simultaneous to another observer suggest that time itself is not absolute but is relative.

An event is a physical occurrence they happens at a specified location at a specified time. The **proper time,** Δt_0, is the time separating two events occurring at the same location. If two events are separated by a proper time Δt_0 occurring in a reference frame that moves with a speed, v, relative to an observer, then the time measured by the observer is given by

$$\Delta t = \frac{\Delta t_0}{\sqrt{1 - v^2/c^2}} \qquad (36\text{–}1)$$

where $c = 3 \times 10^8$ m/s is the speed of light. Or, $\Delta t = \gamma \Delta t_0$,

where
$$\gamma = \frac{1}{\sqrt{1 - v^2/c^2}}$$
(36–2)

where m is called the **rest mass** of the object.

The variation of γ versus speed is shown at the right.

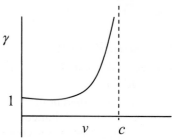

Since γ is greater than 1, $\Delta t > \Delta t_0$. This means that a clock that is moving relative to an observer runs slower than one at rest relative to the observer. This stretching of time is known as the **time dilation**. Time dilation has nothing to do with the mechanics of clocks; it arises from the nature of time itself.

- From Equation (36–1), if $v << c$, $\Delta t \approx \Delta t_0$.

- Also, as $v \rightarrow c$, $\Delta t \rightarrow \infty$. In this limit, the clock will slow down to the point of stopping.

To verify time dilation an atomic clock was carried in a jet plane and an identical clock was left at rest in the laboratory. After flying the moving clock at a high speed, it was found that the clock in the plane ran slower than the clock in the laboratory, in agreement with the time dilation.

A muon is an elementary particle whose lifetime is 2.2 μs when at rest. Experiments had shown that when a muon is traveling at high speeds, its lifetime was longer than when it is at rest in agreement with the time dilation formula.

Space travel?
According to time dilation formula, the time taken by an astronout moving at a relativistic speed, say $v = 0.999c$, to reach a star 100 light-years away is

$(100 \text{ yr}) \sqrt{1 - (0.999)^2} = 4.5$ yr. Thus, time dilation allows such a trip, but there are practical problems of achieving such high speeds.

Twin paradox
Time dilation applies equally to all processes, including biological processes. An astronaut traveling in a spaceship ages slower than one who remains on Earth.

Suppose a 20-year old twin brother takes off in a spaceship traveling at a speed very close to the speed of light to a distant star 20 light-years away and he immediately comes back while his twin brother remains on the earth. When the traveling brother returns, the earthbound twin is forty years old, while according to him time dilates for his traveling brother and his age is less, perhaps he is only 21.

Note that the consequence of special relativity applies only to the inertial frame such as for the twin brother on the earth. For the traveling twin, since he accelerates at the start and decelerates at the end, his predictions from special relativity are not correct and therefore, he cannot claim the reverse of the effects observed by his earthbound twin brother.

*Global positioning system (GPS)
Global positioning system is used to determine accurately the location of cars, airplanes, boats, and hikers. The 24 global positioning system satellites send out precise time signals using atomic clocks. By comparing the time differences with the known satellite positions, the receiver

can determine where he is on the earth. Because of the extremely high precision timing, relativistic effects must be taken into account.

36–6 Length Contraction

Length is not absolute but is relative. The length of an object depends on its speed relative to an observer. The **proper length**, L_0, is the length of an object or distance between two points whose positions are measured at the same time, as measured by an observer at rest with respect to it. An object of proper length L_0 moving with a speed v relative to an observer has a length L as given by

$$L = L_0 \sqrt{1 - v^2/c^2} = L_0/\gamma \qquad (36\text{--}3)$$

This shows that the *length of an object is measured to be shorter when it is moving relative to the observer than when it is at rest.*

- From Equation (26–2), if $v \ll c$, $L \approx L_0$.

- If $v = c$, $L = 0$.

The length contraction occurs only in the direction of motion. Lengths perpendicular to the direction of motion are not affected.

36–7 Four-Dimensional Space–Time

According to Einstein special relativity, space and time are interlinked, giving rise to the concept of **four-dimensional space–time**. Space has three dimensions and time is a fourth dimension. Space contracts but time dilates. Just as when we squeeze a balloon, one dimension gets larger and the other smaller, similarly, when we observe objects or events from different frames, a certain amount of space is exchanged for time, or vice versa. Thus, space and time are not independent of one another.

36–8 Galilean and Lorentz Transformations

Consider two inertial reference frames S and S' as shown below. S' moves at a constant velocity v to the right (x-direction). At $t = 0$ the origins of S and S' coincide. Consider an event that occurs at a point P represented by the coordinates x', y', z', in S' and time t'.

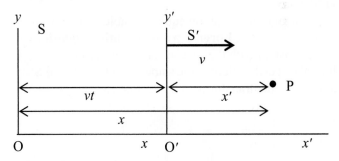

According to **Galilean transformation equations** the corresponding quantities in the S frame are given by:

$$x = x' + vt',$$

$$y = y'$$

$$z = z'$$

$$t = t' \qquad\qquad (36\text{–}4)$$

If the quantities in the S frame are known, then the S' coordinates can be obtained from the equations:

$$x' = x - vt, \, y' = y, \, z' = z \,, \, t' = t.$$

Taking derivative with respect to time on both sides we get the Galilean velocity transformation equations:

$$u_x = u'_x + vt$$

$$u_y = u'_y$$

$$u_z = u'_z \qquad (36\text{--}5)$$

where (u_x, u_y, u_z) and (u'_x, u'_y, u'_z) are x-, y-, and z-components of the velocity of the object in the frames S and S', respectively.

Galilean transformation equations hold when the speed is much less than c. **Lorentz transformation equations** are inconsistent with the special relativity according to which the speed of light is the same in both the S and S' frames. Lorentz transformation equations are:

$$x = \frac{1}{\sqrt{1 - v^2/c^2}} \, (x' + vt'),$$

$$y = y'$$

$$z = z'$$

$$t = \frac{1}{\sqrt{1 - v^2/c^2}} \, (t' + vx'/c^2) \qquad (36\text{--}6)$$

Note:

- The space and time coordinates in the Lorentz transformation equations are mixed as seen from the equations for x and t.

- Lorentz transformation equations reduce to Galilean transformation equations when $v/c \ll 1$.

- The equations for length contraction and time dilation can be derived easily from the Lorentz transformation equations (36-6).

Relativistic Addition of Velocities

The velocity transformation equations can be obtained by taking derivatives of the preceding Lorentz transformation equations with respect to time. The results are:

$$u_x = \frac{dx}{dt} = \frac{u'_x + v}{1 + vu'_x / c^2} \qquad (36\text{--}7a)$$

$$u_y = \frac{dy}{dt} = \frac{u'_y \sqrt{1 - v^2 / c^2}}{1 + vu'_x / c^2} \qquad (36\text{--}7b)$$

$$u_z = \frac{dz}{dt} = \frac{u'_z \sqrt{1 - v^2 / c^2}}{1 + vu'_x / c^2} \qquad (36\text{--}7c)$$

36–9 Relativistic Momentum

In Newtonian mechanics, momentum is defined as $\vec{p} = m\vec{v}$. Classically, momentum is a conserved quantity. Since momentum must be conserved in the collision between particles moving at relativistic speeds, Einstein found that the correct expression for the momentum (called **relativistic momentum**) for an object moving with velocity $\vec{v}$ must be given by

$$\vec{p} = \frac{m\vec{v}}{\sqrt{1 - v^2 / c^2}} = \gamma m\vec{v}. \qquad (36\text{--}8)$$

- For small speeds ($v \ll c$), classical momentum and relativistic momentum are the same.

- As $v \rightarrow c, p \rightarrow \infty$.

The Newton second law of motion in special relativity is given by:

$$\vec{F} = \frac{d\vec{p}}{dt} = \frac{d}{dt}\left(\gamma m \vec{v}\right) = \frac{d}{dt}\left(\frac{m\vec{v}}{\sqrt{1 - v^2/c^2}}\right) \quad (36\text{–}9)$$

*Derivation of relativistic momentum
Derivation of relativistic momentum is done by considering the momentum conservation for relativistic collisions. The result shows that the Newtonian definition of momentum must be multiplied by γ to conserve momentum at such high speeds.

*Relativistic mass
The classical expression for momentum can still be used if one interprets that the **relativistic mass** of an object increases with its speed as

$$m_{\text{rel}} = \frac{m}{\sqrt{1 - v^2/c^2}}$$

where m is called the rest mass of the object.

36–10 The Ultimate Speed
From Equation (36–8) it is seen that as $v \rightarrow c, m_{\text{rel}} \rightarrow \infty$. Since acceleration is $a = F/m$, it appears that an infinite force would be needed to accelerate an object as its speed approached the speed of light c. Also, (36–1) and (36–2)

show that time will be frozen and length would be zero as v approaches c. Thus, the speed of light c is the ultimate speed.

36–11 $E = mc^2$; Mass and Energy

One important conclusion of the special theory of relativity is that mass and energy are equivalent. Energy has mass, and mass has energy.

Work-energy theorem is valid in relativity which states that the net work done on a particle is equal to its change in kinetic energy. Einstein showed that the relativistic kinetic energy of an object is

$$\text{KE} = K = \gamma mc^2 - mc^2 = (\gamma - 1)\, mc^2 \quad (36\text{–}10a)$$

$$= mc^2 \left(\frac{1}{\sqrt{1 - v^2/c^2}} - 1 \right) \quad (36\text{–}10b)$$

When $v \ll c$, this reduces to the classical expression, KE = $\frac{1}{2}\, mv^2$, as expected.

The term γmc^2 is called the total energy E of the object. Thus,

$$E = mc^2 + K, \quad (36\text{–}11a)$$

$$E = \gamma mc^2 = \frac{mc^2}{\sqrt{1 - v^2/c^2}} \quad (36\text{–}11b)$$

For a particle at rest in a reference, $K = 0$. Thus, **rest**

energy, $\qquad\qquad E_0 = mc^2 \qquad\qquad (36\text{–}12)$

This equation represents the equivalence between energy and mass and it also means that mass and energy are interconvertible.

The interconversion of mass and energy is detected in nuclear and elementary particle physics. For example, a neutral pion decays into photons. The mass of the pion disappears in the process and the amount of electromagnetic energy that appears is found to be exactly equal to that predicted by the expression $E_0 = mc^2$. The reverse process is also observed in the laboratory. In a nuclear power plant the energy is produced as a result of loss in mass of the uranium fuel as it undergoes a process called nuclear fission. The energy we receive from the sun is as a result of loss of mass of the sun (which is converted to energy) as it undergoes nuclear fusion.

If the energy of a system changes by an amount ΔE, the mass of the system changes by the amount Δm where $\Delta E = (\Delta m)(c^2)$.

The relationship between energy E and momentum p of a relativistic particle can be obtained from the relations

$$p = \frac{mv}{\sqrt{1 - v^2/c^2}} \text{ and } E = \frac{mc^2}{\sqrt{1 - v^2/c^2}} \,.$$

We can show from these expresstions that

$$E^2 = p^2c^2 + m^2c^4 \qquad (36\text{--}13)$$

Since the rest mass m is the same in any reference frame, $E^2 - p^2c^2$ is **invariant** under Lorentz transformation.

Units in relativity:

Energy	eV (electron volts) or Mev (10^6 MeV)
Momentum	eV/c or MeV/c

| Mass | eV/c^2 or MeV/c^2 |

*When do we use relativistic formulas?
For speeds less than $0.1c$ usually we do not use relativistic equations since the error is less than 1%.

*36–12 Doppler Shift for Light

Doppler effect occurs for light. The shifted frequency or wavelength is given by slightly different equations than from sounds. There are only two equations for each because for light we can make no distinction between motion of the source and motion of the observer.

When the source and the observer move toward each other, the observed wavelength λ and frequency f are given by, respectively:

$$\lambda = \lambda_0 \sqrt{\frac{c-v}{c+v}}, \qquad (36\text{–}14a)$$

$$f = f_0 \sqrt{\frac{c+v}{c-v}}. \qquad (36\text{–}14b)$$

When the source and the observer move away from each other, the observed wavelength λ and frequency f are given by, respectively:

$$\lambda = \lambda_0 \sqrt{\frac{c+v}{c-v}}, \qquad (36\text{–}15a)$$

$$f = f_0 \sqrt{\frac{c-v}{c+v}}. \qquad (36\text{–}15b)$$

Clearly, when a souce of light moves away, it wavelength increases. This effect is called the **redshift**. On the other hand, when a source moves toward the observer, wavelength decreases, which is called the blue shift.

36–13 The Impact of Special Relativity

The predictions of the relativity theory were accurately verified by experiments, and the theory has been accepted as the true description of nature. At speeds of objects much less than the speed of light, the relativistic formulas reduce to the results from classical mechanics. Thus, relativity is a more general theory of which classical mechanics is a limiting case valid for $v \ll c$. The insistence that a general theory gives the same result as a restricted theory is called the **correspondence principle.**

Relativity also changed our view about the world. According to the old concepts, space and time are absolute and separate; mass and energy are different. According to relativity theory, space and time are relative and interlinked; mass and energy are equivalent and interconvertible. Relativity has influenced the other sciences, the world of art and culture, and has entered the general culture.

Tips for Solving the Problems

1. Use the relation $\Delta t = \dfrac{\Delta t_0}{\sqrt{1 - v^2/c^2}}$ for time dilation to solve problems in Section 29–2, and the relation $L = L_0\sqrt{1 - v^2/c^2}$ for contracted length to solve problems in Sections 36–5 and 36–6.

2. Remember that the factor $\gamma = 1/\sqrt{1 - v^2/c^2}$ is always greater than 1 (unless $v = 0$ or $v = c$). Be careful, it is common to forget either to square the factor v/c or to take the square root. The factor tends to 1 as the velocity, v, tends to zero. As a result, the relativistic result will agree with the classical result when v is small.

3. Read the problem carefully several times to find out which observer sees the length contracted (or time dilated), and which observer sees its proper length (or proper time) before making an attempt to solve the problem. The hard part of solving problems on length contraction or time dilation is to figure out which length is proper length (or proper time) and which length is contracted length (or dilated time).

4. Motion in the perpendicular direction does not have any effect on the length.

5. Use Lorentz transformation equations to determine the space coordinates and time of an event in a reference frame when the corresponding values are given in another inertial frame. Use Equation 36–6a for the

addition of two velocities. Note that the result cannot exceed c.

6. Mass of an object is γ times its rest mass. The relativistic momentum γ is times the classical momentum. The energy is γ times rest energy, mc^2.

7. For problems involving relativistic energy and relativistic momentum, such as those in Section 36–11, the equation $E^2 = p^2c^2 + m^2c^4$ is very useful.

8. Note that relativistic kinetic energy $K = (\gamma-1)\,mc^2$ reduces to $\frac{1}{2}\,mv^2$ for $v \ll c$.

9. When choosing the right equation for the Doppler Effect, note that when the source moves away the observed wavelength must increase, and when the source approaches the observed wavelength must decrease.

CHAPTER 37
Early Quantum Theory and Models of the Atom

This chapter introduces the particle-like behavior of radiation (the photoelectric effect and the Compton effect). Wave-particle duality, models of atoms, and atomic line spectra are discussed.

Important Concepts

      Blackbody radiation
      Wien's law
      Quantum of energy
      Photoelectric effect
      Work function
      Cutoff frequency
      Photodiode
      Compton effect
      Compton wavelength
      Pair production
      Wave-particle duality
      Principle of complementarity
      De Broglie wavelength
      Diffraction of electrons
      Electron microscope
      Emission spectrum
      Absorption spectrum
      Balmer series
      The Bohr model
      Stationary states
      The Bohr radius
      Energy level diagram
      Ground state and excited state

37–1 Planck's Quantum Hypothesis; Blackbody Radiation

Hot objects emit electromagnetic radiation, which is proportional to the fourth power of their Kelvin temperatures.

At ordinary temperatures. the dominant radiation is in the infrared, at a temperature of about 2000 K it glows yellowish white, and as the temperature increases, the radiation is strongest at higher and higher frequencies. At any temperature, the emitted radiation forms a continuous spectrum as shown.

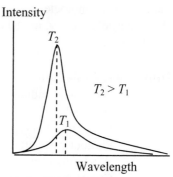

A **blackbody** is a substance that absorbs all radiation incident on it. The radiation such a body would emit is called the **blackbody radiation**.

The distribution of energy in blackbody radiation has these two characteristics:

- As the temperature increases, the area under the curve, and hence, the total energy increases with temperature.

- As the absolute temperature, T, is increased, the peak wavelength (at which the radiant energy is a maximum) moves toward shorter wavelength. The peak wavelength λ_p is related to T by **Wien's law**:

$$\lambda_p T = 2.90 \times 10^{-3} \, \text{m} \cdot \text{K} \qquad (37\text{–}1)$$

According to the classical model, radiation emitted by a hot object is due to the oscillations of electric charges in the molecules in the material. Classical theories given by Wien as well as by Raleigh and Jeans were unsuccessful in satisfactorily explaining the distribution of blackbody radiation. Wien's theory was accurate at short wavelengths but deviated from experiments at longer wavelengths, whereas the reverse was for Raleigh-Jeans theory.

Planck's Quantum Hypothesis

The following empirical formula for the radiated intensity $I(\lambda, T)$ proposed by Planck was found to fit the data well.

$$I(\lambda, T) = \frac{2\pi hc^2 \lambda^{-5}}{e^{hc/\lambda kT} - 1}$$

where h is a universal constant, called **Planck's constant**, and k is Boltzman's constant. The value of h is 6.626×10^{-34} J·s.

To explain blackbody radiation, Planck hypothesized that the energy of the oscillating electric charges of the molecules consists of a finite number of very small discrete amounts each related to the frequency of oscillation, f, by $E_{min} = hf$. Thus, the energy of any molecular vibration, E, in a blackbody radiation is

$$E = nhf \qquad n = 1, 2, 3, \cdots \qquad (37-2)$$

The energy is thus **quantized**. The smallest amount of energy, hf, is called the **quantum of energy**.

37–2 Photon Theory of Light and the Photoelectric Effect

According to Einstein, light is composed of tiny particles called **photons**. For a light of frequency f, the energy of each photon is given by

$$E = hf \qquad (37–3)$$

where h is Planck's constant.

The **photoelectric effect** is the phenomenon where electrons are emitted from a substance when light shines on it. The effect can be observed using a **photocell**, where a photosensitive metal plate and a smaller electrode are placed inside an evacuted tube.

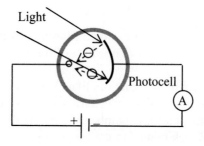

The maximum KE of the emitted electrons can be measured experimentally by applying a negative voltage to the anode. The negative voltage is increased to a value V_0, called *stopping voltage*, until the photoelectric current becomes zero, so that $K_{max} = eV_0$.

The electromagnetic wave theory of light makes the following predictions about the photoelectric effect:

1. If the light intensity is increased, the number of ejected electrons and their maximum KE should be increased.

2. The frequency of the light should not have any effect on the KE of the emitted electrons.

Einstein applied the photon theory of light to explain the photoelectric effect, which occurs when an incident light photon collides elastically with an electron in the material. Electrons are emitted if the photon has sufficient energy.

The minimum energy needed to eject an electron from a material is called its **work function**, W_0. The input energy hf of a photon is related to the maximum kinetic energy of a photoelectron by

$$hf = K_{max} + W_0 \qquad (37\text{–}4)$$

The photon theory makes the following predictions:

1. When light intensity is increased more photons are emitted, but, since energy of each photon does not change, the maximum KE of emitted electrons does not change.

2. The maximum KE increases linearly with light frequency according to the relation $K_{max} = hf - W_0$.

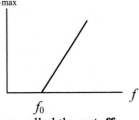

3. There is a minimum frequency, called the **cutoff frequency**, f_0, of light for the photoelectric effect to occur (regardless of light intensity), given by $hf_0 = W_0$.

The experimental results of photoelectric effect were fully in agreement with the photon concept of light.

Applications of the Photoelectric Effect

The photoelectric effect has practical applications such as in automatic door openers, alarms, automatic switches to turn on streetlights, smoke detectors, light meters, and film soundtrack. In many applications, the photocell has been replaced by a semiconductor device called a **photodiode**. Here, an incoming photon liberates an electron which changes the conductivity of the material so that the current through the photodiode is altered.

37–3 Energy, Mass, and Momentum of a Photon

Energy of a photon $E = hf$. The rest mass, m_0, of a photon is zero, but it still has a nonzero momentum. The momentum, p, of a photon for a light of frequency f and wavelength λ is

$$p = \frac{E}{c} = \frac{hf}{c} = \frac{h}{\lambda} \qquad (37\text{--}5)$$

37–4 Compton Effect

When an X-ray photon is scattered from an electron in some material, the scattered photon has a lower frequency (that is, longer wavelength) than the incident photon. This effect is called the **Compton effect**.

The Compton effect is a consequence of energy and momentum conservation in the collision between the incident photon and an electron within the scattered material.

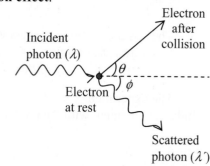

The wavelength of the scattered photon λ' is

$$\lambda' = \lambda + \frac{h}{m_0 c}(1 - \cos\phi) \qquad (37\text{--}6)$$

In this expression λ is the wavelength of the incident photon, λ' is the wavelength of the scattered photon, m_0 is the rest mass of the electron, and θ is the angle between the incident photon and the scattered photon. $\Delta\lambda = \lambda' - \lambda$ is called the **Compton shift**. The quantity $h/m_0 c = 2.43 \times 10^{-12}$ m is called the **Compton wavelength** of an electron.

37–5 Photon Interactions; Pair Production

There are four important types of interactions that a photon can have as it passes through matter:

1. A photon can be scattered by *Compton effect*, losing some of its energy in the process.
2. A photon can eject an electron, showing *photoelectric effect*.
3. A photon can give its energy to an atomic electron and bring the atom to a higher energy state (called *excited state*).
4. A photon disappears and produces an electron and positron. This is called **pair production**.

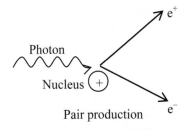

Pair production

In an inverse pair production, an electron collides with a positron and they **annihilate** each other, creating electromagnetic energy.

The intensity of the scattered gamma radiation off bone material is proportional to the density of electrons, and, hence, is proportional to the bone density. This fact is used to determine the changes in bone density which indicate the onset of osteoporosis.

37–6 Wave–Particle Duality; the Principle of Complementarity

Light shows particle properties (as shown in photoelectric effect and Compton effect) and wave properties (as shown in interference and diffraction). This dual nature of light is called **wave-particle duality**. According to the **principle of complementarity**, to understand any given experiment we must use either the wave or the photon theory but not both.

37–7 Wave Nature of Matter

According to Louis de Broglie, since light sometimes behaves as a wave and sometime as a particle, particles of matter (such as the electron and proton) should exhibit wavelike behavior.

According to de Broglie, the relationship between momentum and wavelength should be the same for photons and material particles. Thus, for a particle of mass m and momentum p, the wavelength associated with the particle, called the **de Broglie wavelength**, is

$$\lambda = \frac{h}{p} \qquad (37–7)$$

where h is Planck's constant.

The de Broglie wavelength is extremely small for macroscopic objects. For a microscopic system, such as the motion of an electron, the de Broglie wavelength is of the order of the size of atoms and molecules (10^{-10} m).

Since electrons can have wavelengths of the order of the size of atoms in a crystal, they show interference effects when reflected from the atomic layers of a crystal. Davisson and Germer experimentally observed diffraction patterns produced by low-energy electrons from crystals of nickel in 1927. Using a different experimental arrangement, G. P. Thomson detected the diffraction pattern of electrons, confirming the wave nature of electrons.

The principle of complementarity applies to matter as well.

What is an electron?

We use a wave or particle model for electron to understand what is happening. But we should not say that electron *is a wave or a particle*. Instead we should say that an electron is the set of properties that we can measure.

*37–8 Electron Microscopes

In an *electron microscope*, a beam of electrons is used for imaging instead of light. There are two types, the **transmission electron microscope** and the **scanning electron microscope**. In each case, the objective and eyepiece lenses are magnetic fields that exert forces on the electrons to bring them in focus. In scanning electon microscope, scanning coils move the electrons back and forth across the specimen. When the beam strikes the specimen, secondary electrons are produced, which modulate the intensity of the beam to produce a picture. The wavelength associated with the beam of electrons is typically 1000 times shorter than the wavelength of visible light. Since resolution depends inversely on wavelength, an electron microscope has more resolution and thus can show much finer detail than an optical microscope.

37–9 Early Models of the Atom

In 1897 J. J. Thomson discovered negatively charged electrons from atoms. An atom, being electrically neutral, therefore includes an equal amount of positive charges.

According to the plum-pudding model of Thomson, the positive charges are uniformly distributed within an atom, whereas the electrons are symmetrically situated inside the atom.

Rutherford's α particle scattering experiment by atoms does not support the plum-pudding model. By directing a beam of positively charged α particles on a thin foil of gold, Rutherford found that most α particles went straight without being deflected, and only some α particles were deflected at very large angles. To account for these results, Rutherford argued that all positive charges (containing about 99.9 % of the mass of the atom) are concentrated at the center, called the nucleus, of an atom, whereas the electrons rotate around the nucleus in circular orbits. The most part of an atom is empty. This is called the "planetary" model or "nuclear" model of an atom.

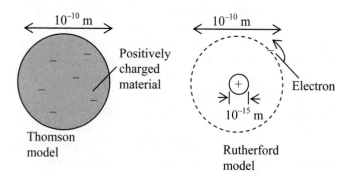

37–10 Atomic Spectra: Key to the Structure of the Atom

Solid objects when heated emit radiation with a continuous spectrum of wavelength. The radiation emitted by an isolated atom (such as from hydrogen gas in a low-pressure discharge tube) consists of a series of bright *discrete* lines of different colors. This type of spectrum is called a **line spectrum**.

When atoms emit light, the observed spectrum is called an **emission spectrum**. The precise wavelength associated with the lines provides a way to identify atoms from their emission spectra. Conversely, if light of all colors passes through a gas, the gas will absorb some wavelengths, giving rise to dark lines (where its atoms absorb colors) against the bright background, producing an **absorption spectrum** of the gas.

The line spectra serve as a key to the structure of atoms since the structure must be able to explain the line spectra and predict what these wavelengths are.

Balmer showed that the wavelength of the visible lines of the spectrum of hydrogen fit the following formula:

$$\frac{1}{\lambda} = R\left(\frac{1}{2^2} - \frac{1}{n^2}\right) \qquad n = 3, 4, 5, \cdots \qquad (37\text{--}8)$$

The constant R is called the *Rydberg constant.* Its value is $1.097 \times 10^7 \text{ m}^{-1}$. The series containing lines calculated using Equation (37–8) is called the **Balmer series**.

There are also other series in the spectrum of hydrogen, such as the **Lyman series** (in the ultraviolet)

$$\frac{1}{\lambda} = R\left(\frac{1}{1^2} - \frac{1}{n^2}\right) \qquad n = 2, 3, \cdots$$

469

and the **Paschen series** (in the infrared)

$$\frac{1}{\lambda} = R\left(\frac{1}{3^2} - \frac{1}{n^2}\right) \qquad n = 4, 5, \cdots .$$

- The Rutherford model cannot explain the line spectra emitted by an atom; the model instead predicts a continuous spectrum.

- Also, an orbiting electron radiates electromagnetic radiation according to Maxwell's electromagnetic theory. As a result, an orbiting electron within an atom would spiral inward and eventually fall into the nucleus. Thus, Rutherford's model is not a model for a stable atom.

37–11 The Bohr Model

Bohr's model of the atom is a modification of Rutherford's model. Bohr introduced the quantum concept of radiation to explain the line spectrum of hydrogen atoms. He made the following assumptions:

- The electron in an atom can rotate only along certain selected orbits around the nucleus without radiating any energy. These orbits are called **stationary states**.

- An atom gives off radiation when an electron jumps from one stationary state to another of lower energy.

E_u

hf

E_l

The frequency, f, of the photon that is emitted when an electron changes its energy from E_u to E_l is given by

$$hf = E_u - E_l \qquad (37\text{–}9)$$

- When an electron is orbiting along any one of these orbits, its angular momentum is quantized. For the nth orbit, the angular momentum, L, is given by the **quantization condition**

$$L = mvr_n = \frac{nh}{2\pi}, \quad n = 1, 2, 3, \cdots, \quad (37\text{–}10)$$

The number n is called the **quantum number** of the orbit.

For an orbiting electron, the electrostatic force between the electron and the nucleus (of charge Ze) provides the centripetal force: $\dfrac{1}{4\pi\varepsilon_0}\dfrac{Ze^2}{r_n^2} = \dfrac{mv^2}{r_n}$. From the condition of angular momentum and the preceding equation for the centripetal force, Bohr showed that

(i) The radius of an electron orbiting in the nth orbit is given by

$$r_n = \frac{n^2 h^2 \varepsilon_0}{\pi m Z e^2} = \frac{n^2}{Z} r_1 \qquad (37\text{–}11)$$

$$\text{where} \quad r_1 = \frac{h^2 \varepsilon_0}{\pi m e^2}. \qquad (37\text{–}12)$$

$$= 0.529 \times 10^{-10} \text{ m} \qquad (37\text{–}13)$$

r_1 is the smallest orbit ($n = 1$) for hydrogen ($Z = 1$), called **Bohr radius**.

(ii)   The allowed energy of the nth orbit is given by

$$E_n = -\frac{Z^2 e^4 m}{8\varepsilon_0^2 h^2}\frac{1}{n^2} \quad n = 1, 2, 3, \cdots \qquad (37\text{–}14)$$

$$= \frac{Z^2}{n^2} E_1$$

where E_1 is the lowest energy state of hydrogen.

$E_1 = \dfrac{e^4 m}{8\varepsilon_0^2 h^2} = -13.6$ eV. Thus, for hydrogen,

$$E_n = \frac{-13.6\,\text{eV}}{n^2}.$$

This gives $E_2 = -3.40$ eV and $E_3 = -1.51$ eV, etc.

The lowest energy state E_1 is called the **ground state**. Higher energy states $E_2, E_2, \cdots$, are known as **excited states**. When $r = \infty$, $E = 0$, which corresponds to $n = \infty$, and in this situation the electron is just barely free from the atom. The minimum energy required to remove an electron from the ground state of the atom is called the **binding energy** or **ionization energy**. An energy level diagram showing the possible energy values as horizontal lines and different series is shown.

Spectral Lines Explained
When an electron in an excited atom jumps down to one of its lower states, it emits a light photon.

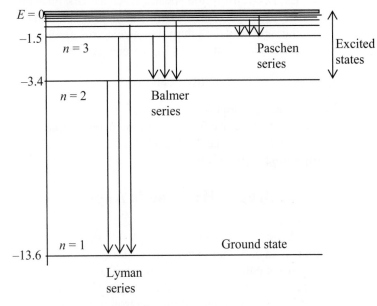

The wavelength of the emitted photon can be predicted from the expression

$$\frac{1}{\lambda} = \frac{1}{hc}(E_n - E_{n'}) = \frac{Z^2 e^4 m}{8\varepsilon_0^2 h^3 c}\left(\frac{1}{n'^2} - \frac{1}{n^2}\right) \qquad (37\text{–}15)$$

where n refers to an upper state and n' to a lower state. The value of the constant $2\pi^2 Z^2 m k^2 e^4 / h^3 c$ agrees with the value of R (the Rydberg constant) that appears in the expression for the Balmer series.

Bohr theory:
- Explains why atoms emit line spectra, and it accurately predicts the wavelengths for hydrogen.

- Explains absorption spectra (such as dark lines in the solar spectrum), where an incoming photon of the right wavelength moves an electron from a lower energy state to a higher energy state.

- Ensures the stability of atoms.

Corresponding principle
When quantum theory overlaps with the macroscopic world, it predicts the classical result. This is the **correspondence principle**.

37–12 de Broglie Hypothesis Applied to Atoms

Louis de Broglie showed that the allowed orbits of electrons in an atom correspond to the formation of standing wave patterns by the electrons. The circumference of the electron orbit is equal to the integer multiple of the wavelength of the electron in that orbit. That is, $$2\pi r_n = n\lambda \qquad n = 1, 2, 3, \ldots.$$

Since $\lambda = h/mv$ (de Broglie wavelength for electron), this gives $mvr_n = \dfrac{nh}{2\pi}$, which is the *quantization condition* proposed by Bohr.

Tips for Solving the Problems

1. According to Wien's law, $\lambda_p T = 2.90 \times 10^{-3}\,\text{m} \cdot \text{K}$. Make sure to convert the temperature to kelvins when using this law to calculate the temperature of a star or a blackbody, such as in the problems in Section 37–1.

2. The energy of a light photon is $E = hf = hc/\lambda$. Use this expression to calculate the frequency or wavelength of a photon.

3. Wavelength is usually given in nanometers (nm). Convert to meters (m) when solving problems.

4. Energy may be given in electron volts (eV) or in joules (J) in numerical problems. If the value of h to be used is in J · s, make sure to convert energy to joules. Remember the conversion $1\,\text{eV} = 1.6 \times 10^{-19}\,\text{J}$.

5. The cutoff frequency or cutoff wavelength for the photoelectric effect can be obtained from the expression $W_0 = hf_0 = hc/\lambda_0$, where W_0 is the work function. Usually it is given in eV. Make sure to convert it to joules (J). The maximum kinetic energy of a photoelectron is $K_{\max} = hf - W_0$. Also for the stopping voltage V_0, $K_{\max} = eV_0$. Use these expressions to solve problems in Section 37–2.

6. Calculation of scattered wavelength of an X-ray by the Compton effect is straightforward. Use the expression $\lambda' = \lambda + \left(h/m_0 c\right)\left(1 - \cos\phi\right)$ for the change in wavelength or the scattered wavelength.

7. The de Broglie wavelength associated with a particle is $\lambda = h/p = h/mv$. Once v is known, the kinetic energy

of the particle can be calculated from its mass, m, and velocity, v.

8. Remember that the formula for the spectral lines of the hydrogen atom (Equation 37–15) gives the inverse of the wavelength. You must take the inverse of the right-hand side of Equation 37–15 to find wavelength, λ.

9. Some useful constants for this chapter are: $h = 6.63 \times 10^{-34}$ J · s; $c = 3 \times 10^8$ m/s, $m = 9.1 \times 10^{-31}$ kg, $e = 1.6 \times 10^{-19}$ C, $\varepsilon_0 = 8.85 \times 10^{-12}$ C^2/ N · m^2, and $R = 1.097 \times 10^7$ m^{-1}. Remember the conversion factor 1 eV $= 1.6 \times 10^{-19}$ J.

10. For a hydrogen-like atom, the radius of its nth orbit is $r_n = (0.529 \times 10^{-10}$ m) n^2/Z, and the energy of its nth orbit is $E_n = -13.6$ eV Z^2/n^2. For hydrogen $Z = 1$. You do not need to memorize Equation 37–15 for the wavelength of the light emitted when an electron jumps from one orbit to another. Use the preceding expression for energy to determine the energies of the initial orbit and final orbit. Set the energy difference to hc/λ to calculate the wavelength. Remember to convert electron volts to joules.

11. To calculate the wavelength of an electron in a Bohr orbit, find the circumference of the orbit, and divide the circumference by n. This is because an integral number of wavelengths must fit in the Bohr orbit, with one wavelength in the ground state ($n = 1$).

CHAPTER 38
Quantum Mechanics

This chapter introduces quantum mechanics, the Heisenberg uncertainty principle, and the Schrödinger wave equation, as well as particles in a rigid box and finite potential.

Important Concepts

> Correspondence principle
> Wave function
> Heisenberg uncertainty principle
> Copenhagen interpretation
> Schrödinger time-independent wave equation
> Schrödinger time-dependent wave equation
> Free-particle wave function
> Probability density function
> Wavepacket
> Rigid box and finite potential well
> Tunneling

38–1 Quantum Mechanics—A New Theory

Quantum mechanics deals with the microscopic world of atoms and light. The theory covers all quantum phenomena from blackbody radiation to atoms and molecules and is accepted as the fundamental theory underlying physical processes.

When applied to macroscopic phenomena, quantum mechanics produces the classical results, as demanded by the **correspondence principle**. Classical theories produce accurate results for motion of ordinary objects. For objects moving at high speeds we must use the theory of relativity, and when dealing with the tiny world of atoms, we must use quantum mechanics.

38–2 The Wave Function and Its Interpretation; the Double-Slit Experiment

For an electromagnetic wave, its amplitude at any point is the strength of the electric (or magnetic) field, and it is related to the intensity of the wave (or the brightness of the light). For material particles such as electrons, the wavelength of the matter wave is related to the momentum by the relation $\lambda = h/mv$, and the amplitude of the matter wave is called the wave function Ψ, which is a function of both position and time. One of the basic tasks of quantum mechanics is to find the wave function Ψ in a given situation. The Schrödinger equation (a differential equation of space and time) determines the behavior of Ψ. Ψ, being the amplitude of a matter wave at any point in space and time, plays the role of the electric field E for an electromagnetic wave. For an electromagnetic wave, the square of the electric field E^2 is a measure of the probability that a light photon will be at a given location. Similarly, for a matter wave, $\left| \Psi \right|^2$, called the **probability density**, at a certain point in space and time represents the *probability of finding the particle (say, an electron) at a given position and time.*

Double-Slit Interference Experiment for Electrons

Young's double-slit experiment is done with electrons where parallel beams of electrons fall on two slit, whose sizes are comparable to the wavelength. An interference pattern is observed; however, the pattern is not evident with only a few electrons. The experiment confirms that when both the slits are open, it is as if both the electrons passed through both the slits, like a wave.Yet, each electron makes a tiny point spot on the screen as if it were a particle.

38–3 The Heisenberg Uncertainty Principle

A natural consequence of the wave properties of a particle is that the position and momentum of a particle cannot be simultaneously determined with infinite accuracy. That is, if the position of a particle is known with a greater accuracy, its momentum becomes uncertain, and vice versa. This consequence is expressed in the **Heisenberg uncertainty principle** (also called **indeterminacy principle**) of *position* and *momentum* as

$$(\Delta x)\,(\Delta p) \geq \frac{h}{2\pi} \qquad (38\text{–}1)$$

where Δx is the uncertainty in the measurement of the position, x, and Δp is the uncertainty in the measurement of the corresponding momentum.

The inaccuracy of simultaneous measurements of energy, E, and time, t, is expressed in the Heisenberg uncertainty principle of energy and time as

$$(\Delta E)\,(\Delta t) \geq \frac{h}{2\pi} \qquad (38\text{–}2)$$

The constant $\dfrac{h}{2\pi} = \hbar = 1.055 \times 10^{-34}$ J · s.

Since h is very small, the uncertainties expressed in the uncertainty principle are negligible at the macroscopic level. But at the atomic level, the uncertainties are important. The uncertainty principle expresses the probabilistic nature of quantum mechanics. If h were large, wavelike properties would be apparent even for macroscopic objects.

38–4 Philosophical Implications; Probability versus Determinism

Classical physics is deterministic. That is, for example, once the position and velocity of an object is known at a particular time, its future position and velocity can be precisely determined if the force acting on the object is known. On the other hand, quantum mechanics is probabilistic. It allows us to calculate only the probability that, say, an electron can be observed at various places. This probabilisitc view of quantum mechanics is called **Copenhagen interpretation**. Quantum mechanics says that there is some inherent unpredictability in nature as expressed by the Heisenberg uncertainty principle.

Quantum mechanics predicts with very high probability that ordinary objects will behave as classical physics predicts. The reason macroscopic objects follow classical laws is because there is a large number of molecules involved since in this case the deviation from the average (or most probable) is negligible.

It is to be noted that there is a difference between the probability imposed by quantum mechanics and probability used in thermodynamics. In thermodynamics, probability is used since there are too many molecules but each molecule is still assumed to obey classical laws.

38–5 The Schrödinger Equation in One Dimension—Time-Independent Form

The Schrödinger equation is the basic equation to determine the wave function ψ. It can be written in time-dependent form and time-independent form.

Let us consider a free particle (with no forces acting) moving along the x-axis. We can write the wave function of the free particle at a specific moment, say $t = 0$, as

$$\psi(x) = A \sin kx + B \cos kx \qquad (38\text{–}3a)$$

where A and B are constants and $k = 2\pi/\lambda$. Since $\lambda = h/p$,

$$k = = \frac{2\pi}{\lambda} = \frac{p}{\hbar} \qquad (38\text{–}3b)$$

From the conservation of energy (E = total energy, U = potential energy)

$$\frac{p^2}{2m} + U = E \quad \text{or}$$

$$\frac{\hbar^2 k^2}{2m} + U = E \qquad (38\text{–}4)$$

The differential equation that satisfies conservation of energy (Equation 38–4) and provides $\psi(x)$ (given by Equation 38–3a) as the solution is:

$$-\frac{\hbar^2}{2m} \frac{d^2\psi(x)}{dx^2} + U(x)\,\psi(x) = E\,\psi(x) \qquad (38\text{–}5)$$

This is the **time-independent Schrödinger equation** in one dimension, where $U = U(x)$. Note that we have constructed Schrödinger's equation but not derived it.

Since $|\Psi|^2$ represents the probability of finding a particle at a given position, $|\Psi|^2\,dV$ represents the probability of finding the particle within a volume dV. In

481

one-dimensional case, $dV = dx$, and since the probabilities summed over all space must be unity (that is, the particle must be somewhere in space), we get

$$\int_{all\ space} |\psi|^2 dV = \int |\psi|^2 dx = 1. \qquad (38\text{--}6)$$

This is called the **normalization condition**. In general, x ranges from $x = -\infty$ to $x = +\infty$.

*38–6 Time-Dependent Schrödinger Equation

The general form of the Schrödinger equation, called the **time-dependent Schrödinger equation**, for the motion of a particle of mass m in one dimension is

$$-\frac{\hbar^2}{2m}\frac{\partial^2\psi(x,t)}{\partial x^2} + U(x)\ \psi(x,\ t) = i\hbar\frac{\partial\psi(x,t)}{\partial t}. \quad (38\text{--}7)$$

where $i = \sqrt{-1}$.

In many situations in quantum mechanics it is possible to write the wave function as a product of two separate functions of space and time:

$$\psi(x,\ t) = \psi(x)f(t)$$

Substituting this in Equation (38–7) we can separate the variables and get:

$$-\frac{\hbar^2}{2m}\frac{d^2\psi(x)}{dx^2} + U(x)\ \psi(x) = C\psi(x) \qquad (38\text{--}8a)$$

$$i\hbar \frac{1}{f(t)} \frac{\partial f(t)}{\partial t} = C. \qquad (38\text{–}8b)$$

It is evident that the constant C in Equation (38-8a) must be the total energy E [compare Equations (38–5) and (38–08a)]. Also, from Equation (38–8b) we get:

$$f(t) = e^{-i\left(E/\hbar\right)t}$$

Thus, the total wave function is

$$\psi(x,\,t) = \psi(x)\; e^{-i\left(E/\hbar\right)t} \qquad (38\text{–}9)$$

Note that, in general, $\psi(x,\,t)$ is a complex function. Since $\psi(x,\,t)$ is not purely real, it cannot be physically measureable. We can measure only $\left| \psi \right|^2$, which is real. Also note that

$$\left| f(t) \right|^2 = \left| e^{-i\left(E/\hbar\right)t} \right|^2 = 1^2 = 1.$$

Thus, the probability density in space does not depend on time:

$$\left| \psi(x,\,t) \right|^2 = \left| \psi(x) \right|^2.$$

38–7 Free Particles; Plane Waves and Wave Packets

For a free particle $U(x) = 0$. Thus, from Schrödinger's equation:

$$-\frac{\hbar^2}{2m} \frac{d^2\psi(x)}{dx^2} = E\psi(x)$$

or $\qquad -\dfrac{d^2\psi(x)}{dx^2} + k^2\psi(x) = 0$ where

$$k^2 = \frac{2mE}{\hbar^2} = \frac{p^2}{\hbar^2} \text{ (since } E = \tfrac{1}{2}\, mv^2) \qquad (38\text{--}10)$$

This is the equation of a simple harmonic motion. The solution is

$$\psi(x) = A \sin kx + B \cos kx \qquad (38\text{--}11)$$

Thus, a free particle is represented by a plane wave that varies sinusoidally. For a free particle, momentum p is given ($\Delta p = 0$) and the particle can be anywhere between $x = -\infty$ and $x = \infty$.

To describe a particle whose position is well localized, we use the concept of **wave packet**. A wave packet can be represented as the sum of many plane waves of slightly different wavelengths. A localized particle moving through space can be represented by a moving wavepacket.

38–8 Particle in an Infinitely Deep Square Well Potential (a Rigid Box)

For an **infinitely deep square well potential** or **rigid box** (one-dimensional box of width L)

$$U(x) = 0 \qquad 0 < x < L$$

$$U(x) = \infty \qquad x \le 0 \ \text{ and } x \ge L.$$

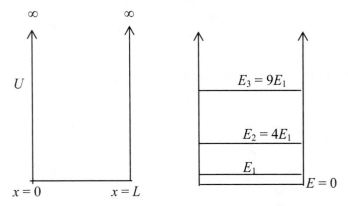

If a particle of mass m is confined to the potential well, the wave function is $\psi(x) = A \sin kx + B \cos kx$, for $0 < x < L$, where $k^2 = \dfrac{2mE}{\hbar^2}$.

For $x \leq 0$ and $x \geq L$ (outside the well), $\psi(x) = 0$.

Since $\psi(x)$ is continuous, $\psi(0) = 0$ and $\psi(L) = 0$. From these boundary conditions, we get

$$kL = n\pi \qquad n = 1, 2, 3, \ldots, \qquad (38\text{–}12)$$

and

$$E_n = n^2 \frac{h^2}{8mL^2} = n^2 E_1 \qquad (38\text{–}13)$$

where E_n represents the energy of the nth state and n is called the **quantum number** of the state. The lowest energy is E_1, called the **zero point energy**. The wave function for the nth-state is:

$$\psi_n = A \sin \left(\frac{n\pi}{L} x \right) \qquad (38\text{--}14)$$

From the normalization condition, $\int_{-\infty}^{\infty} |\psi|^2 dx = 1$, we get

$A = \sqrt{\dfrac{2}{L}}$. Thus,

$$\psi_n = \sqrt{\frac{2}{L}} \sin \left(\frac{n\pi}{L} x \right) \qquad (38\text{--}15)$$

The approximate sketches of the wave functions and probability distribution functions are shown below.

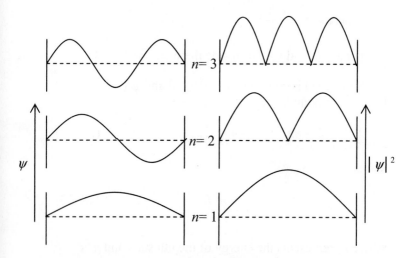

38–9 Finite Potential Well

For a **finite potential well**, the potential energy outside the box rises to some level, say U_0, as shown. This model serves as an approximation, for example, for a neutron in a nucleus.

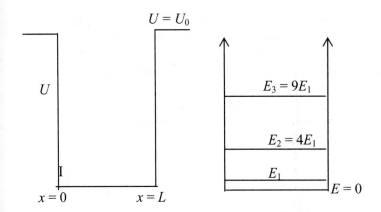

Inside the well (region II) $U = 0$, and the solution is

$$\psi_{II} = A \sin kx + B \cos kx, \qquad 0 < x < L$$

In regions I and III, $U = U_0$, and the Schrödinger equation is:

$$-\frac{\hbar^2}{2m}\frac{d^2\psi}{dx^2} + U_0\psi = E\psi$$

The solution provides the wave functions :

$$\psi_I = Ce^{Gx}, \qquad x < 0 \text{ (region I)}$$

and $\qquad \psi_{III} = De^{-Gx}, \qquad x > L \text{ (region III)}$

where C and D are constants and $G^2 = \dfrac{2m(U_0 - E)}{\hbar^2}$.

Thus, in regions I and III, the wave functions decrease exponentially with distance from the wall. The form of the wave function inside the well is different (sinusoidal) but the wave functions and their first derivatives are

487

continuous at the walls. Since ψ is not zero outside the walls, there is a finite probability that the particle can be found outside the walls, which is not possible classically. Also, outside the walls U_0 (potential energy) > E (total energy), violating conservation of energy.

The approximate sketches of the wave functions and probability distribution functions are shown below.

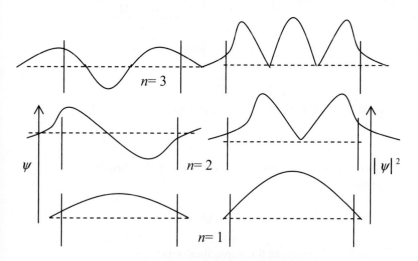

38–10 Tunneling Through a Barrier

Classically, when a particle encounters a narrow potential barrier of height U_o and width l with an energy $E < U_o$), it is reflected and returns in the opposite direction. However, quantum mechanically, thee is a nonzero probability that the particle penetrates the barier. This is called **tunneling** through a barrier or **barrier penetration**. The tunneling probability is described by *transmission coefficient*, T, and *reflection coefficient*, R, where $R + T = 1$. In this case, $T \approx e^{-2Gl}$.

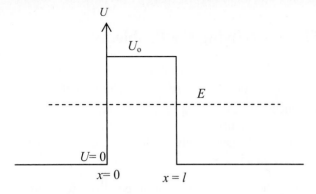

Applications of Tunneling

Tunneling is a result of wave properties of particles and also occurs in classical waves. Some unstable nuclei undergo radioactive decay by emitting alpha particles.

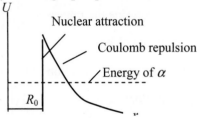

This can be explained from quantum theory. The potential energy for an alpha particle in a nucleus of radius R_0 is shown.

Classically the α particle would escape the nucleus only if there were an input of energy equal to the height of the barrier. According to quantum mechanics, it actually passes through the barrier by **tunneling**.

In a **tunnel diode** that is made of two types of semiconductors carrying opposite charge carriers and separated by a very thin neutral region, the current tunnels through this barrier. In a **scanning tunneling microscope**, the tip of a tiny probe remains very close to the specimen as it scans. A voltage applied between the probe and the surface causes the electron to tunnel through the vacuum between them.

Tips for Solving the Problems

1. Use the Heisenberg uncertainty relation $\Delta p_y \, \Delta y \geq h/2\pi$ or $\Delta E \, \Delta t \geq h/2\pi$ to determine the uncertainties relating to the simultaneous measurements of position and momentum or energy and time, respectively, such as in the problems in Section 38–3.

2. Use the normalization condition $\int |\psi|^2 \, dx = 1$ to determine the constant A in the wave function. In general, x ranges from $x = -\infty$ to $x = +\infty$. To determine the probability of finding a particle in a certain region, determine the integral $\int |\psi|^2 \, dx$ between the two limits of x.

3. The average position of the particle can be obtained by evaluating the integral $\int x |\psi|^2 \, dx$ between the two limits of x. The most likely position(s) of a particle can be determined by finding $|\psi|^2$, then taking its derivative and setting it to zero.

4. Recall that k, λ, and p are related by the equation $k = \dfrac{2\pi}{\lambda} = \dfrac{p}{\hbar}$, which is often useful to determine λ or p.

5. The probability of tunneling through a finite barrier of width l is given by $T \approx e^{-2Gl}$.

CHAPTER 39
Quantum Mechanics of Atoms

This chapter describes the quantum mechanical description of atoms, the Pauli exclusion principle, multielectron atoms, and the periodic table. X-rays, principles of operation of a laser, fluorescence, and phosphorescence are also described.

Important Concepts

> Principle quantum number
> Orbital quantum number
> Magnetic quantum number
> Zeeman effect
> Spin quantum number
> Selection rule
> Pauli exclusion principle
> Periodic table
> Moseley plot
> Bremsstrahlung
> Space quantization
> Total angular momentum
> Fine structure
> Fluorescence and phosphoresence
> Laser
> Stimulated emission
> Helium-neon laser
> Holography

39–1 Quantum-Mechanical View of Atoms

Bohr model does not provide an accurate description of nature, but quantum mechanics reaffirms that (i) atomic energy levels are discrete and (ii) a photon of light is emitted or absorbed when an electron makes a transition from one state to another.

According to quantum mechanics, electrons do not exist in well-defined circular orbits, but they can be thought of as spread out in space in the form of a "cloud." The electron cloud roughly indicates the size of an atom, but an atom does not have a precise boundary or a well-defined shape since a cloud does not have a distinct border. The electron cloud is a result of the wave nature of electrons and can be interpreted as the **probability distributions** for an electron. The size and shape of the electron cloud can be determined for a given state of an atom. For the ground state in hydrogen atom, the electron cloud is spherically symmetric.

39–2 Hydrogen Atom: Schrödinger Equation and Quantum Numbers

A hydrogen atom consists of a single electron of charge $-e$ moving around a central nucleus containing one proton of charge $+e$. The potential energy due to Coulomb force between electron and proton is

$$U = -\frac{1}{4\pi\varepsilon_0}\frac{e^2}{r}$$

where r is the radial distance from the proton to the electron. The Schrödinger equation of a hydrogen atom is then

$$-\frac{\hbar^2}{2m}\left(\frac{\partial^2\psi}{\partial x^2}+\frac{\partial^2\psi}{\partial y^2}+\frac{\partial^2\psi}{\partial z^2}\right) - \frac{1}{4\pi\varepsilon_0}\frac{e^2}{r}\,\psi = E\psi \quad (39\text{–}1)$$

To solve, the equation is written in spherical coordinates because of the spherical symmetry. The solutions are characterized by four quantum numbers. The energy levels are the same as given by Bohr's theory:

$$E_n = \frac{-13.6\,\text{eV}}{n^2}. \qquad n = 1, 2, 3,\ldots \quad (39\text{--}2)$$

where n is called the principal quantum number. The following four quantum numbers are required for the quantum mechanical description of the hydrogen atom:

(i) The **principal quantum number**, n, which is the same as n in Bohr theory,

(ii) The **orbital quantum number**, l, which can have the values $l = 0, 1, 2, \ldots(n-1)$. The orbital angular momentum, L, is given by the expression

$$L = \sqrt{l(l+1)}\,\hbar \qquad\qquad (39\text{--}3)$$

(iii) The **magnetic quantum number**, m_l. The allowed values of m_l are

$$m_l = -l, -l+1, -l+2, \cdots, -1, 0, +1, +2, \ldots, l-1, l$$

The z-component of the angular momentum is

$$L_z = m_l \hbar \qquad\qquad (39\text{--}4)$$

When a gas discharge tube is placed in a magnetic field, the spectral lines are split into closely spaced lines. This is known as the **Zeeman effect**. This shows that energy of a state depends not only on n but also m_l when a magnetic field is applied.

(iv) The **electron spin quantum number**, m_s, which can have the values $m_s = -\frac{1}{2}, \frac{1}{2}$. These two values correspond to the electron's spin being "up" or "down" with respect to the z-axis, respectively.

A careful observation of the spectral lines of hydrogen showed that each line consisted of two or more very closely spaced lines even in the absence of a magnetic field. This tiny splitting of lines is called **fine structure** and is due to the angular momentum associated with the spinning of the electron.

Selection Rules: Allowed and Forbidden Transitions
Quantum mechanics predicts that when a photon is emitted or absorbed, the transition can occur only if it follows the **selection rule**: $\Delta L = \pm 1$. For example, if an electron is $l = 2$ state, it can jump to a state with $l = 1$ or 3. These are called **allowed transitions**. But a transition such as $l = 2$ to $l = 0$ is called a **forbidden transition** because the probability of occuring this transition is very low.

39–3 Hydrogen Atom Wave Functions

For the ground state $n = 1$, $l = 0$ and $m_l = 0$. The ground state wave function (that serves for both $m_s = \frac{1}{2}$ and $m_s = -\frac{1}{2}$) is given by

$$\psi_{100} = \frac{1}{\sqrt{\pi r_0^3}} e^{-\frac{r}{r_0}} \qquad (39\text{–}5a)$$

where $r_0 = h^2\varepsilon_0/\pi me^2 = 0.0529$ nm, called the Bohr radius. The probability density function for the ground state is:

$$|\psi_{100}|^2 = \frac{1}{\pi r_0^3} e^{-\frac{2r}{r_0}} \qquad (39\text{–}5b)$$

The radial probability distribution function is given by

$$P_r = 4\pi r^2 \left| \psi \right|^2 \qquad (39\text{–}6)$$

For the ground state,

$$P_r = 4\frac{r^2}{r_0^3} e^{-\frac{2r}{r_0}} \qquad (39\text{–}7)$$

The most probable distance can be obtained by setting $dP_r/dr = 0$. The result shows that $r = r_0$ as the most probable distance, in coincidence with the Bohr theory. The plot of P_r as a function of r is shown.

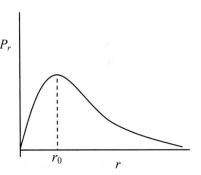

For the first excited state $n = 2$, $l = 0$. The wave function is spherical symmetric and is given by

$$\psi_{200} = \frac{1}{\sqrt{32\pi r_0^3}}\left(2 - \frac{r}{r_0}\right) e^{-\frac{r}{2r_0}} \qquad (39\text{–}8)$$

and the corresponding P_r is given by

$$P_r = \frac{r^2}{8r_0^3}\left(2 - \frac{r}{r_0}\right)^2 e^{-\frac{r}{r_0}}. \qquad (39\text{–}9)$$

P_r shows two peaks; the second, at $r \approx 5\, r_o$, is higher and corresponds to the most probable value of r in $n = 2$.

39–4 Complex Atoms; the Exclusion Principle

In multielectron atoms, the electrons experience mutual repulsive electrostatic forces in addition to their attractive interactions with the nucleus. Quantum mechanics shows that the energy level of a multielectron atom depends on the principal quantum number, n, and the orbital quantum number, l. The energy value increases with n for a given l and also with l for a given n. The number of electrons (which is also the same as the number of positive charges) is called the atomic number, Z.

Each particular state in an atom is characterized by the quantum numbers n, l, m_l, and m_s. The possible arrangement of electrons in an atom can be understood using the **Pauli exclusion principle** which states: **No two electrons in an atom can occupy the same quantum state**. That is, only one electron at a time may have a particular set of quantum numbers n, l, m_l, and m_s. Once a particular state is occupied by an electron, other electrons are excluded from that state.

39–5 The Periodic Table of Elements

The Russian chemist Mendeleev arranged all the chemical elements in a table, called the **periodic table**, by grouping the elements based on their chemical properties. Electrons having the same value of n are said to be in the same **shell**. For $n = 1$, an electron is in the K shell, for $n = 2$ it is in the L shell, and for $n = 3$ it is in the M shell, and so on. Electrons in a given shell with the same value of l are said to be in the same **subshell**. The subshells are named s, p, d, ... for $l = 0, 1, 2, \ldots$, respectively. As electrons fill subshells of increasing l, they produce different elements of the periodic table. Atoms with the same configuration of the outermost electrons have similar chemical properties. For example, lithium, sodium, potassium, rubidium, cesium, and francium are called **alkali metals**. They have similar chemical properties, and all have only one electron in the outermost subshell. Column VII contains the **halogens**, which lack one electron from a filled shell. Group VIII consists of noble gases. These elements all have completely filled subshells, and they are relatively inert. The **transition elements** have incomplete inner shells. Most of the chemical properties of these transition elements are governed by the relatively loosely held $4s$ electron and they usually have valence of $+1$ or $+2$. For the rare earths, chemical properties are determined by their two outer $6s$ or $7s$ electrons.

39–6 X-Ray Spectra and Atomic Number

X-rays are produced when electrons accelerated by a high voltage hit a metal target in an X-ray tube. The spectrum of wavelengths emitted by an X-ray tube consists of a continuous spectrum with a cut-off at some λ_0 which depends on the voltage across the tube and a series of peaks superimposed on the continuous spectrum.

X-ray radiation is created by two separate physical mechanisms. The first mechanism is called **bremsstrahlung**, in which a continuous spectrum is emitted when electrons, accelerated through a high potential difference, undergo a rapid deceleration as they collide with a metal target made of elements such as molybdenum or platinum. From the conservation of energy, the energy of the emitted photon, hf, is equal to the loss of kinetic energy of the electron: $hf = \Delta K$. The shortest-wavelength X-ray is produced when it gives up all its kinetic energy, in which case $hf_0 = eV$, where V is the accelerating voltage. This gives

$$\lambda_0 = hc/eV, \tag{39–10}$$

in agreement with the value observed experimentally. The sharp peaks in the X-ray spectrum are called **characteristic X-rays**. These are created when the energetic incident electrons knock electrons out from one of the inner shells of the target atoms, and other electrons of the target atoms drop to fill the vacancy in the inner shell. The approximate energy needed to knock an electron out of a K orbit can be obtained from the Bohr formula by replacing Z with $(Z–1)$ to take into account the "shielding" of the nucleus by the second electron in the K shell. Moseley found that a plot of $(1/\lambda)^{1/2}$ versus Z for K_α (strongest peak) X-ray lines produced a straight line, a plot known as a **Moseley plot**. The plot can be used to determine an unknown Z.

*39–7 Magnetic Dipole Moments; Total Angular Momentum

Magnetic Dipole Moment and the Bohr Magneton

An electron orbiting in an atom can be considered as a current loop and possesses a magnetic dipole moment. The magnetic dipole moment vector $\vec{\mu}$ is related to the amgular momentum vector $\vec{L}$ by

$$\vec{\mu} = -\frac{1}{2}\frac{e}{m}\vec{L}. \tag{39–11}$$

When the electron is placed in a magnetic field $\vec{B}$, its potential energy is given by

$$U = -\vec{\mu} \cdot \vec{B}.$$

If the magnetic field is in the z-direction, $U = \mu_z B_z$, which gives

$$\mu_z = -\mu_B m_l, \qquad (39\text{--}12)$$

where

$$\mu_B = \frac{e\hbar}{2m} \qquad (39\text{--}13)$$

is called the **Bohr magneton**.

*Stern-Gerlach Experiment and the *g*-Factor for Electron Spin

The Stern-Gerlach experiment provides the evidence of *space quantization*. In the experiment, a collimated beam of silver atoms was passed through a nonhomogeneous magnetic field along the z direction, giving a force in the z direction. Two distinct lines were observed for silver in the experiment. The two states for silver in the experiment are due to the spin of its one valence electron. With the spin of $\frac{1}{2}$, the electron spin can have only two orientations in space.

μ_z for spin is found to be given by

$$\mu_z = -g\mu_B m_s \qquad (39\text{--}14)$$

where g is called the g-factor or gyromagnetic ratio. $g = 2.0033\ldots$ for a free electron.

* Total Angular Momentum $\vec{J}$

The total angular momentum vector $\vec{J}$ is given by

$$\vec{J} = \vec{L} + \vec{S}$$

where $\vec{L}$ and $\vec{S}$ represent the orbital angular momentum and spin, respectively.

The magnitude of the total angular momentum vector is

$$J = \sqrt{j(j+1)}\,\hbar \qquad (39\text{--}15)$$

For a single electron in a H atom, j can be $l+1/2$ or $l+1/2$ whereas $m_j = j, j-1, \ldots, -j$.

*Spectroscopic Notation
For a single electron state we write as nL_j, where the value of L is specified as follows:

$L =$	0	1	2	3	4
	S	P	D	F	G

*Fine Structure; Spin-Orbit Interaction
A **fine structure** splitting occurs in an atom even in the absence of an external field because of the magnetic field produced by the atom itself. This is so because the electron itself has an intrinsic magnetic dipole moment μ_s, and from the reference frame of an electron, a revolving charged nucleus produces a magnetic field. This interaction is called the **spin-orbit interaction**. The resulting fine structure is proportional to a dimensionless constant called the **fine structure constant**, $\alpha = \dfrac{e^2}{2\varepsilon_0 hc} \approx \dfrac{1}{137}$.

39–8 Fluorescence and Phosphorescence
When electrons in an atom are excited to higher-energy states, they may return to the ground state by a series of lower-energy jumps. These jumps produce lower-frequency (and, hence, longer-wavelength) photons than the photons that caused the original excitation. This

process of emission of light of longer wavelength after illumination by shorter-wavelength light is called **fluorescence**. The wavelength for which fluorescence occurs depends on the energy levels of the particular atoms. In **fluorescent light bulbs**, the ultraviolet light emitted by mercury atoms is absorbed by a phosphor, which then emits longer-wavelength light in the visible.

When the emitted glow occurs over an extended period of time, the process is called **phosphorescence**. In phosphorescent substances, atoms are excited to **metastable** states, where they can stay longer before they jump into lower energy levels.

39–9 Lasers

A laser is a device that produces light that is intense, highly directional, monochromatic, and coherent. (By *coherent* we mean that there is a constant phase relationship between different beams.)

When electrons drop from higher-energy orbits to lower-energy orbits, photons are emitted in random directions with random phases. This process is called spontaneous emission. In contrast, if an incident photon with energy equal to the energy difference between two orbits *stimulates* an electron in an atom to move to a lower-energy orbit, the resulting emission of a photon is called a **stimulated emission**. In a stimulated emission, the emitted photon has the same energy and phase as the incident photon and both photons travel in the same direction.

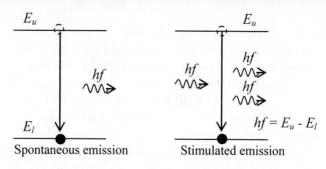

Spontaneous emission   Stimulated emission

$hf = E_u - E_l$

Repeated stimulated emissions increase the number of photons with the same energy and phase and traveling in the same direction, thus causing amplification of light. In order for light amplification to occur, there must be more atoms with electrons in the higher-energy orbit (excited state) than in the lower-energy orbit. This situation is known as **inverted population** in the system. Also, the electrons must spend a longer time in the excited state than they would ordinarily so that incident photons continue to encounter excited atoms to produce a stimulated emission. A long-lived excited state is called a **metastable state**. Thus, the excited state must be metastable for light amplification.

The excitation of atoms can be done in several different ways to obtain an inverted population. In a ruby laser, where lasing material is a ruby rod consisting of Al_2O_3 with a small percentage of Al replaced by Cr (chromium), the atoms are excited by strong flashes of light of wavelength 550 nm. This is called **optical pumping**.

In a helium–neon (He-Ne) laser, the lasing material is a mixture of 15 % He gas and 85 % Ne gas. Neon atoms are excited to a metastable state E_3 of energy 20.66 eV by resonant energy transfers from excited helium atoms. When the excited electrons of the neon atoms drop to energy level E_2 of energy 18.70 eV, they emit light photons of the same energy as the energy difference, 1.96 eV, between the levels, corresponding to the 632.8 nm red light of the laser.

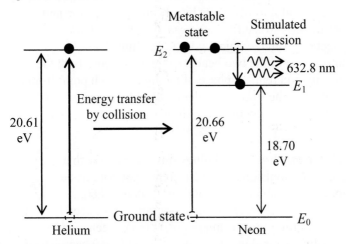

In a **pulsed laser**, the atoms are excited by periodic inputs of energy and the output is discrete pulses. In a **continuous laser**, energy is supplied continuously so that output is also a continuous beam.

*Applications
Lasers that emit high-energy light photons are used in laser eye surgery to correct for nearsightedness and farsightedness. Laser pulses vaporize corneal material until the cornea is flattened just enough to correct the nearsightedness. Laser beams are used to destroy cancerous and precancerous cells, used to destroy tissue in a localized area and to break gallstones and kidney stones. Because of the heat produced, a laser beam is used to weld broken tissue such as a detached retina. Lasers are used as bar-code readers at store checkouts and in compact disk (CD) players.

*39–10 Holography

Lasers are used for three-dimensional imaging, called holography. To make a laser **hologram**, a broadened laser beam is split into two parts by a beam splitter. One part goes directly to the film whereas the other part after reflection from the object goes to the film. The interference of the two beams enables to record both the intensity and relative phase of the light at each point from the object. The film is developed and placed again in a laser beam to view the three-dimensional image of the object.

Volume or **white-light holograms** are made with a laser on a thick emulsion. The interference pattern can be thought of as an array of bands or ribbons where interference occurred. If the hologram is produced by lasers emitting three primary colors (red, green, and blue), three-dimensional image is seen in full color when viewed with white light.

Tips for Solving the Problems

1. For a given value of the principal quantum number, n, the values of the orbital angular momentum quantum number, l, are 0, 1, 2, ..., $(n - 1)$. The allowed values of the magnetic quantum numbers, m_l, are $-l$, $-l + 1$, $-l + 2$, ..., -1, 0, $+1$, $+2$, ..., $l - 1$, l. The spin quantum number, m_s, can have two values: $-\frac{1}{2}$, $\frac{1}{2}$.

2. To find the probability of an electron at a certain distance from the nucleus use the radial probability distribution function given by $P_r = 4\pi r^2 \left| \psi \right|^2$ (not just $\left| \psi \right|^2$). To get the most probable distance take the

derivative of P_r with respect to r and set the derivative to zero to solve for r.

3. For $l = 0$, $|\psi|^2$ is spherically symmetric. For $l > 0$, $|\psi|^2$ depends on the angle θ.

4. Note that the radial probability density is maximum at distances that correspond roughly to the Bohr orbits.

5. The magnitude of the total angular momentum vector is $J = \sqrt{j(j+1)}\,\hbar$. For a single electron in H atom, j can be $l + 1/2$ or $l + 1/2$, whereas $m_j = j, j - 1, \ldots, -j$.

6. To determine the energy of a K-shell electron, replace the nuclear charge Ze by $(Z-1)e$ because of the screening of the nuclear charge by the remaining K-shell electron. As a result, the energy of the K-shell electron is $E_K = -13.6$ eV $(Z-1)^2/1^2$. Similarly, the energy of an L-shell electron is $E_L = -13.6$ eV $(Z-1)^2/2^2$. The wavelength of the radiation emitted (K_α X-ray) when an electron drops from the L shell to the K shell can be obtained from $hc/\lambda = E_L - E_K$.

7. The wavelength of a laser light can be obtained from the equation $hc/\lambda = E_u - E_l$ where E_u and E_l are the energies of the upper state (metastable state) and the lower state.

CHAPTER 40
Molecules and Solids

Important Concepts

        Covalent and ionic bonds
        Weak (van der Waals) bonds
        Rotational and vibrational energies of molecules
        Fermi energy
        Fermi-Dirac probability distribution
        Band theory of solids
        n-type semiconductor
        p-type semiconductor
        p-n junction diodes
        Half-wave and full-wave rectifiers
        Junction transistor

40–1 Bonding in Molecules

A molecule is a group of two or more atoms strongly held together forming chemical bonds. There are two main types of strong chemical bond: covalent and ionic.

Covalent Bonds

In a *covalent bond*, such as that formed in a hydrogen molecule (H_2), as two H atoms approach each other, the electron clouds overlap and the electrons from each atom can orbit both nuclei. This is also called "sharing" of electrons. If both the electrons are in the ground state of their respective atoms, a bond is formed only when the spins of the electrons are opposite and the total spin $S = 0$. In this case the two electrons are in different quantum states and they can come close together. The bond can be said to be the result of constructive interference of the electronic wave functions in the space between the two atoms and of the electrostatic attraction of the two positive nuclei to the concentration of negative charges between them. In this configuration the molecule has less energy than two separate atoms and is stable, as can be explained from uncertainty principle. The energy needed to break a bond is called the **bond energy**, the **binding energy**, or the **dissociation energy**.

Ionic Bonds

In an ionic bond, such as exists in sodium chloride, the outer electron of sodium atom spends nearly all its time around the chlorine. In neutral sodium, the 10 inner electrons shield the nucleus, so the single outer electron is attracted by a net charge of $+e$. In a neutral chlorine atom, the $+17e$ of the nucleus is shielded by the 12 electrons in the inner shells and subshells. Four of the five $3p$ electrons form donut-shaped clouds symmetric about the z-axis, and the fifth can have a barbell-shaped distribution, which is half empty. If an extra electron from a Na atom comes to be in the vicinity, it can occupy that state. As a result, the chlorine acquires a negative charge and sodium is left with a net positive charge. A bond is formed because of the electrostatic attraction between these two charged atoms.

Partial Ionic Character of Covalent Bonds

In a covalent bond when the atoms involved are different, the shared electrons are found to be more likely in the vicinity of one atom than the other. For example, in the water molecule, the shared electrons are more likely to be found around the oxygen atom than around the two hydrogens. The bond has a partial ionic character and the molecule is polar. That is, one part of the molecule has a net positive charge and the other part a net negative charge.

40–2 Potential Energy Diagrams for Molecules

The force between charges of the same sign is repulsive and the force between the two opposite charges is attractive. The potential energy (PE) U for two point charges are shown below as a function of their separation.

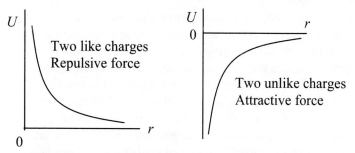

When a covalent bond is formed, such as in H_2 molecule, the PE of one H atom in the presence of the other has a shape that is shown below.

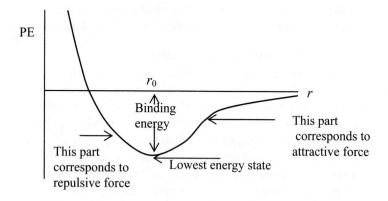

Starting at large r (separation) the PE decreases as the atoms approach each other. There is an optimum separation r_0 for which the energy is lowest. The energy difference between PE = 0 and the lowest energy state near the bottom of the well is called the *binding energy*. A reasonable approximation of binding energy is

$$U = -\frac{A}{r^m} + \frac{B}{r^n} \qquad (40\text{--}1)$$

where A and B are constants associated with the attractive and repulsive parts of the potential energy and m and n are small integers.

For many bonds the PE curve has a shape different from the above as shown on the next page. In this case as the atoms approach from large distance, the force is initially repulsive. Some additional energy, called the **activation energy**, must be injected into the system to get over the barrier. For example, to form H_2O from H_2 and O_2, activation energy is needed to break H_2 and O_2 molecules first into H and O atoms before they combine to form H_2O.

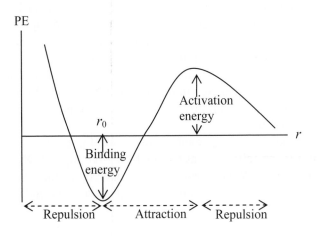

Potential energy diagram for the formation of ATP (adenosine triphosphate) from ADP (adenosine diphosphate) and phosphate (P) is shown to the right. In this case, an energy input is needed to make the bond (and hence the binding energy is negative) and there is energy release when the bond is broken.

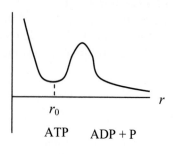

40–3 Weak (van der Waals) Bonds

The binding energy of covalent and ionic bonds is about 2 to 5 eV. These bonds are called **strong bonds**. An attachment between molecules due to simple electrostatic attraction (such as between polar molecules) is called a **weak bond**. Binding energy for weak bonds is about 0.04 eV to 0.3 eV. Weak bonds are generally a result of attraction between dipoles, also called the **dipole-dipole bonds**. There are also **dipole-induced dipole bonds** in which a polar molecule with permanent dipole moment can induce a dipole moment in an otherwise nonpolar molecule. All these weak bonds are called **van der Waals bonds** and the forces involved are called **van der Waals forces**. When one of the atoms in a dipole-dipole bond is hydrogen, it is called a **hydrogen bond**.

Weak bonds are important in liquids and for understanding the activities of cells, such as the double helix shape of DNA and DNA replication. Weak bonds can be broken by a molecular collision and because of this they are important in the cell.

*Protein Synthesis

Weak bonds (especially hydrogen bonds) are important to the process of protein synthesis. A protein molecule consists of one or more chains of amino acids. There are 20 different amino acids and a single protein chain may contains hundreds of amino acids in a certain order.

40–4 Molecular Spectra

A molecule can absorb energy of the right frequency to excite one of its electrons from a lower electronic state to a higher electronic state. The excited electrons then return to a lower state by emitting a photon. In addition, a molecule can rotate as a whole and the atoms in the molecule can vibrate. The rotational and vibrational energy levels are quantized and are generally spaced much more closely (10^{-3} to 10^{-1} eV) than electronic levels (1 to 10 eV). The emitted lines are not always distinguishable (that is, very closely spaced) and the lines form a **band spectrum**.

Rotational Energy Levels in Molecules

For a diatomic molecule rotating about a vertical axis, the rotational energy levels are given by

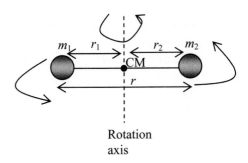

Rotation
axis

$$E_{rot} = \frac{1}{2} I\omega^2 = \frac{(I\omega)^2}{2I} = l(l+1)\frac{\hbar^2}{2I}, \, l = 0, 1, 2, \ldots \quad (40\text{--}2)$$

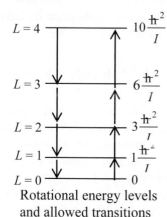

where l is called the rotational angular momentum quantum number. The transitions between rotational energy levels are given by the *selection rule*:

$$\Delta l = \pm 1$$

Rotational energy levels and allowed transitions

The energy of the emitted and absorbed photon for a transition between two rotational states with angular momentum quantum number l and $l - 1$ is given by

$$\Delta E_{rot} = \frac{\hbar^2}{I} l. \quad (40\text{--}3)$$

The measured absorption lines fall in the microwave or far-infrared regions of the spectrum.

The moment of inertia of the molecule of masses m_1 and m_2 (as shown in the sketch) rotating about an axis passing through the center of mass is given by

$$I = m_1 r_1^2 + m_2 r_2^2$$

$$= \mu r^2 \quad (40\text{--}4)$$

where $\mu = m_1 m_2/(m_1 + m_2)$, called the **reduced mass**, and r is the distance between the atoms.

Vibrational Energy Levels in Molecules

The PE of the two atoms in a diatomic molecule can be approximated as the PE of a simple harmonic oscillator (SHO) by the equation, $PE = \frac{1}{2} kx^2$. The classical frequency of vibration is

$$ f = \frac{1}{2\pi} \sqrt{\frac{k}{\mu}}, \qquad (40\text{--}5) $$

where k is the stiffness constant.

The solution of the Schrödinger equation for the SHO shows that the energy levels are quantized and are given by

$$ E_{\text{vib}} = (v + \frac{1}{2})\, hf, \quad v = 0, 1, 2, \dots \qquad (40\text{--}6) $$

where f is the classical frequency of the vibration and the integer v is called the **vibrational quantum number**. The selection rule for the transitions is $\Delta v = \pm 1$.

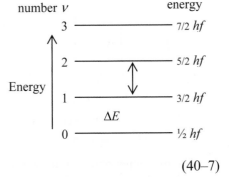

$$ \Delta E_{\text{vib}} = hf \qquad (40\text{--}7) $$

Typical transition energies are about 10^{-1} eV and the wavelengths are in the infrared region (10^{-5} m) of the spectrum.

Rotational plus Vibrational Levels

When energy is given to a molecule, both the rotational and vibrational modes can be excited. Transitions follow the selection rules:

$$\Delta v = \pm 1 \text{ and}$$

$$\Delta l = \pm 1.$$

$$\Delta E = \Delta E_{\text{vib}} + \Delta E_{\text{rot}}$$

$$= hf + (l+1)\,\frac{\hbar^2}{I}\,[l \to l+1],\, l = 0, 1, 2, \dots$$

$$(40\text{–}8)$$

$$= hf - l\,\frac{\hbar^2}{I}\,\,[l \to l-1],\,\, l = 1, 2, 3, \dots$$

Expected spectrum for transitions between combined rotational and vibrational states

The molecular absorption spectrum of HCL molecules follows the pattern well.

40–5 Bonding in Solids

In *amorphous* substances (such as glass) the atoms and molecules show no long-range order. In *crystalline* substances, the atoms, ions, or molecules are accepted to form an orderly array in a geometrical arrangement, called the *lattice*. The three common arrangements of atoms in a crystal are: simple cubic (atoms at each corner of cube), face-centered cubic (atoms at each corner and at the center of each face), and body-centered cubic (atoms are at each corner and at the center of the cube).

NaCl crystal is face-centered cubic with one Na^+ ion or one Cl^- ion at each lattice point. Each Na^+ ion has six nearest neighbor Cl^- ions that provide attractive Coulomb potential. Each Na^+ ion also feels a weaker repulsive Coulomb potential from other Na^+ ions that are far away. The net potential is given by

$$U = -\frac{\alpha}{4\pi\varepsilon_0}\frac{e^2}{r} + \frac{B}{r^m}, \qquad (40–9)$$

where the constant α is called the *Madelung constant* ($\alpha = 1.75$ for NaCL crystal).

Metallic bonds hold a solid. In metal atoms there are loosely held outer electrons. These outer electrons move freely among the metal atoms that without their outer electrons act like positive charges. The electrostatic attraction between the metal ions and this electron gas is responsible for holding the solids together. The binding energy of metal bonds is about 1 to 3 eV. The free electrons are responsible for providing the high electrical and thermal conductivity of metals. The free electrons account for the shininess of smooth metal surfaces. The free electrons can vibrate at any frequency and when a white light falls on a metal the electrons can vibrate and re-emit those same frequencies.

40–6 Free-Electron Theory of Metals; Fermi Energy

In a metal, electrons are free to move inside the metal and at high temperatures they can leave the metal (such as by thermionic emission). Thus, the electrons inside a metal can be modeled by a potential well whose height is infinite at ordinary temperatures but finite at higher temperatures. The energy is quantized but the spacing between energy levels is very small since L (width of the well) is very large. To deal with such a large number of very closely spaced states we define the **density of states** $g(E)$ (so that $g(E)\,dE$ represents the number of states per unit volume having energy between E and $E + dE$), which is given by in this situation:

$$g(E) = \frac{8\sqrt{2}\,\pi m^{\frac{3}{2}}}{h^3} E^{\frac{1}{2}}. \qquad (40\text{--}10)$$

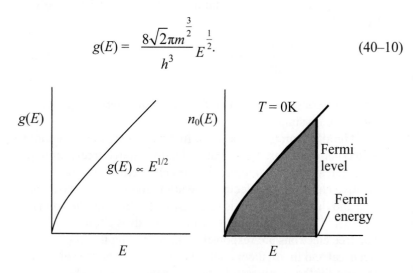

Electrons have spin ½, follow exclusion principle, and obey **Fermi-Dirac statistics**. The electron gas inside a metal is often called a **Fermi gas**. At $T = 0$ K, the possible energy levels will be filled, two electrons each (spins up and down), up to a maximum level called the **Fermi level**, as shown in the sketch above where vertical axis is labeled $n_0(E)$ for density of occupied states. If N/V represents the number of conduction electrons per unit volume, then

$$\frac{N}{V} = \int_0^{E_F} g(E)\,dE. \tag{40-11}$$

This gives

$$E_F = \frac{h^2}{8m}\left(\frac{3}{\pi}\frac{N}{V}\right)^{\frac{2}{3}}. \tag{40-12}$$

The average energy is:

$$\overline{E} = \frac{3}{5}E_F. \tag{40-13}$$

For copper, $E_F = 7.0$ eV and $\overline{E} = 4.2$ eV, which is much greater than the energy of thermal motion at room temperature.

At temperature $T > 0$ K, at least some of the electrons will increase in energy due to thermal motion. The probability of a given state of energy E being occupied is given by the Fermi-Dirac probability function:

$$f(E) = \frac{1}{e^{(E - E_F)/kT} + 1}, \tag{40-14}$$

which is plotted at two different temperatures.

Note that

$$f(E) = \begin{cases} 1 & E < E_F \\ 0 & E > E_F \end{cases} \text{ at } T = 0.$$

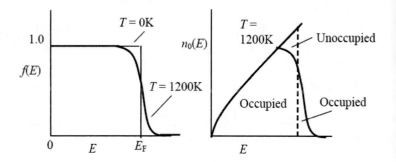

At higher temperature (say $T = 1200$ K for copper, just below its melting point) the Fermi factor changes only a little. When $E = E_F$, $f(E) = 0.50$. This means that there is a 50% chance that the state is occupied.

The density of occupied states is given by the following product:

$$n_0(E) = g(E)\,f(E)$$

$$= \frac{8\sqrt{2}\pi m^{\frac{3}{2}}}{h^3} \frac{E^{\frac{1}{2}}}{e^{(E - E_F)/kT} + 1}. \quad (40\text{–}15)$$

A plot at a higher temperature (say 1200 K for copper) shows that the distribution differs very little from that at $T = 0$. Small changes that occur are near Fermi level, showing a few electrons slightly below the Fermi level move to energy states slightly above it. The average energy increases only very slightly as the temperature is increased from absolute zero. This behavior is very different from that of an ideal gas for which kinetic energy increases directly with T.

This simple model of uniform potential well explains the electrical and thermal properties of conductors but cannot explain why some materials are good conductors and others are good insulators. The model needs to be refined to include the effect of the crystal lattice. The outcome of the existing periodic potential (that takes into account the attraction of electrons for each atomic ion in the lattice) into the Schrödinger equation is that the allowed energy levels are divided into *bands* with the energy gaps in between.

40–7 Band Theory of Solids

When two atoms approach each other, the wave functions overlap and the two same energy states (one for each atom) split into two states of different energy. In a solid, where there are a large number of atoms close together, each of the original energy levels becomes a **band**. Each band seems to be continuous since the energy levels are very close together in the band.

In a good **conductor**, the highest energy band containing electrons is only partially filled. For example, in sodium the outermost 3s band is half full. When a potential difference is applied, electrons can respond by accelerating and increasing their energy since there are many unoccupied states of slightly higher energy available. Thus, current flows and sodium is a good conductor. In magnesium, its 3s band is filled, but, unfilled 3p band overlaps the 3s band in energy. Thus there are many available states for the electrons to move into. As a result, magnesium is a good conductor.

In an **insulator**, the outermost band containing electrons, called the **valence band**, is completely filled. The next higher energy band, called the **conduction band**, is separated by an energy gap (Eg) of about 5 to 10 eV. At room temperature no electrons can move into the conduction band. When a potential difference is applied, no current flows since no states are available for electrons.

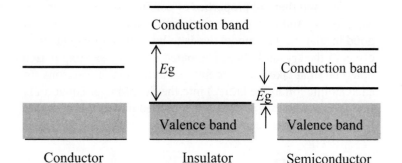

Conductor   Insulator   Semiconductor

In a pure or intrinsic semiconductor, such as silicon and germanium, energy bands are like those of an insulator except the energy gap between the valence and conduction bands is much less, typically about 1 eV. At room temperature a few electrons can acquire enough thermal energy to move into the conduction band where that can move. At higher temperatures more electrons can jump the gap and thus the resistivity of semiconductors decreases with increasing temperature. When a potential difference is applied, the few electrons in the conduction band move toward the positive electrode. This creates a few unoccupied states, called the **holes**, that were left empty by the electrons reaching the conduction band. Each electron in the valence band tries to fill a hole as it moves toward a positive electrode leaving behind a hole. Thus, the holes move toward the negative electrode.

40–8 Semiconductors and Doping

Silicon (Si) and germanium (Ge) are common semiconductors where four outer electrons of the atoms hold the atoms in the regular lattice structure. Adding a tiny amount of impurity in the crystal can enhance the electrical conductivity of a semiconductor. This process is called **doping**. Two types of doped semiconductors are made. In an *n*-type **semiconductor**, the impurity element (such as arsenic) has five outer electrons. Only four of the impurity electrons fit into the bonding with silicon (or germanium) and the remaining free electron can move relatively freely. Thus, the conductivity is higher. Since the charge carriers are negatively charged electrons, it is called an *n*-type semiconductor. In a **p-type semiconductor**, the impurity element (such as gallium) has only three outer electrons. The three impurity electrons fit into the bonding with silicon (or germanium) and there is an empty spot or *hole*. Since silicon atoms need four electrons to be electrically neutral, the hole has a positive charge in the structure. Electrons from nearby silicon atoms can jump into this hole, creating a positive hole where it had previously been. Positive holes in this case seem to carry the electric current, and it is called a *p*-type semiconductor.

According to the band theory in an *n*-type semiconductor the impurity provides an energy level that is just below (about 0.05 eV below for silicon) the conduction band. This energy level supplies electrons to the conduction band and is called a **donor**. In a *p*-type semiconductor, the impurity provides an energy state just above the valence band. Electrons from the valence band can easily jump into it and it is called an **acceptor** energy level.

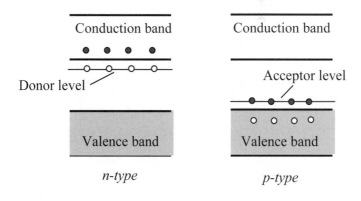

n-type   *p-type*

40–9 Semiconductor Diodes

When an *n*-type semiconductor is joined to a *p*-type, a *p-n* junction diode is developed. A few electrons near the junction diffuse from the *n*-type to the *p*-type, where they fill a few of the holes. Thus a potential difference is established at the junction with the *n* side positive relative to the *p* side, preventing further diffusion of electrons.

When a potential difference is applied across the junction with the positive terminal to the *p* type and the negative terminal to the *n* type, called the **forward bias**, the applied potential difference opposes the internal potential difference and a current flows through the junction. The current becomes very large as the applied potential difference is increased. If the applied potential difference is reverse, called **reverse bias**, the holes in the *p* end are attracted to the negative terminal of the battery and the electrons in the *n* end are attracted to the positive terminal of the battery. No current flows through the junction. A diode is a *nonlinear* device since the graph of current versus voltage is not linear.

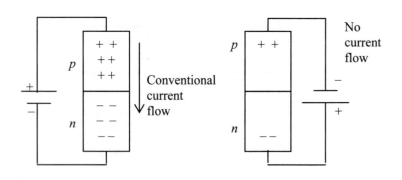

I

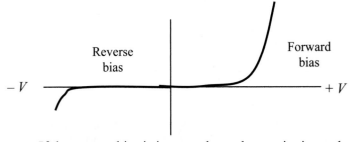

Reverse bias

Forward bias

$-V$

$+V$

If the reverse bias is increased greatly, a point is reached where breakdown occurs. The electric field across the junction becomes so large that electrons are pulled off the atoms and contribute to a larger and larger current as breakdown continues. The voltage remains constant over a wide range of currents and this situation can be used to regulate a power supply. A diode used for this reason is called a **Zener diode**.

A *p-n* junction allows current to flow only in one direction (so long as the voltage is not too large). It can be used as a **rectifier**, that is, to change ac to dc. The ac source applies a voltage across the diode alternately positive and negative. Only during the half cycle the junction will be forward biased and the current will flow through the circuit. This is **half-wave rectification** where the current is unidirectional but still not constant. A **full-wave rectifier** uses two diodes (sometimes four) so that at any instant either one diode or the other will conduct. The current flows for both the cycles in the same direction. A capacitor is used in parallel to the load R to smooth out the current.

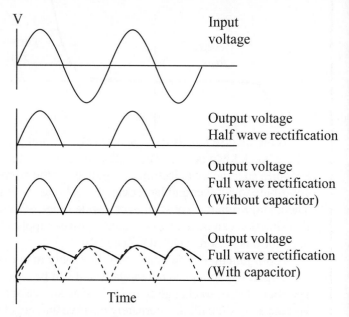

In a **light-emitting diode**, when a *p-n* junction is forward biased, electrons cross from the *n* region to the *p* region. When they combine with the holes, a light photon is emitted with an energy about the same as the band gap, E_g. **Photodiodes** and **solar cells** are *p-n* junctions used in a reverse way. Photons are absorbed, creating electron–hole pairs if the photon energy is greater than the band gap E_g.

40–10 Transistors and Integrated Circuits (Chips)

A **junction transistor** consists of one type of doped semiconductor sandwiched between two doped semiconductors of opposite type. Both *npn* and *pnp* transistors are made. The three semiconductors are called *emitter*, *base*, and *collector*.

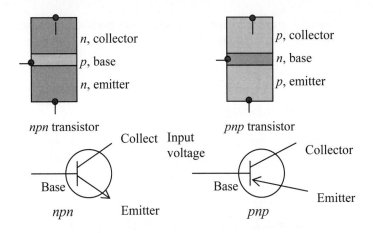

npn transistor   *pnp* transistor

A transistor can amplify a small ac signal to a large one and is the basic element in modern electronic amplifiers. A small change in the base voltage due to the input signal causes a large change in the collector current and, hence, a large change in the voltage drop across the output resistor R_C.

Current gain, $\beta_I = \dfrac{\text{output (collector) ac current} \left(i_C \right)}{\text{input (base) ac current} \left(i_B \right)}$,

Voltage gain, $\beta_V = \dfrac{\text{output (collector) ac voltage}}{\text{input (base) ac voltage}}$.

Transistors can be connected in many different ways and many new and innovative uses have been found for them. An **integrated circuit** consists of a tiny semiconductor crystal or chip on which many diodes, transistors, resistors, and other circuit elements are constructed using careful placement of impurities. The integrated circuit not only allows extremely complicated circuits to be placed in a small place, but has allowed a great increase in the speed of operation because the distances the electrical signals travel are tiny.

Tips for Solving the Problems

1. The force between two atoms in a diatomic molecule $F = -dU/dr$, where U is the potential energy. Set $F = 0$ in order to calculate the equilibrium separation between the atoms.

2. Remember the expression $E_{rot} = l(l+1)\dfrac{\hbar^2}{2I}$, $l = 0, 1, 2,$ ... for the rotational energy of a diatomic molecule and the selection rule $\Delta l = \pm 1$ to calculate the rotational transition frequencies or wavelengths.

3. The moment of inertia for a diatomic molecule rotating about a vertical axis passing through its center of mass is given by $I = m_1r_1^2 + m_2r_2^2 = \mu r^2$, where m_1 and m_2 are the masses of the atoms and r_1 and r_2 are the distances of the masses, respectively, from the center of mass, and μ is the reduced mass [$\mu = m_1m_2/(m_1 + m_2)$].

4. Remember the expression for the vibrational energy $E_{vib} = (v + \dfrac{1}{2})\,hf$, $v = 0, 1, 2, \ldots$ for a diatomic molecule and the selection rule $\Delta v = \pm 1$ to determine the vibrational transition.

5. The constant k can be obtained from the expression $f = \dfrac{1}{2\pi}\sqrt{\dfrac{k}{\mu}}$.

6. Use the Fermi-Dirac probability distribution $f(E)$ to determine the order of magnitude of the number of free electrons in the conduction band of a solid. For example, if the value of the function is 10^{-9}, then 1 atom in 10^9 can contribute an electron to the conductivity.

7. At $T = 0$, the probability that a state being occupied, $f(E) = 1$ $(E < E_F)$ and 0 $(E > E_F)$. At $T = 0$, average energy $\overline{E} = \dfrac{3}{5}E_F$.

8. The dependence of E_F on number density of electrons (N/V) is given by electron $E_F = \dfrac{h^2}{8m}\left(\dfrac{3}{\pi}\dfrac{N}{V}\right)^{\frac{2}{3}}$. The Fermi speed is given by $v_F = (2E_F/m)^{1/2}$.

9. Use the relation $E = hc/\lambda$ for the energy of a light photon for calculating the wavelength of light from a LED. In this case E is the energy gap, E_g.

CHAPTER 41
Nuclear Physics and Radioactivity

This chapter focuses on the nucleus of an atom, its composition and stability. The binding energy, forces inside a nucleus, radioactive decay, half-life, radioactive dating, and detection of radiation are described.

Important Concepts

Atomic number
Atomic mass number
Isotope
Atomic mass unit
Binding energy
Strong nuclear force
Radioactivity
α, β, and γ decays
Disintegration energy
Half-life
Radioactive decay law
Decay series
Radioactive dating
Carbon-14 dating
Tunneling
Geiger counter
Scintillation counter
Semiconductor detector
Photographic emulsion
Cloud chamber
Bubble chamber
Wire chamber

41–1 Structure and Properties of the Nucleus

The nucleus of an atom is composed of two types of particles: protons and neutrons. A neutron is electrically neutral and has a mass (m_n) slightly greater than that of a proton (m_p). These particles are called **nucleons**.

$$m_p = 1.6726 \times 10^{-27} \text{ kg}$$
$$m_n = 1.6749 \times 10^{-27} \text{ kg}$$

The different types of nuclei are called **nuclides**. The **atomic number**, Z, is the number of protons in a nucleus. The **atomic mass number**, A, is the number of nucleons in the nucleus. Thus, $A = Z + N$, where N is the number of neutrons in the nucleus. Clearly, $N = A - Z$.

A nucleus X with atomic number Z and mass number A is written as $_Z^A X$. Thus, the nucleus of carbon-14 (which has 6 protons and 8 neutrons) is $_6^{14}C$. Nuclei with the same atomic number (Z) but different numbers of neutrons (N) are called **isotopes**.

The average radius r of a nucleus of mass number A is approximately given by

$$r = (1.2 \times 10^{-15}\text{m}) \, A^{1/3} \qquad (41\text{--}1)$$

Since volume of a sphere is proportional to r^3, the volume of a nucleus is proportional to A.

The masses of nuclei and nucleons are often expressed in **unified atomic mass units**, u, given by

$$1 \text{ u} = 1.6605 \times 10^{-27} \text{ kg}$$

The mass of a proton is 1.007276 u, and the mass of a neutron is 1.008665 u. The atomic mass unit can be expressed in terms of Einstein's mass–energy relation $E = mc^2$ as

$$1 \text{ u} = 931.494 \text{ MeV}/c^2$$

where 1 MeV $= 10^6$ eV. In terms of this unit (MeV/c^2) the masses of a proton and neutron are 938.27 and 939.57, respectively.

A nucleus has a **nuclear spin**, I, that can be integer or half-integer, depending on whether it is made up of an even or an odd number of nucleons. The *nuclear angular momentum* of a nucleus is given by $\sqrt{I(I+1)}\,\hbar$. The nuclear magnetic moments are measured in terms of nuclear magnetor

$$\mu_N = \frac{e\hbar}{2m_p}. \tag{41–2}$$

41–2 Binding Energy and Nuclear Forces

The mass of a nucleus is found to be less than the total mass of its component nucleons. For example, the mass of a neutral ^4_2He is 4.002602 u. But the mass of the constituents in its nucleus (that is mass of two neutrons, two protons, and two electrons) is 4.032980 u. The reduction in energy, E, due to the reduction of the mass, Δm, (according to $E = \Delta mc^2$) is the **total binding energy** of the nucleus. The binding energy of a nucleus represents the minimum energy that is needed to separate a nucleus into its component nucleons.

The binding energy per nucleon increases rapidly with mass number A and becomes nearly constant at about 8.7

MeV per nucleon above $A \approx 40$. Beyond $A \approx 80$, binding energy per nucleon decreases slowly demonstrating that larger nuclei are held together a little less tightly than those in the middle of the periodic table.

Nuclear Forces
The protons inside a nucleus repel one another because of the Coulomb force. There is also a large attractive force, called the **strong nuclear force**, among the nucleons that holds the nucleus together.

Nuclear force is more complicated than gravitational and electromagnetic forces. The strong force is **short-range** and extends only a few fermis (1 fermi = 10^{-15}m). But the gravitational and electromagnetic forces act over long distances and are called **long-range** forces.

For a nucleus to be stable, the attractive strong nuclear force among its nucleons must be more than the repulsive force among the protons. Small nuclei are most stable when the number of protons and neutrons are approximately equal ($Z = N$). Larger stable nuclei require more neutrons than protons to be stable. It is found that if the atomic number (that is, the number of protons) of a nucleus is more than 82, the strong force cannot overcome the repulsive force among the protons.

There is another type of nuclear force called **weak nuclear force**, which shows itself in certain types of radioactive decay. The strong and weak nuclear forces together with gravitational and electromagnetic forces are four known types of forces that exist in nature.

41–3 Radioactivity
Radioactivity refers to the emissions from an unstable nucleus. In radioactivity, the nucleus changes its

composition by emitting particles, or an excited nucleus decays to a lower-energy state.

Many unstable isotopes occur in nature and such radioactivity is called natural radioactivity. Also, unstable isotopes can be produced in the laboratory by nuclear reactions and these are said to have artificial radioactivity.

Radioactive radiations are classified into three types: α rays, β rays, and γ rays.

- α rays are positively charged and actually consist of the nuclei of $^4_2 \text{He}$.

- β rays are negatively charged and consist of electrons.

- γ rays are electrically neutral and consist of photons.

41–4 Alpha Decay

An α particle consists of two protons and two neutrons. A nucleus that emits an α particle decreases its mass by 4 and its atomic number by 2. This decay is represented as

$$ ^A_Z N \longrightarrow \quad ^{A-4}_{Z-2} N' + {}^4_2 \text{He} $$

where N and N' are the **parent** nucleus and **daughter** nucleus, respectively. For example,

$$ ^{226}_{88} \text{Ra} \longrightarrow \quad ^{222}_{86} \text{Rn} + {}^4_2 \text{He} $$

Alpha decay occurs because strong nuclear force is overpowered by the Coulomb repulsive force and is unable to hold the nucleus together. The mass of the parent nucleus is greater than the mass of the daughter nucleus plus the mass of the α particle. The total released energy, called the **disintegration energy** (Q), is given by

$$Q = (M_P - M_D - m_\alpha)c^2 \qquad (41\text{--}3)$$

α-Decay Theory— Tunneling

We can understand alpha decay using the model of a nucleus inside which there is an alpha particle bouncing around. The following sketch shows the potential energy for alpha particle and (daughter) nucleus, showing the Coulomb barrier through which the alpha particle must tunnel to escape.

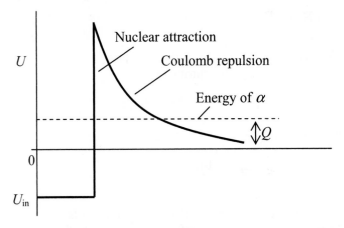

Classically the α particle would escape the nucleus only if there were an input of energy equal to the height of the barrier. According to quantum mechanics, it actually passes through the barrier by **tunneling**. Classically, this would violate conservation of energy, but according to uncertainty principle, energy conservation can be violated by an amount ΔE for a length of time Δt given by $(\Delta E)(\Delta t) = h/2\pi$. That is, quantum mechanics allows conservation

533

of energy to be violated for brief periods that may be long enough for an α particle to "tunnel" through the barrier.

Smoke Detectors—An Application

A **smoke detector** contains 0.2 mg of americium isotope $^{241}_{95}\text{Am}$ in the form of AmO_2, which is radioactive. The radiation ionizes the air molecules between two oppositely charged plates, resulting in a small steady current. But, if smoke enters, the radiation is absorbed by the smoke particles. This reduces current, which is sensed by the electronics, setting off the alarm.

41–5 Beta Decay

β^- decay

In β^- decay, a neutron is converted to a proton (p), an electron (e^-), and an antineutrino ($\bar{\nu}$):

$$\text{n} \longrightarrow \text{p} + e^- + \bar{\nu}$$

For example,

$$^{14}_{6}\text{C} \longrightarrow {}^{14}_{7}\text{N} + e^- + \bar{\nu}$$

Thus, in β^- decay, the mass number of a nucleus is unchanged but its atomic number increases by 1. The emitted electron is not the orbital electron; instead the electron is created within the nucleus from the decay of a neutron.

β^+ decay

Many isotopes decay by electron emission. There are unstable isotopes that have too few neutrons compared to their number of protons, which decay by emitting a

positron instead of an electron. In positron decay (β^+ decay), a positron (e^+) and a neutrino (ν) are emitted. For example,

$$^{19}_{10}\text{Ne} \quad \rightarrow \quad ^{19}_{9}\text{F} + e^+ + \nu$$

positron is called the **antiparticle** of the electron.

In general, for β^- and β^+ decay, respectively,

$$^{A}_{Z}\text{N} \quad \rightarrow \quad ^{A}_{Z+1}\text{N}' + e^- + \bar{\nu}$$

$$^{A}_{Z}\text{N} \quad \rightarrow \quad ^{A}_{Z-1}\text{N}' + e^+ + \nu$$

Electron Capture

In **electron capture**, a nucleus absorbs one of its orbiting electrons. The electron disappears in this process, a proton in the nucleus becomes a neutron, and a neutrino is emitted. For example,

$$^{7}_{4}\text{Be} + e^- \rightarrow \quad ^{7}_{3}\text{Li} + \nu$$

In general,

$$^{A}_{Z}\text{N} + e^- \rightarrow \quad ^{A}_{Z-1}\text{N}' + \nu$$

Usually, an electron in the innermost (K) shell is captured, in which case it is called "K-capture."

41–6 Gamma Decay

In a γ decay, an excited nucleus decays to a lower-energy state by emitting high-energy photons. In this process, neither the mass number nor the atomic number of the nucleus is changed. γ decay can be written as:

$$^{A}_{Z}\mathrm{N}^{*} \quad \longrightarrow \quad ^{A}_{Z}\mathrm{N} + \gamma$$

Isomers; Internal Conversion
In some cases, the nucleus may remain in the excited state before it emits a γ ray. The nucleus is said to be in a **metastable state** and is called an **isomer**.

An excited nucleus sometimes returns to the ground state by **internal conversion** without emitting a γ ray, in which case the excited nucleus interacts with one of the orbital electrons and ejects the electron with the same KE as the emitted γ ray would have had.

41–7 Conservation of Nucleon Number and Other Conservation Laws
In any radioactive decay, three conservation laws (energy, momentum, and angular momentum) and conservation of electric charge are obeyed. Also, the **law of conservation of nuclear number** is obeyed, according to which the total number of nucleons remains constant.

41–8 Half-Life and Rate of Decay
We can determine approximately how many nuclei in a sample will decay over a given time. The number of decay ΔN over a short time Δt is given by

$$\Delta N = -\lambda N \Delta t \qquad (41\text{--}4)$$

where N is the total number of radioactive nuclei present and λ is a constant called the decay constant. In the limit $\Delta t \rightarrow 0$, this gives

$$dN = -\lambda N dt \qquad (41\text{–}5)$$

If N_0 is the number of nuclei at $t = 0$, the number of nuclei, N, remaining at time t as obtained from the solution of the preceding equation is

$$N = N_0 e^{-\lambda t} \qquad (41\text{–}6)$$

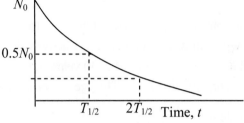

This is called the **radioactive decay law**. dN/dt (number of decays per second) is called the **activity** of the sample and since it is proportional to N,

$$\frac{dN}{dt} = \left(\frac{dN}{dt} \right)_0 e^{-\lambda t} \qquad (41\text{–}7)$$

The **half-life** ($T_{1/2}$) of a radioactive material is the time required for a given number of its nuclei to decrease to half their original number. In terms of the decay constant,

$$T_{1/2} = \frac{\ln 2}{\lambda} = \frac{0.693}{\lambda} \qquad (41\text{–}8)$$

41–9 Decay Series

If the daughter nucleus of an unstable nucleus is also unstable, it will decay to produce its own daughter nucleus. As a result, an unstable parent nucleus will produce a series of radioactive nuclei, called a radioactive **decay series**. For example ^{238}U after a series of α and β decays produces ^{206}Pb.

41–10 Radioactive Dating

Radioactive dating is a technique by which the age of ancient materials can be determined.

$^{14}_{6}C$ is unstable. The $^{14}_{6}C$ activity of an organism is constant until it dies, at which point the activity decreases exponentially with a half-life of 5730 years. $^{14}_{6}C$ can be used to date organic materials. The age can be determined from the equation, $t = \dfrac{1}{\lambda} \ln \dfrac{\lambda N_0}{\Delta N / \Delta t}$.

Geological Time Scale Dating
Carbon dating is useful for determining the age of objects less than 60,000 years old; the amount of remaining $^{14}_{6}C$ in older objects is too small to measure accurately. To measure dates on different time scales, different radioactive isotopes need to be used. $^{238}_{92}U$ (half-life 4.5 × 10^9 years) is useful in determining the ages of rocks on a geologic time scale. It found the age of the oldest earth rocks to be about 4 × 10^9 yrs.

41–11 Detection of Radiation
Different instruments are used for the detection of individual particles such as electrons, protons, α particles, neutrons, and γ rays.

In a **Geiger counter**, the incoming charged particles ionize a noble gas atom of the counter, thus freeing an electron that is accelerated toward the central positive electrode. On the way, the electron produces additional electrons through ionization of atoms of the gas. This results in a current pulse that is detected as a voltage across an external resistance.

In a **scintillation counter**, an incoming particle excites a **phosphor** atom and causes a photon to be emitted. This causes the emission of a photoelectron from the *photocathode* of a **photomultiplier**. Accelerated through a potential difference in the photomultiplier, the photoelectrons emit secondary electrons as they collide with the successive electrodes at higher potentials. After several steps, weak scintillations are converted into sizable electric pulses that are counted electronically. In tracer work and other biological experiments, **liquid scintillators** are used.

A **semiconductor detector** consists of a reverse-biased *p-n* junction diode. An incoming particle passing through the junction excites electrons into the conduction band, leaving holes in the valence band. The charges, which are free, produce a short electric pulse that can be detected.

In a **photographic emulsion**, a particle passing through a layer of photographic emulsion ionizes the atoms along its path. This results in a chemical change at these points. When the emulsion is developed, the particle's path is seen.

In a **cloud chamber**, super cooled vapor condenses into droplets on ionized molecules created along the path of the particle. The path becomes visible as the droplets scatter light when illuminated.

In a **bubble chamber** a reduction in pressure causes a liquid to be superheated and boil. Ions produced along the path of an energetic particle are sites for bubble formation and a trail of bubbles is created. A bubble chamber uses a liquid, and the tracks are readily visible.

A **wire chamber** consists of closely spaced, fine parallel wires among which alternate planes of wire are grounded and the ones in between are held at high voltage. As a charged particle passes through, the ions produced in the gas travel through the wires, creating an avalanche of ions as they further ionize the gas. A large current results, producing either a visible spark of an electrical pulse.

Tips for Solving the Problems

1. Useful constants and conversions for this chapter are:
   $1\ u = 1.660540 \times 10^{-27}$ kg, $1\ u = 931.494$ MeV/c^2, mass of electron $= 9.109390 \times 10^{-31}$ kg $= 0.511$ MeV/c^2, mass of proton $= 1.672623 \times 10^{-27}$ kg $= 938.28$ MeV/c^2, mass of neutron $= 1.674929 \times 10^{-27}$ kg $= 939.57$ MeV/c^2, 1 fermi (fm) $= 10^{-15}$ m.

2. Remember that $A = Z + N$. A nucleus X with atomic number Z and mass number A is written as $_Z^A X$.

3. To find the radius of a nucleus from its mass number A, use the relation $r = (1.2 \times 10^{-15}\ \text{m})\ A^{1/3}$. To calculate $A^{1/3}$, you need to use the x^y key of your calculator. Once you know the radius, you can calculate volume from $V = 4/3\ \pi r^3$.

4. The nuclear binding energy is the difference in mass between the nucleus and the total mass of its component nucleons times the square of the speed of light. Use this to solve problems in Section 41–2.

5. The total number of protons and neutrons is the same before and after an α decay. Remember that α particles are the nuclei of ^4_2He.

6. When a nucleus emits an α particle, it loses two protons and two neutrons in a β^- decay, the number of protons is increased by one, and the number of neutrons is decreased by one. In a β^+ decay, the number of protons is decreased by one, and the number of neutrons is increased by one.

7. If N_0 is the number of nuclei at $t = 0$, the number of nuclei, N, remaining at time t is given by $N = N_0 e^{-\lambda t}$, where λ is the decay constant. From this equation, the half-life is $T_{1/2} = 0.693/\lambda$. Use the preceding equation to determine the half-life from the decay constant, or vice versa. When calculating N, be sure to use the proper units of λ and t so that the product λt is dimensionless (that is if t is in years, λ must be in 1/years). The age of a sample can be determined from the equation, $t = \dfrac{1}{\lambda} \ln \dfrac{\lambda N_0}{\Delta N / \Delta t}$ using radioactive-dating. The preceding equations will be useful for solving problems in Section 41–8 to 41–10.

8. Remember that both N and $\Delta N/\Delta T$ decrease exponentially.

9. It is not hard to estimate an answer using an integer number of half-lives. If you want N after 2.5 half-lives, make sure the answer is between 1/4 (two half-lives) and 1/8 (three half-lives) of N_0.

CHAPTER 42
Nuclear Energy: Effects and Uses of Radiation

This chapter describes nuclear reactions, nuclear fission and fusion, and measurement of radiation, as well as applications of radiation in medicine.

Important Concepts

>Nuclear reaction
>Nuclear fission
>Liquid-drop model
>Chain reaction
>Nuclear reactor
>Breeder reactor
>Nuclear fusion
>Proton–proton cycle
>Magnetic confinement of plasma
>Inertial confinement
>Source activity
>Curie
>Becquerel
>Roentgen
>Rad
>Gray
>Rem
>Tomography image (CT)
>Nuclear magnetic resonance
>Magnetic resonance imaging

42–1 Nuclear Reactions and the Transmutation of Elements

A **nuclear reaction** is said to occur when a given nucleus is struck by another nucleus, or by particles (such as a γ ray or neutron) so that an interaction takes place with the first nucleus. For example,

$$^{4}_{2}\text{He} + ^{14}_{7}\text{N} \longrightarrow ^{17}_{8}\text{O} + ^{1}_{1}\text{H}$$

A general reaction is:

$$a + \text{X} \longrightarrow \text{Y} + b, \qquad (42\text{–}1)$$

where a is a projectile particle (or small nucleus) that strikes a nucleus X and produces nucleus Y and a particle b (such as p, n, α, γ). The **reaction energy** or **Q-value** is given by

$$Q = (M_a + M_X - M_b - M_Y)c^2, \qquad (42\text{–}2a)$$

$$= K_b + K_Y - K_a - K_X \qquad (42\text{–}2b)$$

If $Q > 0$, the reaction is *exothermic* or *exoergic*, and if $Q < 0$, the reaction is *endothermic* or *endoergic*.

The minimum energy for the projectile for a nuclear reaction to occur is known as the **threshold energy**.

Neutron Physics

Because neutrons do not have electric charge, they are not repelled by positively charged nuclei; they are found to be most effective projectiles for nuclear reactions. Bombarding uranium with neutrons, "transuranic" elements such as neptunium ($Z = 93$) and plutonium ($Z = 94$) were produced.

42–2 Cross Section

The *cross section* specifies the probability of a collision or of a nuclear reaction. It is given by

$$\sigma = \frac{R}{R_0 nt},$$ (42–3)

where n is the number of nuclei per unit volume and t is the thickness of the target. R and R_0 are the rate at which collision occurs and the rate at which the projectile particles strike the target, respectively. The concept of σ is useful since it depends only on the properties of the interacting particles.

For each possible kind of collision (such as elastic or inelastic) and each possible reaction, there is a cross section. The total cross section σ_T is given by

$$\sigma_T = \sigma_{el} + \sigma_{inel} + \sigma_R$$

where σ_R is the sum of the separate reaction cross sections.

σ is measured in barns. 1 barn $= 10^{-28}$ m^2. For example, the neutron cross section for cadmium is very large for low energies (between 100 and 10000 barns for $K \leq 1$ eV). The neutrons that have been slowed down and have reached equilibrium with matter at room temperature are called **thermal neutrons**.

42–3 Nuclear Fission; Nuclear Reactors
Nuclear Fission and Chain Reactions

The splitting of a heavy nucleus into two less-massive nuclei is known as **nuclear fission**. A typical nuclear fission reaction generates two to three neutrons together with the daughter nuclei. The reaction can be written as:

$$n + {}^{235}_{92}U \longrightarrow {}^{236}_{92}U \longrightarrow N_1 + N_2 + \text{neutrons} \quad (42\text{--}4)$$

The resulting nuclei N_1 and N_2 are called **fission fragments**. A typical fission reaction is:

$$n + {}^{235}_{92}U \longrightarrow {}^{141}_{56}Ba + {}^{92}_{36}Kr + 3n \quad (42\text{--}5)$$

According to the **liquid-drop model**, the neutron absorbed by the ${}^{235}_{92}U$ nucleus forms an intermediate compound nucleus ${}^{236}_{92}U$, which is in an elongated state because of the increased motion of the individual nucleons. When the nucleus is elongated like a dumbbell, the nucleus splits into two because of the dominant electric repulsive force.

Nuclear fission releases an amount of energy because the mass of the nucleus is considerably greater than the mass of the fission fragments and neutrons. It can be shown that total energy released per fission is (0.9 Mev/nucleon)(236 nucleons) $\approx$ 200 MeV.

The neutrons released in one fission reaction may initiate additional fission reactions in other nuclei. A fission reaction that proceeds from one nucleus to another in this fashion is called a **chain reaction**.

Nuclear Reactors
In a **nuclear reactor, self-sustaining chain reactions** occur, releasing energy in a controlled manner. Note the following:

- Since the probability that the ${}^{235}_{92}U$ nucleus will absorb a neutron is large only for slow neutrons, a substance known as a **moderator** is used to slow down the neutrons. **Heavy water** and **graphite** are found to be effective moderators.

- To increase the probability of fission of $^{235}_{92}U$ nuclei, natural uranium, which contains only 0.7% $^{235}_{92}U$, is enriched to increase the percentage of $^{235}_{92}U$.

- The minimum mass of uranium needed for a self-sustaining chain reaction is called the **critical mass**. If the mass is less than the critical, most neutrons escape before additional fissions occur, and the chain reaction is not sustained. The average number of neutrons per fission that produces further fissions is called the **multiplication factor**, f. If $f < 1$, the reactor is "subcritical.".If $f > 1$, it is "supercritical." Reactors are equipped with **control rods** to absorb neutrons and maintain the reactor at just barely "critical," $f = 1$.

A "power reactor" is used to produce electric power. The heat energy that is produced by nuclear fission in the fuel rods is carried off by water or liquid sodium and is used to boil water to produce steam. The steam drives a turbine to generate electricity and is then cooled in a condenser.

Besides thermal pollution, there is the problem of disposal of the radioactive fission fragments produced in the reactor plus radioactive nuclides produced by neutrons interacting with the structural parts of the reactor. The accidental release of radioactive fission fragments into the atmosphere poses a serious threat to human health.

$^{238}_{92}$ U has an appreciable fission cross section for fast neutrons. In a **breeder reactor**, by reduced moderation one or more neutrons from $^{235}_{92}$ U fission are absorbed by $^{238}_{92}$ U nuclei to produce $^{239}_{94}$ Pu . $^{239}_{94}$ Pu is fissionable and serves as additional fuel.

Atom Bomb
An atom bomb works on the principle of nuclear fission. Besides its great destructive power, when an atom bomb explodes, highly radio active isotopes are released. These are known as *radioactive fallout*.

42–4 Nuclear Fusion
Nuclear Fusion; Stars
When two light nuclei combine to form a relatively massive nucleus, the process is called **nuclear fusion**. In a nuclear fusion, the larger nucleus formed in the fusion reaction has less mass than the sum of the masses of the original less massive nuclei. The mass difference appears as the liberated energy (according to $E = mc^2$) in the fusion.

The sun (and many other stars) releases energy by nuclear fission, which is believed to be due to the following sequence of reactions, called **proton–proton cycle**:

$$^1_1H + ^1_1H \longrightarrow ^2_1H + e^+ + \nu \quad (0.42 \text{ MeV}) \quad (42\text{–}6a)$$

$$^1_1H + ^2_1H \longrightarrow ^3_2He + \gamma \quad (5.49 \text{ MeV}) \quad (42\text{–}6b)$$

$$^3_2He + ^3_2He \longrightarrow ^4_2He + ^1_1H + ^1_1H \quad (12.86 \text{ MeV}) \quad (42\text{–}6c)$$

where the energy released (Q-value) for each reaction is mentioned in the parenthesis. The overall reaction is

$$4{}_1^1\text{H} \longrightarrow {}_2^4\text{He} + 2e^+ + 2\nu + 2\gamma \qquad (42\text{--}7)$$

The total energy released is 26.7 MeV. In stars hotter than the Sun, the energy output comes from a different cycle, called the **carbon cycle**. ${}_6^{12}\text{C}$ is used as a catalyst in the cycle.

Possible Fusion Reactors

Fusion reactions for possible application using deuterium (which is very plentiful in the water of the oceans) are given as follows:

$$\begin{aligned}
{}_1^2\text{H} + {}_1^2\text{H} &\longrightarrow {}_1^3\text{H} + {}_1^1\text{H} \quad (4.03 \text{ MeV}) \quad (42\text{--}8a) \\
{}_1^2\text{H} + {}_1^2\text{H} &\longrightarrow {}_2^3\text{He} + \text{n} \quad (3.27 \text{ MeV}) \quad (42\text{--}8b) \\
{}_1^2\text{H} + {}_1^3\text{H} &\longrightarrow {}_2^4\text{He} + \text{n} \quad (17.59 \text{ MeV}) \quad (42\text{--}8c)
\end{aligned}$$

Energy released in fusion reactions can be greater than in fission for a given mass of fuel.

To start a nuclear fusion, the initial lighter nuclei must have enough energy to overcome their mutual Coulomb repulsion. The temperature required to give the needed kinetic energy is about 2 to 4×10^8 K for a useable fusion reactor. At this temperature, the atoms are ionized, and the resulting collection of nuclei and electrons is called plasma. Also for fusion, there must be a high density of nuclei to obtain a sufficiently high collision rate.

In **magnetic confinement**, magnetic fields are used in an attempt to contain the hot plasma. A more complex device, called a **tokamak**, is toroid-shaped and utilizes a combination of two magnetic fields (one directed along the axis of the toroid and the other produced by the current that passes through the plasma), producing a helical field. The **Larson criterion**, $n\tau \geq 3 \times 10^{20}$ s/m^3 (n is ion density and τ is confinement time), must be reached to produce **ignition**.

In **inertial confinement**, a small pellet containing the fuel for fusion is struck simultaneously from several directions by intense laser beams. This creates the density about 10^3 times the normal and heats up the core to a temperature at which fusion occurs. The confinement time is short, about 10^{-10} s, during which time the ions do not move appreciably because of their inertia.

42–5 Passage of Radiation Through Matter; Radiation Damage

Charged particles, such as α and β rays and protons, when they pass through a material, can attract or repel electrons strongly enough to ionize the atoms of the material. Neutral particles such as X-rays and γ-ray photons can ionize atoms by ejecting electrons by photoelectric and Compton effects. Energetic γ-rays can produce electrons and protons by pair production. Neutrons can cause a nuclear reaction altering the nucleus and the fragments produced can cause ionization.

*Biological Damage

Radiation damage in biological organisms is primarily due to the ionization produced in cells. The produced ions or radicals are highly reactive and take part in chemical reactions, thus interfering with the normal operation of the cell. A living cell exposed to radiation may no longer function or behave normally, or it may soon die. Radiation damage to a biological organism is classified as "somatic" or "genetic." *Somatic damage* refers to damage to any part of the body except the reproductive organs. *Genetic damage* refers to damage to reproductive cells.

42–6 Measurement of Radiation—Dosimetry

Radiation can be used to treat diseases, particularly, in a proper dose, to destroy cancerous tumors.

The rate at which nuclear decay occurs (that is, the number of disintegrations per second) is called the **source activity**. The units of activity are the curie (Ci) and the **becquerel** (Bq). 1 Ci = 3.7 × 10^{10} disintegrations/s,
1 Bq = 1 disintegration/s

The first radiation unit, the **roentgen** (R), is the dosage of X-rays or γ rays that deposits 0.878 × 10^{-2} J of energy per kilogram of air.

The **rad** is that amount of radiation that deposits energy at a rate of 1.00 × 10^{-2} J/kg in any absorbing material. The SI unit of absorbed dose is **gray** (Gy):

$$1 \text{ Gy} = 1 \text{ J/kg} = 100 \text{ rad} \qquad (42\text{–}9)$$

The **relative biological effectiveness** (RBE) or **quality factor** (QF) of a given type of radiation is defined as the number of rads of X or γ radiation that produces the same biological effect as 1 rad of the given radiation. Combining the absorbed dose in rad and QF gives the effective dose in **rem** (*rad equivalent man*)

$$\text{dose in rem} = \text{dose in rad} \times \text{QF} \qquad (42\text{–}10)$$

Defined in this fashion, 1 rem of radiation produces the same amount of biological damage regardless of the radiation involved.

Human Exposure to Radiation

We are constantly exposed to low level radiation from natural sources including cosmic rays, natural radioactivity in rocks and soils, and naturally occurring isotopes in our food. The average natural radioactive background is about 0.36 rem (360 mrem) per year per person and from medical X-rays the average is 40 mrem per person per year. The recommended upper limit of allowed radiation per person is 500 mrem per year, exclusive of natural sources.

Large dose of radiation causes **radiation sickness** including reddening of skin, drop in white blood cell count, nausea, etc., and can be fatal. People working around radiation generally carry **radiation film badges** to monitor the dose of incident radiation.

*42–7 Radiation Therapy

In **radiation therapy**, radiation is used for the treatment of diseases such as cancer. A well localized cancerous tumor can be killed by a narrow beam of γ or X-rays of large doses. Radiation from a radioactive source such as $^{60}_{27}Co$, high energetic photons, as well as protons, neutrons, electrons, and pions produced in a particle accelerator are used in cancer therapy. In some cases a radioactive isotope is inserted directly inside a tumor to destroy a majority of the cells. The radioactive isotope $^{131}_{53}I$ is inserted into blood to determine the condition of a patient's thyroid gland, which concentrates iodine present in the blood.

Radiation is also used for sterilizing bandages, surgical equipment, and packaged food since bacteria and viruses can be killed by large doses of radiation.

*42–8 Tracers and Imaging in Research and Medicine

Radioactive isotopes are used in biological and medical research as **tracers**. A given compound is artificially synthesized using a radioactive isotope such as $^{14}_{6}C$ or $^{3}_{1}H$, which can then be traced as it moves through an organism or undergoes a chemical reaction. Geiger or scintillation counters can be used to detect these tagged molecules. Using this technique, the mechanism of how food molecules are digested and to what parts of the body they are diverted can be traced. Tracers can be used to determine how amino acids and other essential compounds are synthesized by organisms.

In **autoradiography**, the position of a radioactive isotope is determined on film. This technique is useful in the following patterns of nutrient transport in plants.

For medical diagnosis, $^{99m}_{43}$Tc (m stands for metastable state) is the commonly used radio nuclide. It is formed when $^{99}_{42}$Mo decays. It has a half-life of 6 h and it can combine with a large variety of compounds. A detector outside the body records or forms image of the distribution of the radioactively labeled compound.

*42–9 Imaging by Tomography: CAT Scan, and Emission Tomography
*Norm X-ray Image
In this case, X-rays pass through the body and are detected on a photographic film.

Tomography Images (CT)
In tomographic imaging, the X-ray source and the detector move together across the body. The transmitted intensity is measured at a large number of points. The process is repeated for more scans by rotating the source-detector assembly (say, by 1° until 180°). The computer reconstructs the image of a slice and it is displayed on a TV or computer monitor.

Emission Tomography

Gamma rays are used for imaging purposes. In **single emission tomography** (SPET), a gamma detector is moved around the patient to measure gamma emission at different angles. **Positron-emission tomography** (PET) scans are used to locate tumors in the brain and also to find which regions of the brain are most active when a person performs a specific mental task. In a PET scan, a patient is given a radiopharmaceutical that decays by positron emission. The positrons encounter electrons in the body and they annihilate one another producing energetic γ rays that move in opposite directions. Detectors determine the origin of the γ rays and thus locate the areas where glucose metabolism is most intense. One advantage of PET is that no collimator is needed and thus fewer photons are wasted and lower doses can be administered.

*42–10 Nuclear Magnetic Resonance (NMR) and Magnetic Resonance Imaging (MRI)

*Nuclear Magnetic Resonance (NMR)

The atomic electrons have an intrinsic angular momentum called *spin*. When atoms are placed in a magnetic field, the atomic levels are split into several closely spaced levels. Nuclei also have magnetic moments. For example, the hydrogen nucleus consists of a single proton. Its spin angular momentum and its magnetic moment, like that of an electron, can have only two values when placed in a magnetic field (spin up or parallel to the field and spin down or antiparallel to the field). In a magnetic field B_T the energy level of the nucleus splits into two levels and the difference in energy $\Delta E = 2\mu_p B_T$ where μ_p is the magnetic moment of the proton.

In a nuclear magnetic resonance setup, the sample to be examined is placed in a static magnetic field and a radio frequency pulse of electromagnetic radiation (that is, photon) is applied to the sample. If the frequency f of the pulse is same as the energy difference between the two energy levels, that is

$$hf = \Delta E = 2\mu_p B_T$$

then photons will be absorbed, thus exciting many nuclei from the lower state to the upper state. Thus, there will be absorption for frequency close to $f = 2\mu_p B_T/h$. This is called **nuclear magnetic resonance** (NMR).

*Magenetic Resonance Imaging (MRI)
Magnetic resonance imaging (MRI) produces NMR images using hydrogen, which is the most common element of the human body. In the experimental apparatus, large coils set up the static magnetic field, and RF coils generate RF pulse of electromagnetic wave that cause the nuclei to jump from the lower state to the upper state. These same coils (or another coil) detect the absorption of energy or the emitted radiation when nuclei jump back to the lower state. To create an image, the intensity of absorbed or reemitted radiation from many different points of the body is measured. This is a measure of the density of H atoms at each point. Usually instead of applying a uniform magnetic field, a field gradient is applied and by varying f, absorption by different planes is measured. A reconstructed image based on the density of H atoms is not very interesting. More useful images are based on the rate at which the nuclei decay back to the ground state. Such images produce resolution of 1 mm or better. This NMR technique, also called **spin-echo**, produces images of great

diagnostic value, both in the delineation of structure and in the study of metabolic processes.

Tips for Solving the Problems

1. To determine the energy released in a nuclear reactions determine the mass difference before and after the reaction, then use Einstein's relation $E = |\Delta m|c^2$. If mass is in kg, the result is in Joules (J), and likewise, the energy has to be in joules in order to get the mass in kg. Usually, the energy released is expressed in MeV. Use the necessary conversion factor to convert the energy to MeV. Note that for energy to be released in a nuclear reaction, the mass difference (final mass minus initial mass) has to be negative.

2. Note the following definitions:
    1 curie (Ci) = 3.7×10^{10} disintegrations/s
    1 becquerel (Bq) = 1 disintegration/s
    1 roentgen (R) = 0.878×10^{-2} J/kg
    1 rad = 1.00×10^{-2} J/kg
    1 Gy = 1 J/kg = 100 rad
    Dose in rem = dose in rad $\times$ QF

3. Energies released in nuclear fusion and nuclear fission can be obtained from Einstein's relation, $E = |\Delta m|c^2$.

4. Useful constants and conversions are:
    1 u = 1.660540×10^{-27} kg
    1 u = 931.494 MeV/c^2
    Mass of electron = 9.109390×10^{-31} kg = 0.511 MeV/c^2, mass of proton = 1.672623×10^{-27} kg = 938.28 MeV/c^2, mass of neutron = 1.674929×10^{-27} kg = 939.57 MeV/c^2, 1 fermi (fm) = 10^{-15} m.

CHAPTER 43
Elementary Particles

This chapter focuses on particle accelerators, elementary particles, antiparticles, particle classification, quarks, and particle interactions.

Important Concepts
> Cyclotron
> Cyclotron frequency
> Synchrotron
> Linear accelerator
> Feynman diagram
> Lepton
> Baryon
> Strange particles
> Quarks
> Quantum chromodynamics (QCD)
> Electroweak theory
> Grand unified theory

43–1 High-Energy Particles and Accelerators

Particles accelerated to high energy are used as projectiles to probe the interior of nuclei and nucleons they strike. Because the projectile particles have high energy, the field is called **high-energy physics**.

Wavelength and Resolution

The greater the momentum p of a particle, the shorter is its wavelength, which is given by the relation

$$\lambda = \frac{h}{p}, \qquad (43\text{--}1)$$

Because the resolution is limited by the wavelength, shorter wavelengths can obtain finer detail.

Cyclotron

In a cyclotron, charged particles (usually protons) move within two D-shaped cavities. The source of the ion is at the center of the gap between the cavities. A magnetic field is applied to maintain the ions in nearly circular orbits. Each time the ions pass into the gap between the "dees," a voltage is applied that accelerates the ions, thus increasing the speed of the ions and the radius of curvature of their paths. After many revolutions, the ions attain high kinetic energy and reach the outer edge of the cyclotron. The speed of the ion of charge q and mass m when moving in a circle of radius r can be obtained from the magnetic force relation

$$qvB = \frac{mv^2}{r} \text{ or } v = \frac{qBr}{m}$$

This gives the frequency f of revolution, known as the **cyclotron frequency**, as

$$f = \frac{qB}{2\pi m} \qquad (43\text{--}2)$$

The frequency f does not depend on the radius r. But at higher speeds, the mass of the ion increases (relativistic increase of mass) and the frequency of the applied voltage must be reduced. Such a modified cyclotron is called a **synchrocyclotron**, where using complex electronics the frequency is decreased as protons increase their speeds and reach larger orbits.

Synchrotron

This can accelerate protons to about 1000 GeV or 1 TeV (1 TeV $= 10^{12}$ eV). This uses a narrow ring of magnets with each magnet being placed at the same radius from the center of the circle. The magnets are interrupted by gaps where high voltage accelerates the particles. Once particles are injected, they move in a circle of constant radius. This is done by providing them considerable energy initially in a much smaller accelerator, and then slowly increasing the magnetic field as their speed increases in the large synchrotron.

Since charge particles are moving in a circle, they are constantly accelerating toward the center. Accelerating charged particles produce radiation. In a synchrotron, the radiation is known as **synchrotron radiation**. Synchrotons can be used to produce intense beams of X-rays, UV light, or visible light.

Linear Accelerators

In this arrangement, the ions pass through a series of tubular conductors. The frequency of the applied voltage is such that when positive ions (say) reach a gap, the tube in front of them is negative and the one just left is positive. Thus, they are accelerated in each gap and increase their speed. Linear accelerators are used to accelerate electrons since because of their small mass they reach high speed very quickly. The largest electron linear accelerator is that at Stanford Linear Accelerator Center (SLAC), which is over 2 miles long and can accelerate electrons to 50 GeV (1 GeV $= 10^9$ eV).

Colliding Beams

In the **colliding beams** method, the target particles and the projectile particles are both moving and they are allowed to collide head-on. To accomplish this in one accelerator, **storage rings** are used where one type of particles, after they are accelerated to a maximum energy, continue to circulate for many hours. It then accelerates the second type of particles and the beams are then directed to collide. For example, the large electron–positron collider at CERN produces oppositely revolving beams of e^+ and e^-, each of energy 93 GeV, for a total interaction energy of 186 GeV.

43–2 Beginnings of Elementary Particle Physics—Particle Exchange

The elementary particles are the fundamental building blocks of all atoms. Among the three subatomic particles (electrons, protons, and neutrons), only the electron is considered to be elementary, whereas protons and neutrons are composed of smaller elementary particles. About 300 new particles have been discovered so far, most of which are unstable.

Elementary particle physics began in 1935 when Yukawa predicted the existence of a new particle that would

mediate the strong nuclear force. For the electromagnetic force that exists between two charged particles, photons are exchanged between the particles, giving rise to the force. A **Feynman diagram** illustrating a photon acting as a carrier of the electromagnetic force between two electrons is shown on the right. Starting at the bottom, the two electrons approach each other.

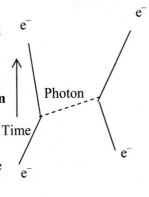

As they get close, momentum and energy are transferred from one to the other by a photon and the electrons bounce apart. Since the photon is absorbed by the second particle very shortly after it is emitted by the first, it is not observable and is called a *virtual* photon. A photon is called the quantum of the electromagnetic field or force.

According to Yukawa, a particle, called a **meson,** mediates the strong nuclear force, the force that holds the nucleus together. A meson is exchanged when a proton and neutron interact via strong nuclear force as shown. The mass m of the meson can be obtained from

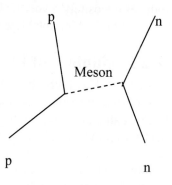

$$mc^2 \approx \frac{hc}{2\pi d} \qquad (43\text{–}3)$$

where d is the maximum distance that can separate the interacting nucleons. This results in a mass m of about 130 MeV/c^2, which is 250 times the electron mass of 0.51 MeV/c^2.

In 1937 a particle, called a **muon**, was observed whose mass is 106 MeV/c^2. The particle does not interact strongly and can hardly mediate the strong nuclear force. In 1947, particle predicted by Yukawa was observed in cosmic rays. This is called the "π" or pi meson or simply **pion**, which can have three charge states +, −, or 0. The π^+ and π^- have mass of 139.6 MeV/c^2 and π^0 a mass of 135.0 MeV/c^2. All three particles interact strongly with matter. The reactions observed in the laboratory include

$$p + p \rightarrow p + p + \pi^0$$

$$p + p \rightarrow p + n + \pi^{+} \qquad (43\text{--}4)$$

Other mesons were also subsequently observed that were considered to mediate strong force. The recent theory of quantum chromodynamics replaced mesons with *gluons* as the basic carrier of strong force.

It is believed that the weak nuclear force and the force of gravity are also mediated by particles. The particles that mediate the weak force are called W^{+}, W^{-}, and Z^{0} and were detected in 1983. The carrier (or quantum) for the gravitational force is called the **graviton**, which has not been observed yet.

43–3 Particles and Antiparticles

A positron has the same mass and the same magnitude of charge as an electron, but the sign of the charge is opposite. A positron is called the **antiparticle** to the electron. An **antiproton**, which carries a negative charge, is the antiparticle to the proton. Other antiparticles have also been found. The photons, π^{0}, and a few other particles do not have antiparticles. A particle and its antiparticle annihilate each other, producing energy of γ rays or of other particles.

43–4 Particle Interactions and Conservation Laws

In addition to the conservation laws of physics, two other conservation laws are found to be true in particle reactions. These conservation laws are helpful in understanding why some particle reactions do not occur.

Baryon number is conserved in particle reactions. All nucleons have baryon number $B = +1$; all antinucleons (antiprotons, antineutrons) have $B = -1$.

The following reaction does occur if the incoming proton has sufficient energy since it conserves baryon number.

$$p + n \rightarrow p + n + \bar{p} + p$$

$$B = 1 + 1 = 1 + 1 - 1 + 1$$

Three **lepton numbers** are conserved with weak interactions including decays. Electron (e^-) and the electron neutrino (ν_e) are assigned $L_e = +1$; e^+ and $\bar{\nu}_e$ are assigned $L_e = -1$; whereas all other particles have $L_e = 0$. Electron lepton number is conserved in all interactions including β decays.

In a decay involving muons, such as $\pi^+ \rightarrow \mu^+ + \nu_\mu$, muon lepton number ($L_\mu$) is conserved. μ^- and ν_μ have $L_\mu = +1$; μ^+ and $\bar{\nu}_\mu$ have $L_\mu = -1$; all other particles have $L_\mu = 0$. L_μ is believed to be conserved in all interactions or decays. A third lepton number has been associated with τ lepton and its neutrino ν_τ.

Antiparticles have opposite charge, baryon number, and lepton number.

43–5 Neutrinos—Recent Results

There are three types of neutrinos: ν_e, ν_μ, ν_T. These can change into one another, a phenomenon called **neutrino flavor oscillation**. This has been confirmed by experiments and solves the solar neutrino problem. Astrophysical experiments show that the sum of all three neutrino masses is less than $0.14 \text{ eV}/c^2$. Because of flavor oscillations, it seems likely that at least one neutrino has a mass of at least $0.04 \text{ eV}/c^2$.

Experimental measurements of neutrinos emitted in a supernova explosion shows an upper limit for the mass of electron anti-neutrino about 4 eV/c^2. Even more recent measurements provide evidence that mass is even smaller and that it is not zero.

43–6 Particle Classification

About 300 new particles have been discovered so far, most of which are unstable. Particles are classified according to their interactions.

Gauge bosons include photons and the W and Z particles that carry the electromagnetic and weak interactions, respectively. The particles that experience the weak nuclear force but not the strong nuclear force are called **leptons**. There are only six leptons: electron, muon, and the tau (τ) and three types of neutrino (ν_e, ν_μ and ν_τ). **Hadrons** are particles that experience both the strong and the weak nuclear force. Hundreds of hadrons are known to exist, including the proton and the neutron. They are divided into two subgroups: **baryons** having baryon number +1 (or -1 for their antiparticles) and **mesons** having baryon number 0.

43–7 Particle Stability and Resonances

The lifetime of an unstable particle depends on which force is most active in causing the decay. When a strong force causes decay, the decay occurs more quickly. Decays that occur via weak force have lifetimes of 10^{-13} s or longer. Particles that decay via electromagnetic forces have lifetimes about 10^{-16} to 10^{-19} s. W and Z particles are exceptions to this scheme. Many particles decay via strong interaction into other strongly interacting particles and their lifetimes are very short (about 10^{-23} s). Their decay products can be detected and from them the existence of such short-lived particles is inferred. Fermi studied the number of π^+ particles scattered by a proton target as a function of the incident π^+ kinetic energy. The large **resonance** peak around 200 MeV of width about 100 MeV led him to conclude that the π^+ and proton combined momentarily to form a short-lived particle before coming apart again. This new particle is called the Δ. In fact, the lifetime of about 10^{-23} s for such resonances can be obtained from the measured width of resonance using the uncertainty principle.

43–8 Strange Particles? Charm? Toward a New Model

Particles, namely the K, Λ, and Σ, are called the **strange particles**. They are always produced in pairs. For example, $\pi^- + p \rightarrow K^0 + \Lambda^0$ occurs with high probability, but the reaction $\pi^- + p \rightarrow K^0 + n$ is never found to occur. They are called strange particles because the unobserved reaction would not have violated any known conservation laws.

Another feature of these strange particles was that although they were clearly produced via strong interaction, they did not decay at a rate characteristic of the strong interaction. A new quantum number, called **strangeness**, was introduced and a new conservation law, called the conservation of strangeness, was introduced to explain the production of strange particles in pairs. Antiparticles have opposite strangeness from their particles; one of each pair has S = +1 and the other S = −1.

Note that strangeness is conserved in strong interactions but is not conserved in weak interactions.

43–9 Quarks

No internal structure has been detected in any of the leptons and they are considered to be elementary particles.

All hadrons are composed of a number of truly elementary particles called **quarks**. There are six types of quarks. They have been named *up* (u), *down* (d), *strange* (s), *charm* (c), *top* or *truth* (t), and *bottom* or *beauty* (b). There are also antiparticles to the quarks. All quarks have charges that are fractions of the charge of the electron. The charges for u, d, s, c, b, and t quarks are, respectively, $+2/3$ e, $-1/3$ e, $-1/3$ e, $+2/3$ e, $-1/3$ e, and $+2/3$ e. The charges of antiquarks are equal and opposite to the correspding quarks.

For example, the proton is composed of two up quarks and one down quark (uud). The neutron is composed of one up quark and two down quarks (udd). Mesons are composed of bound pairs of quarks and antiquarks. For example, a π^+ meson is considered a u$\overline{\text{d}}$ pair ($Q = 2/3$ e + $1/3$ $e = +1e$, $B = 1/3 - 1/3 = 0$, $S = 0 + 0 = 0$).

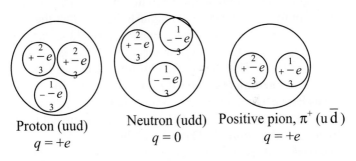

Proton (uud)
$q = +e$

Neutron (udd)
$q = 0$

Positive pion, π^+ (u$\overline{\text{d}}$)
$q = +e$

No free quark has been observed. It is believed that a free, independent quark cannot exist.

Today it is believed that the true elementary particles are the six quarks, the six leptons, and the gauge bosons.

43–10 The "Standard Model": Quantum Chromodynamics (QCD) and the Electroweak Theory

Quarks have a property called **color**. The distinction between the six quarks is referred to as **flavor**. Each flavor of quarks (u, d, s, c, b, t) has three colors: red, green, and blue. The colors of antiquarks are antired, antigreen, and antiblue. Baryons consist of three quarks, one of each color. Mesons consist of a quark–antiquark pair of a particular color and its anticolor.

Particles that follow the exclusion principle are called **fermions** and they have half-integer spin, usually $\frac{1}{2}$. Those that do not follow the principle are called **bosons** and they have integer spins (0, 1 etc.). The quarks have spin $\frac{1}{2}$ and should obey the exclusion principle. Quarks were given an additional quantum number (color) that is different for each quark to distinguish them so that the exclusion principle holds. Each quark is thought to carry a *color charge* and the strong force between them is called the color force. This theory of color force is called **quantum chromodynamics** (QCD). The strong force that exists between two hadrons is a force between the quarks that make them up. The particles that mediate the color force are called **gluons**. There are eight gluons, all massless, and six of them have color charge.

The strength of the color force increases with increasing distance and as two quarks approach very closely, the force between them becomes very small. This is called **asymptotic freedom**.

The weak force is mediated by W^+, W^-, and Z^0 particles. It acts between the weak charges each particle has.

The true elementary particles are the leptons, the quarks, and the gauge bosons (photon, W and Z, and the gluons).

A **gauge theory** unifies the weak and electromagnetic interactions. In this **electroweak theory**, the weak and electromagnetic forces are seen as two different manifestations of a single *electroweak* interaction. The electroweak theory plus QCD for the strong interaction are known as **standard model**.

43–11 Grand Unified Theories

A **grand unified theory** (GUT) is an attempt to unify electroweak theory and QCD for strong force into one theory. One type of such a grand unified theory of the electromagnetic, weak, and strong forces has been worked out in which there is only one class of particle—leptons and quarks belonging to the same family. This unity is predicted to occur within a scale of less than 10^{-30} m, called the **unification scale**. The theory predicts the existence of particles, called X bosons, that can be exchanged between a quark and a lepton, allowing one to change to the other. The mass of the X boson would be about 10^{14} GeV/c^2.

At 10^{-30} m the force between elementary particles is believed to be a single force —it is symmetrical and does not single out one type of charge over another. But, at large distances we find three distinct forces. This is known as **symmetry breaking**. During the first 10^{-35} s after the big bang that created the universe, the temperature was so high that the particles had energies corresponding to this unification scale. Baryon numbers were not conserved, then allowing an imbalance that accounts for the observed predominance of matter ($B > 0$) over antimatter ($B < 0$) in the universe.

43–11 Grand Unified Theories

A theory that attempts to unify all four forces is called
string theory. In the theory, the elementary particles are
imagined not as points but as one-dimensional strings of
about 10^{-35} m long. A super string theory includes a related
idea, called **super symmetry**, into the string theory. Super
symmetry predicts that interactions exist that would
change fermions into bosons and vice versa and that all
fermions have supersymmetric boson partners.

Tips for Solving the Problems

1. Useful constants and conversions are:
    $1 \text{ u} = 1.660540 \times 10^{-27} \text{ kg}$
    $1 \text{ u} = 931.494 \text{ MeV}/c^2$
    Mass of electron = $9.109390 \times 10^{-31} \text{ kg} = 0.511$
    MeV/c^2, mass of proton = $1.672623 \times 10^{-27} \text{ kg} =$
    938.28 MeV/c^2, mass of neutron = 1.674929×10^{-27}
    kg = 939.57 MeV/c^2.
    $1 \text{ GeV} = 10^9 \text{ eV}$.

2. Note that the cyclotron frequency is given by
    $f = qB/2\pi m$, which does not depend on the radius r.

3. Remember that baryon number is conserved in particle
    reactions. All nucleons have baryon number $B = +1$;
    all antinucleons (antiprotons, antineutrons) have
    $B = -1$.

4. Electron (e^-) and the electron neutrino (ν_e) are
    assigned $L_e = +1$; e^+ and $\overline{\nu}_e$ are assigned $L_e = -1$;
    whereas all other particles have $L_e = 0$. Electron lepton

number is conserved in all interactions including β decays. In a decay involving muons, muon lepton number (L_μ) is conserved. μ^- and ν_μ have $L_\mu = +1$; μ^+ and $\overline{\nu}_\mu$ have $L_\mu = -1$; all other particles have $L_\mu = 0$.

CHAPTER 44
Astrophysics and Cosmology

This chapter deals with the basics of astrophysics and cosmology. Different types of stars, stellar evolution, expanding universe, and the standard cosmological model are described.

Important Concepts

Luminosity
Apparent brightness
Main sequence stars
Red giant
White dwarf
Neutron star
Black hole
General theory of relativity
Principle of equivalence
Inertial mass and gravitational mass
Schwartzschild radius
Doppler effect
Red shift
Hubble's law
Cosmological principle
Big bang
Cosmic microwave background
Dark matter and dark energy

The study of objects such as stars and galaxies is called **astrophysics**. **Cosmology** is the study of the universe, such as its origin, evolution, and fate.

44–1 Stars and Galaxies

The Earth–Moon distance = 3.84×10^5 km.
The Earth–Sun distance = 1.50×10^{11} m.

The distance light travels in one year is called a **light-year** (ly). 1 ly = 9.46×10^{15} m. Light-yeara are used to specify distances of stars and galaxies. The distance of the nearest star (Proxima Centauri) = 4.3 ly. The distance to the Andromeda galaxy is about 2.25×10^6 ly.

Our **galaxy** (called the Milky Way galaxy) has a bulging central nucleus and spiral arms, and it contains about 10^{11} stars. The galaxy has a diameter of about 100,000 ly and a thickness about 2000 ly. The Sun is located at about 28,000 ly from the center of the galaxy and it orbits the galactic center once in about 200 million years with a speed of about 250 km/s relative to the galaxy. The mass of our galaxy is about 3×10^{41} kg.

Star clusters are groups of stars that are so numerous that they appear as a cloud. Galaxies are grouped in **galaxy clusters** containing from a few to many thousands of galaxies in the cluster. Clusters of clusters of galaxies are called **superclusters**.

In addition to usual stars, clusters of stars, galaxies, and their clusters, the universe contains other stars (such as *red giants*, *white dwarfs*, *neutron stars*, and *black holes*), exploding stars (called *novae* and *supernovae*) and *quasars*, which are a thousand times brighter than ordinary galaxies.

44–2 Stellar Evolution: Nucleosynthesis, the Birth and Death of Stars
Luminosity and Brightness of Stars

The **intrinsic luminosity**, L, of a star is the total power radiated by the star in watts. The **apparent brightness**, b, is the power crossing a unit area held perpendicular to the path of light at the Earth. If d is the distance from the star to the earth, then

$$b = \frac{L}{4\pi d^2} \qquad (44-1)$$

The absolute luminosity depends on the mass of a star: *the greater its mass, the greater its luminosity.* The peak wavelength λ_p in the spectrum of light emitted depends on the temperature T of the star and is given by Wien's law: $\lambda_p T = 2.90 \times 10^{-3}$ m · k.

H–R Diagram

A plot of absolute intensity versus surface temperature of different stars is known as Hertzsprung–Russell (H–R) diagram. Most stars lie along the diagonal band in the diagram and are called *main sequence stars*. Above and to the right there are large stars with high luminosities but with low temperature. These are called *red giants*. At the lower left there are stars of low luminosity but with high temperatures. These are called *white dwarfs*.

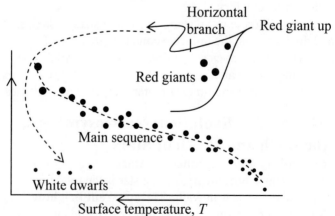

Hertzsprung–Russell (H–R) diagram and evolution of a star like the Sun

Stellar Evolution; Nucleosynthesis

When gaseous clouds (mostly hydrogen) contract due to the force of gravity, stars are born. At sufficiently high energy (when the temperature is more than 10 million degrees), the kinetic energy of the hydrogen nuclei becomes enough to overcome the Coulomb repulsion barrier and the hydrogen nuclei fuse to form helium nuclei. This is called nuclear fusion, and it liberates a tremendous amount of energy. The release of energy produces an outward pressure sufficient to halt the inward gravitational contraction and the star stabilizes in the main sequence.

As the core of helium grows at the center because of hydrogen fusion, hydrogen fuses in the shell around it. The production of energy decreases at the center and the core contracts because of the inward force of gravity and the core heats up. The hydrogen at the shell fuses more fiercely, causing the shell to expand and cool. The star has then entered the **red giant** stage. When the temperature at the center reaches about 100 million degrees, helium nuclei fuse. The reactions are:

$$\begin{align} {}^{4}_{2}\text{He} + {}^{4}_{2}\text{He} &\rightarrow {}^{8}_{4}\text{Be} + \gamma \quad (Q = -95 \text{ keV}) \end{align}$$

$$\text{(44–2)}$$

$$\begin{align} {}^{4}_{2}\text{He} + {}^{8}_{4}\text{Be} &\rightarrow {}^{12}_{6}\text{C} + \gamma \quad (Q = 7.4 \text{ MeV}) \end{align}$$

Further nuclear reactions, called **nucleosynthesis**, are possible, creating elements of higher and higher Z depending on the mass of the star until iron is formed.

If the mass of a residual star is less than 1.4 solar mass (called the *Chandrasekhar limit*), the core of the star collapses because of gravity until the point at which the electron clouds start to overlap. The star becomes a **white**

dwarf, which continues to radiate energy, decreasing in temperature until it becomes a dark cold chunk of ash. If the residual mass is greater than the Chandrasekhar limit, the stars can contract under gravity, reaching extremely high temperatures. The fusion of elements heavier than iron is possible even though the reactions require energy input. The size of the star is not limited by the exclusion principle applied to electrons but that applied to neutrons. The star contracts rapidly, forming an enormously dense **neutron star**. An explosion, called a **supernova explosion**, accompanies the final core collapse. The **pulsars**, which are astronomical objects that emit sharp pulses of radiation at regular intervals, are neutron stars. The core of a neutron star contracts to the point at which all neutrons are as close together as they are in the nucleus of an atom and the density of the neutron star becomes about 10^{14} times greater than that of solids and liquids on the earth. If the mass of a neutron star is greater than two to three times the mass of the sun, the gravity becomes so strong that even light emitted from it would not escape. This is known as a **black hole**.

Novae and Supernovae

Novae are stars that have suddenly increased in brightness and last for about a month or two before fading. Novae are believed to be white dwarfs that have pulled mass from a nearby star forming a binary system. **Supernovae** are brief explosive events of a massive star that release a lot more energy, roughly 10^{10} times more than the Sun.

44–3 Distance Measurements

The distance D of a star can be measured by the method of *parallax*. This includes measuring the angle of parallax ϕ of the star measured from the earth at an interval of six months. Since the distance d from the earth to the Sun is known, D can be obtained from the relation $D = d/\phi$. The distance of stars that are very far can be determined by measuring the "red shift" in the line spectra of the elements of the star.

The distance of a star is also expressed in **parsec**, which is defined as $1/\phi$ where ϕ is in seconds of arc (1 second = 1/3600 of a degree). 1 pc = 3.26 ly.

44–4 General Relativity: Gravity and the Curvature of Space

The general theory of relativity deals with accelerated frames of reference, and it provides an accurate interpretation of gravity. It is based on the **principle of equivalence**, which states: *No observer can determine by experiment whether he or she is accelerating or is rather in a gravitational field.*

Newton's second law, $F = ma$, provides the definition of **inertial mass**. The more the inertial mass a body has, the less acceleration it produces. Another kind of mass is **gravitational mass**. The gravitational force is proportional to the product of the gravitational masses of two objects (Newton's law of gravitation). No experiment has been able to find any difference between inertial mass and gravitational mass. Another way to state the equivalence principle is: *Gravitational mass is equivalent to inertial mass.*

One important conclusion from the Einstein general theory of relativity is that light is deflected in a gravitational field of a massive body. The theory predicted that the light from a star would be deflected by 1.75 seconds of arc as it passes near the Sun. Such prediction was first verified during a solar eclipse of the sun in 1919 and then by other subsequent experiments. When a large galaxy or a cluster of galaxies lies between Earth and a more distant galaxy, the intermediate galaxy produces a significant amount of bending of light resulting in multiple images of the distant galaxy. This is called *gravitational lensing*.

Light travels by the shortest, most direct path between two points. If the light beam follows a curved path (as mentioned above), then the *space must be curved* in the presence of gravity. Thus, according to general relativity, gravity causes a curvature in the four-dimensional space-time.

For a plane geometry (Euclidean), the shortest distance between two points is a straight line and the sum of the angles in a triangle is 180°. If the surface is curved, the shortest distance between two points (which is not a straight line) is called a **geodesic**. On a sphere, the geodesic is an arc of a great circle and the sum of the angles in a large triangle is more than 180°. Also, on a plane surface, $C = 2\pi r$, where C is the circumference of a large circle of radius r. For a curved surface (such as a sphere) with a *positive curvature* $C < 2\pi r$, and for a surface (such as a saddle-shaped surface) with a *negative curvature* $C > 2\pi r$.

Curvature of the Universe

If the universe has a positive curvature, then the universe is *finite*, or *closed*. In such a universe, the galaxies would be spread through the space and the space would fall back and close on itself. If the universe has a zero or negative curvature, the universe would be *open*. An open universe would be infinite. The mass in the universe determines whether it is open or closed; if the mass were great enough, it would bend space into a positively curved, closed, and finite space. The evidence is very strong that the universe on a large scale is very close to being flat.

Black Holes
According to Einstein general theory, objects and light rays move along geodesics in curved space-time. The extreme curvature of space-time is produced by a **black hole**. To become a black hole, an object of mass M must undergo **gravitational collapse** to within the **Schwarzschild radius** R, given by $R = 2GM/c^2$, where G is the gravitational constant and c is the speed of light.

44–5 The Expanding Universe: Red Shift and Hubble's Law

Hubble in 1929 put forward the idea that the universe is expanding. His idea was based on the observed **Doppler-shifted** wavelength (λ_{obs}) of the light emitted from stars and galaxies, which is given by the formula

$$\lambda_{obs} = \lambda_{rest} \sqrt{\frac{1 + v/c}{1 - v/c}} \qquad (44\text{–}3)$$

where v is the speed of the source. According to this, if the source moves away from the observer, the wavelength will be **red-shifted.**

Hubble found that the galaxies are receding from us and the greater the distance of a galaxy, the greater is its speed of recession. That is, the velocity, v, of a galaxy moving away from us is proportional to its distance, d, from us, as expressed by **Hubble's law**:

$$v = Hd \qquad (44\text{–}4)$$

where the constant H is called the **Hubble parameter**, which is 71 km/s/Mpc (that is, 22km/s/Mly).

Redshift Origins

More distant galaxies have higher recession velocity and a larger redshift and we call their redshift a **cosmological redshift**, which is interpreted due to the expansion of space itself. Another kind of redshift, called the **gravitational redshift**, occurs when light travels from a place where gravity is stronger to a place where gravity is weaker, since energy of the light photon decreases as it travels. The *redshift parameter* is defined as

$$z = \frac{\Delta\lambda}{\lambda_{rest}} = \frac{\lambda_{obs} - \lambda_{rest}}{\lambda_{rest}} \approx \frac{v}{c} \text{ (from Equation 44–3)}$$

For large z, Equation (44–3) does not apply because the redshift is due to the expansion of space, not the Doppler effect.

Scale Factor

If two galaxies are at a distance d_0 apart at some initial time, then at time t their distance $d(t)$ is given by

$$\frac{d(t) - d_0}{d_0} = 1 + z$$

Expansion and the Cosmological Principle

The expansion of the universe means that there is no center of expansion and all galaxies are moving away from *each other* at an average rate of 71 km/s per megaparsec of distance between them.

A basic assumption in cosmology is that the universe is isotropic (looks the same in all directions) and homogeneous (looks the same from any place); this is called the **cosmological principle.** On a local scale this does not apply, but the expansion of the universe is consistent with the cosmological principle. But matter (that is, galaxies) even on a large scale seems not to be homogeneous. This might be because over 90 percent of the universe may be nonluminous dark matter, which is uniformly distributed.

The expansion of the universe suggests that the galaxies must have been closer together in the past, which is the basis of the *Big Bang* theory of the origin of the universe. With $H = 22$ km/s/Mly, the age of the universe, $t = d/v = d/Hd = 1/H \approx 14$ billion years.

*Steady-State Model
The steady-state model assumes that the universe is infinitely old and on the average looks the same now as it always had.

44–6 The Big Bang and the Cosmic Microwave Background
According to the **Big Bang** theory, the universe was created by a huge explosion about 14 billion years ago. The observed expansion of the universe supports the theory. Another strong support of the theory came from the discovery of the **cosmic microwave radiation** (CMB) by Penzias and Wilson in 1964. They found that the universe is bathed in a microwave radiation of wavelength, $\lambda = 7.35$ cm. The intensity of this radiation as measured at $\lambda = 7.35$ cm corresponds to blackbody radiation at a temperature of 2.725 ± 0.002 K.

Measuremnets of the cosmic background radiation over the entire sky represent slightly hotter and colder spots from the average temperature of 2.725 K. This discovery of anisotropy of CMB ranks with the discovery of CMB itself in cosmology.

Immediately after the big bang, there was a tremendous release of energy, temperature was extremely high, and the universe consisted solely of radiation (photons) and elementary particles. As the universe expanded, the temperature dropped. Also, as the universe expanded, the wavelengths of the radiation also expanded, redshifting to a longer wavelength corresponding to a lower temperature (since $\lambda_{max} T$ = constant, Wien's law) until it reached the 2.7 K corresponding to the microwave background radiation that we observe today.

Looking Back toward the Big Bang —Lookback Time
We look back in time as we look out from the Earth. The farther an object is from the Earth, the longer back in time the light had left it. We cannot see as far as the Big Bang. We can see only as far as the **surface of last scattering** that radiated the CMB.

The Observable Universe
The entire universe is greater than the **observable universe**, which is a sphere of radius $r_0 = ct_0$ centered on the observer where t_0 is the age of the universe. The edge of our observable universe is called the **horizon**. Two observers on two widely separated galaxies would have different observable universes with different horizons.

44–7 The Standard Cosmological Model: The Early History of the Universe
The model that describes the origin and evolution of the universe is called the **standard cosmological model**.

- At the time of the Big Bang, the four fundamental forces were combined into a single force called the unified force. Gravity became a separate force after about 10^{-43} s of the Big Bang, when the temperature of the universe was about 10^{32} K. The symmetry of the four forces was broken, but the strong, weak, and electromagnetic forces were united (**grand unified era**). There was no distinction between quarks and leptons; baryons and lepton number were not conserved.

- The strong nuclear force became a separate force after 10^{-35} s, when the temperature of the universe was about 10^{27} K. The universe was then filled with leptons (electrons, muons, taus, neutrinos, and their antiparticles) and quarks. The quarks began to condense to form nucleons, other hadrons, and their antiparticles (**confinement of quarks** and **hadron era**).

- When the universe was about 10^{-6} μs old and the temperature dropped to 10^{13} K, the vast majority of the hadrons disappeared. Lighter particles including electrons, positrons, neutrinos, and photons (in roughly equal magnitude) dominated (**lepton era**).

- In about 1 s, the universe had cooled to 10^{10} K. In a few more seconds, because of annihilation of electrons and positrons, only a slight excess of electrons (later to join nuclei to form atoms) remained and in about 10 s after the big bang a majority of the components were photons and neutrinos (**radiation era**). As the temperature dropped to about 10^9 K, nucleons struck each other and fused to form nuclei. The matter mainly consisted of nuclei of hydrogen, nuclei of helium, and electrons, and the radiation (photons) continued to dominate.

- About 300,000 years later when the temperature dropped to about 3000 K, electrons could orbit the nuclei forming atoms (matter dominated). After the birth of atoms, stars and galaxies were formed about a million years after the big bang. The present temperature of the universe is 2.7 K.

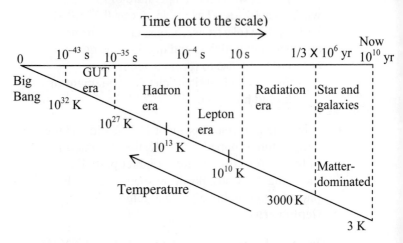

Very recently it was discovered that when the universe was about half as old as it is today, its expansion began to accelerate. Another important recent discovery is that ordinary matter makes up little of the total mass-energy of the universe.

44–8 Inflation: Explaining Flatness, Uniformity, and Structure

The universe underwent a period of exponential inflation early in its life.

Flatness

Best measurements suggest that the universe is flat with zero curvature. Inflation explains flatness of the universe.

CMB Uniformity

Inflation explains why CMB is uniform. By making the early universe very small, inflation shows that it could have been in thermal equilibrium. After inflation the universe would be large enough to provide us with the observable universe.

Galaxy Seeds, Fluctuations, Magnetic Monopoles

Forces undergo tiny **quantum fluctuations** according to the quantum theory. The inflation could have magnified these fluctuations, which would give the density irregularities seen in the cosmic microwave background (CMB). These density variations that are observed in CMB are believed to have coalesced under gravity into galaxies and glactic clusters, including stars and planets.

Inflation explains why **magnetic monopoles** were never observed. After inflation, they would be so far apart that we would never stumble on one.

44–9 Dark Matter and Dark Energy

Until late 1990, the universe was thought to be dominated by matter and the question whether the universe will continue to expand was connected to the curvature of space-time. If the curvature was *negative*, the universe would be open and infinite and the expansion would continue forever. If the universe is *flat* (no curvature), the universe is open and infinite but the expansion would slowly approach zero rate. If the curvature was *positive*, the universe would be closed and finite, the expansion would eventually stop and the universe would contract, collapsing back onto itself in a **big crunch**.

Critical Density

The fate of the universe depends on the **critical density** ρ_c $\approx 10^{-26}$ kg/m^3. If the density of the universe were $\rho > \rho_c$, the curvature would be positive, if $\rho = \rho_c$, the universe is flat. If $\rho < \rho_c$ the curvature would be negative. Today we believe the universe is flat and open and is expanding at an accelerating rate.

Dark Matter

There is evidence for a significant amount of nonluminous matter, called the *dark matter*, in the universe. Observation of rotation of galaxies and the motion of galaxies within the clusters suggest that they have considerably more mass than can be seen.

Dark Energy—Cosmic Acceleration

Recently astrophysicists received an incredible shock with the evidence that the universe is accelerating in spite of attracting gravitational force. We do not know the reason and there are several speculations. Clearly it must have a repulsive effect on matter, causing the objects to speed away from each other ever faster. It has been given the name *dark energy*. Today the best estimate of the mass-energy of the universe is: 75% dark energy and 25% of normal matter. Of this 25%, 20% is dark matter and 5% is baryons (what atoms are made of) and only 1/10 of this is visible matter, stars, and galaxies.

44–10 Large-Scale Structure of the Universe

As mentioned, small but significant inhomogeneities in the temperature of the CMB were observed recently. The anisotropies reflect compressions and expansions in the primordial plasma just before decoupling from which the stars, galaxies, and clusters of galaxies formed. Recent computer simulations predicted the distribution of clusters of galaxies and superclusters of galaxies similar to what is observed today. These simulations are successful if they contain dark energy and if the dark matter is cold.

44–11 Finally...

Based on the observations, interpretations, computer modeling, and theory, we now have a better idea about the universe: it is flat, it is about 14 billion years old, and contains only about 5% normal baryonic matter, and so on. The universe is exquisitely tuned, because if it were even a little different, life could not exist. This is called the **anthropic principle.**

Tips for Solving the Problems

1. Some important conversions in this chapter are
   1 ly = 9.46×10^{15} m
   1 light second = 3.8×10^8 m
   1 pc = 3.26 lys
   1 arc second = 1/3600 of a degree.

2. Use the relation $D = d/\phi$ to determine the distance of a star. Note that ϕ is in radians.

3. Use the relation $b = L/4\pi d^2$ to determine the apparent brightness of a star where L is the absolute luminosity and d is the distance. Clearly, $b_1/b_2 = d_2^2/d_1^2$.

4. The temperature T of a star can be determined by the Wien law: $\lambda_p T = 2.90 \times 10^{-3}$ m · k where λ_p is the wavelength for which the radiated intensity is a maximum. The Stefan law $E = \sigma T^4$ provides the energy radiated in one second by one square meter area of the star.

5. The Schwartzschild radius can be obtained from the relation $R = 2GM/c^2$.

6. The amount of red shift from a galaxy is given by $\Delta\lambda/\lambda \approx v/c$, where v is the speed of the galaxy and c is the speed of light.

7. The distance d of a galaxy can be determined from the Hubble law $v = Hd$, where v is the speed of the galaxy and H is called the Hubble constant. Note the value and the unit of the Hubble constant. $H = 71$ km/s/Mpc or 22 km/s/Mly. The age of the universe is the reciproal of the Hubble constant.

Notes

Notes

Notes

Notes

<u>Notes</u>

<u>Notes</u>

<u>Notes</u>

Notes

<u>Notes</u>

Notes